T0192774

Design Principles of Ships and Marine Structures

Design Principles of Ships and Marine Structures

S. C. Misra

CRC Press
Taylor & Francis Group
Boca Raton London New York

CRC Press is an imprint of the
Taylor & Francis Group, an **informa** business

CRC Press
Taylor & Francis Group
6000 Broken Sound Parkway NW, Suite 300
Boca Raton, FL 33487-2742

First issued in paperback 2020

© 2016 by Taylor & Francis Group, LLC
CRC Press is an imprint of Taylor & Francis Group, an Informa business

No claim to original U.S. Government works

Version Date: 20151020

ISBN 13: 978-0-367-57526-7 (pbk)
ISBN 13: 978-1-4822-5446-4 (hbk)

Visit the Taylor & Francis Web site at
http://www.taylorandfrancis.com

and the CRC Press Web site at
http://www.crcpress.com

Dedicated to

My teacher, Professor R. P. Gokarn

Contents

Preface

In the last 40 years, the author has been involved in teaching ship design and associated subjects, stability and hydrodynamics and industrial activity related to marine design. During this period, design methodology underwent tremendous changes and improvements. With the availability of high-speed computing and many sophisticated software, advanced numerical techniques are being applied to maritime design procedure at the concept design stage. The use of empirical formulations has reduced. Also, in recent years, concerns of energy consumption, environment, safety and reliability have increased. Designers are increasingly incorporating such aspects at the design stage to reduce these concerns. With increasing competition, it has become necessary to look at the effectiveness of design from the economic point of view such as project profitability and shipbuilding cost. With these varied requirements, the system design aspects in large complex products have become important and it is necessary to evaluate alternatives for each system or unit, as well as the final product. Systems engineering applications try to integrate technical, environmental or societal and economic systems, making a final complex marine platform. The management of design activities has also become increasingly important and complex, providing space for creativity and innovation. This book intends to make the reader aware of all these aspects of design and their integration.

As the title of the book indicates, the subjects deal with the many and varied principles of design. The chapters of the book are fairly varied in nature, the intention being to focus on the importance of each of the topics covering the entire sphere of marine design. For each chapter, there are enough books written by renowned authors and the reader is advised to refer to those books if needed, to produce good design work. With the author's experience of more than 40 years in marine design applications, the book is written focussing on marine design with examples of ships wherever necessary. He believes that the principles described in this book can be applied to successful designs of any other marine structure or vehicle, the details being different in each case.

The book has 16 chapters starting with an introduction of marine design which includes a description of various marine products which are used for transportation, defence and exploitation of marine resources. Chapter 2 introduces the reader to marine environment in which the product has to work. Chapter 3 discusses various design methodologies such as sequential design process with the application of concurrent engineering. Set-based design, which has been successfully implemented in the automobile industry, has been introduced in the book, and it is expected that this will be useful in marine applications also. Chapter 4 discusses applications of engineering economics to marine design, highlighting the effect of design parameters on the profitability over the life of the ship and building cost based on shipyard facilities. Chapter 5 addresses the issue of parameter estimation using different techniques such as statistical data and empirical formulae for parameter estimation, as well as the performance prediction at the concept design stage. Chapter 6 discusses intact and damage stability issues applied to ship design. Hydrodynamic issues of resistance, propulsion, sea keeping and manoeuvring and their effects on design are discussed in Chapter 7. We do not go into the fundamental or advanced details of these subjects in this chapter, but the application of computational fluid dynamics (CFD) and experimental fluid dynamics (EFD) in these areas have been highlighted. Chapter 8 discusses hull form design and is purely based on vehicles, particularly ships. The use of computer-aided design techniques has been highlighted. Machinery systems consisting of main and auxiliary machinery, redundancies, piping systems and energy consumption patterns have been discussed in Chapter 9. Structural design is a subject by itself. But without the fundamentals of materials, loads and design techniques, a marine platform design remains incomplete. Chapter 10 briefly discusses structural design, including materials. Space layout for payload, equipment and machinery and accommodation for personnel on board on the platform, commonly known as general arrangement, is discussed in Chapter 11. Safety has become an important aspect of design of complex products and systems these days. In Chapter 12, design aspects related to safety, including risk assessment, are discussed. Design for sustainability includes protection of air and water environment and protection against invasive species. These issues have brought to the fore a number of innovative ideas. These are discussed in Chapter 13. Chapter 14 discusses designs for production to reduce construction cost and time. Standardisation and modularisation are the main issues discussed here. Chapter 15 states the principles of numerical optimisation for decision-making. The importance of heuristic optimisation and multi-objective decision-making processes are highlighted. Chapter 16 is on design management, the crucial factors being the encouragement of creativity and innovation in marine design.

S. C. Misra

Acknowledgements

A book of this kind could not have been written without the direct and indirect help and support of a large number of well-wishers. My teacher Professor R. P. Gokarn has been a constant source of inspiration for me to start and complete the book. His untiring support, information on various topics covered in the book and numerous reviews of the draft manuscript are gratefully acknowledged. My colleagues at the Indian Institute of Technology, Kharagpur, Professors O. P. Sha and N. Vishwanath, have patiently listened to my exposition of the book from the early days and have given me valuable inputs in the form of lecture notes and students' calculation results, particularly on resistance and manoeuvrability applications to design and decision-making processes. My friends A.R. Kar and Karan Doshi of the Indian Register of Shipping have helped me by supplying relevant information and reviewing my work on structural design and design for safety. I have received a lot of help from scientists and faculty members of the Indian Maritime University at Visakhapatnam Campus in the form of information, reviews of chapters and preparation of drawings, which are gratefully acknowledged. A special mention must be made of the director in charge U. S. Ramesh, for his unstinting support and for providing some materials on ship costing; Arun Kishore Eswara for the supply of materials and drawings on machinery system design, design for safety and design for sustainability; G.V.V. Pavan Kumar for his support for the chapters on layout design and hull form design and Dr. K. V. R. K. Pattanayak for reviewing the chapter on marine environment and suggestions for modifications. I gratefully acknowledge the help rendered by Avinash Godey and Jaswant Samal for their untiring and continuous contribution in manuscript preparation and finalisation of all diagrams for days. I also acknowledge the help rendered by N. Madhu Kumar, V. Sunitha, Madhu Joshi, Dr. A. Mukherjee and D. S. P. Vidyasagar for their support during the progress of manuscript preparation. My friends P. P. Singh, Bijit Sarkar and A. Otta have encouraged me to write a book on design which prompted me to start writing this book and motivated me through to its completion.

Some material in the book has been taken from other publications. I am thankful to the publishers for their permission.

My publishers CRC Press of the Taylor & Francis Group, particularly the coordinators of this project, Gagandeep Singh and Ashley Weinstein, have shown a tremendous amount of patience and have encouraged me during this pressing period; and I am indeed thankful to them.

More than 35 years of interaction with students of naval architecture and ocean engineering at the Indian Institute of Technology, Kharagpur, and the Indian Maritime University at Visakhapatnam campus has been the prime motivating factor for undertaking the project of writing this book. I am grateful to all my students for that.

My wife Rachita has shown endless patience and given constant encouragement during the long period of this project. I sincerely thank her and my sons, Kunal and Amrut, and their families for their love and support.

Author

S. C. Misra earned a B.Tech. (Hons.) degree in naval architecture from IIT Kharagpur, India, in 1970, and his PhD from the University of Newcastle upon Tyne, United Kingdom, in 1976. After serving for a few years as design engineer in Hindustan Shipyard Ltd., Visakhapatnam, he joined the Indian Institute of Technology, Kharagpur, as assistant professor in naval architecture. He became professor there and also served three years as head of the Department of Ocean Engineering and Naval Architecture and finally retired from active service in 2013. During his service period, he spent six months at Glasgow University, United Kingdom, two years as visiting professor at IIT Madras and five years as the director of the Indian Maritime University, Visakhapatnam Campus, where he initiated undergraduate and postgraduate programs in naval architecture and ocean engineering and dredging and harbour engineering. Subsequent to his retirement, he has been involved in a number of research and consultancy projects in the areas of design of ships, marine structures and inland water transportation.

Nomenclature

A	Annual repayment/attained subdivision index
B	Breadth
D	Depth/propeller diameter
E	Energy/Young's modulus
F	Future sum of money
I	Moment of inertia
J	Advance coefficient/polar moment of inertia
L	Length/load
M	Bending moment
N	Revolutions per minute/number of years
P	Present sum of money/principal amount of investment/engine power
Q	Torque
R	Required subdivision index/resistance
S	Wetted surface
T	Draught/wave period in seconds/thrust
U	Free stream velocity in x-direction
V	Speed
W	Weight in tonnes
c	Wave celerity $= \omega/k = \lambda/T$
cgt	Compensated gross tonnes
dwt	Deadweight
f	Frequency in Hz
g	Acceleration due to gravity $= 9.81$ m/s^2/Limit state function
g	Limit state function
h	Water depth below surface/height above water
i	Interest rate/effective rate of return/discount rate
k	Wave number $= 2\pi/\lambda$/form factor
k_s	Surface roughness
n	Revolutions per second
p	Pressure
q	Shear flow
t	Thrust deduction fraction
u, v and w	Perturbation velocities in x, y and z directions, respectively
w	Wake fraction
z	Number of propeller blades
Δ	Displacement in tonnes
∇	Volume of displacement
ε	Phase angle
ζ	Wave elevation
$\zeta_{a1/3}$	Significant wave height
ζ_ω or ζ_a	Wave amplitude with frequency ω
η	Efficiency
θ	Trim angle
λ	Wave length/model scale

λ_c	Average wavelength in a seaway
ν	Kinematic coefficient of viscosity
ξ	Motion response
ρ	Density
σ	Cavitation number/bending or flexural stress
σ_ζ	Standard deviation of ζ distribution
$\sigma_\zeta{}^2$	Variance of ζ distribution
τ	Shear stress
τ_c	Thrust loading coefficient
φ	Velocity potential/heel angle
ω	Circular wave frequency in rad/s $= 2\pi/T$
ω_e	Encounter frequency
A_P, A_E, A_D, A_0	Propeller projected, expanded, developed and disc area, respectively
A_x	Midship area
A_{WP}	Water plane area
BAR	Blade area ratio $= A_E/A_0$
C_B	Block coefficient
C_F	Frictional resistance coefficient
C_P	Prismatic coefficient
C_{PV}	Vertical prismatic coefficient
C_R	Residuary resistance coefficient
C_T	Total resistance coefficient
C_{WP}	Water plane area coefficient
C_x	Midship area coefficient
Fn	Froude number $= V/\sqrt{(g \cdot L)}$
$F_{n\nabla}$	Displacement Froude number $= V/\sqrt{(g \cdot \nabla^{1/3})}$
$F(x)$	Probability distribution function
$f(x)$	Probability density function
H_ω	Wave height of wave frequency ω
h_0	Standard height above water
i_e	Half angle of entrance
K_Q	Torque coefficient
K_T	Thrust coefficient
L_{BP}	Length between perpendiculars
L_{OA}	Length overall
L_{WL}	Length on water line
p_v	Vapour pressure
P/D	Pitch ratio
R_F	Frictional resistance
R_n	Reynolds number
R_P	Pressure resistance
R_R	Residuary resistance
R_T	Total resistance
R_V	Viscous resistance
R_{VP}	Viscous pressure resistance
R_W	Wave making resistance
S_ζ	Energy spectrum
$S_\zeta(\omega)$	Energy spectrum ordinate at circular frequency ω
T_c	Average zero-crossing period

V_0	Standard speed
V_A	Speed of advance
V_c	Critical speed
AAC	Average annual cost
AAW	Anti-aircraft warfare
AIS	Automatic identification system
ALARP	As low as possible
AP	After perpendicular
ASD	Allowable stress design
ASW	Anti-submarine warfare
AUV	Automatic underwater vehicle
BM	Metacentric radius
CA	Compound amount factor
CB	Centre of buoyancy with coordinates LCB, TCB and VCB or KB
CAD	Computer-aided design
CAM	Computer-aided manufacturing
CESA	Community of European Shipbuilders Association
CFC	Chlorofluorocarbon
CFD	Computational fluid dynamics
CG	Centre of gravity with coordinates LCG, TCG and VCG or KG
CIM	Computer-instructed manufacturing
CR	Capital recovery factor
CRV	Coastal research vessel
CSD	Cutter suction dredger
CSR	Common structural rules
DCF	Discounted cash flow
DE	Diesel engine
ECA	Emission control areas
EEDI	Energy efficiency design index
EEZ	Exclusive economic zone
FEM	Finite element methods
FFA	Fire-fighting appliance
FORM	First-order reliability method
FP	Forward perpendicular
FRP	Fibre-reinforced plastic
FPSO	Floating production storage and offloading unit
FSA	Formal safety assessment
FSO	Floating storage and offloading unit
FSU	Floating storage unit
GA	General arrangement
GM	Metacentric height
GT	Gross tonnage
GZ	Statistical stability lever or arm
HFO	Heavy fuel oil
IACS	International Association of Classification Societies
IBC	International Regulations for Carriage of Dangerous Chemicals in Bulk
IGC	The International Code for the Construction and Equipment of Ships Carrying Liquefied Gases in Bulk
ILO	International Labour Organisation

IMDG	International Regulations for Carriage of Dangerous Goods
IMO	International Maritime Organisation
INCOSE	International Council on System Engineering
IRR	Internal rate of return
ITTC	International Towing Tank Conference
KN	Perpendicular distance from keel to the perpendicular through CB to LWL
KSA	Korean Shipbuilders Association
LCF	Longitudinal centre of floatation
LNG	Liquefied natural gas
LPG	Liquefied petroleum gas
LSA	Life-saving appliances
LRFD	Load and resistance factor design
LWL	Load water line
MARPOL	International Convention on Prevention of Pollution from Ships
MCT 1 cm	Moment to change trim 1 centimetre
MDO	Marine diesel oil
MSI	Motion sickness index
MSL	Mean sea level
NCCV	Non-cargo carrying vessels
NESDIS	National Environmental Satellite, Data and Information Service
NOAA	National Oceanic and Atmospheric Administration
NPV	Net present value
NPVI	Net present value index
NT	Net tonnage
OECD	Organisation of Economic Co-operation and Development
ODS	Ozone depleting substances
ORV	Ocean research vessel
OTEC	Ocean thermal energy conversion
OWC	Oscillating water column
QPC	Quasi propulsive coefficient
PSF	Partial safety factor
PSSA	Particularly sensitive sea areas
PW	Present worth factor
RAO	Response amplitude operator
RFR	Required freight rate
ROI	Return on investment
ROLO	Roll on load off
RORO	Roll on roll off
SAJ	Shipbuilders Association of Japan
SCA	Series compound amount factor
SOFAR	Sound fixing and ranging
SOLAS	Safety of life at sea
SPW	Series present worth factor
ST	Steam turbine
SWATH	Small water plane area twin hull vessel
TBT	Tri-butyl tin
TEU	Twenty feet equivalent unit
TLP	Tension leg platform
TP 1 cm	Tonnes per centimetre immersion

TSHD	Trailing suction hopper dredger
UI	Unmanned installation
ULCC	Ultra-large crude carrier
UNCLOS	United Nations Conference on Law of the Sea
UNCTAD	United Nations Conference on Trade and Development
UNEP	United Nations Environment Program
VDR	Voyage data recorder
VLCC	Very large crude carrier
VLOC	Very large ore carrier
WSD	Working stress design

1

Introduction

Over the years, human beings have developed systems, processes and products which have served people in many ways. Their approach to such developments has been based on the design of these products and systems. As a form of human activity, design has evolved in a number of ways from the earliest times until today. In ancient times, design was understood as the aesthetic beauty of a structure. The designer built into the design the functional aspects based on intuition and the understanding of the physical world around him or her rather than any scientific analysis. Creativity played a major role in such designs. Major townships built by ancient civilizations across the globe, large monuments such as the pyramids, palaces, forts and places of worship such as temples, cathedrals and mosques bear testimony to this fact. Engineering inventions such as wheels found their application in artefacts which were designed more for beauty than for technical considerations, e.g. chariots for sports and warfare. Similarly, in the water world, humans could move across rivers and oceans in boats, brave the high seas and explore new shores without the scientific knowledge of the mechanics of floating bodies. Boats were designed primarily for their looks. Even the boats used for battles were designed having sleek and slender bodies with figures of mermaids in front. Thus, design was more of an art than science. Due to the understanding of the physical world (or lack of it!), the artefacts designed and built in ancient times have stood the test of time. We do not know if these were designed to stand for such long periods or were simply over-designed due to the lack of scientific knowledge.

With the advent of Industrial Revolution, a large number of inventions took place. These inventions could be used in new structures for obtaining better performance using machines instead of human labour. Engineering emerged as a discipline. Scientific knowledge increased in the physical world of solid and fluid mechanics, light, heat, electrical and magnetic energies due to both theoretical and experimental advancement. As a result, engineering design slowly moved from an art to a science. Most of the last century saw engineering design evolving as a closed system where the loads on structures were precisely known and the science of designing was well defined. This found application in component design such as a gear to transmit a known torque at a given value of revolutions per minute, a structure to withstand a given load or an electrical machine to produce a given power with inputs known a priori. Larger artefacts were designed using the same principles of isolated systems without considering interaction with the outside environment. Slowly but surely the design environment has changed to take care of multiple systems working simultaneously and efficiently in an environment where the variables may not be known or well defined. The complexity of the design process can be understood if the variety of marine vehicles, structures and facilities can be appreciated.

1.1 Development of Marine Vehicles, Structures and Facilities

In ancient times, the first marine structure was a log of wood used to move people around in water. Gradually, wood was carved, shaped and joined together to create ships and catamarans to have a means of transportation in water, across seas and to win wars. In ancient times, human muscle power was the only source of power for ships in the form of movement of oars in water for forward motion. Later, paddle wheels were developed, which used mechanical effort for movement of paddles in water. In later times, wind, the only source of renewable and sustainable energy, was harnessed by using sails to provide motive force to ships to move forward. In the early nineteenth century, steam engines slowly replaced sails and mechanical power was available for ship powering. As a result, coal became the fuel for ships. Simultaneously, screw propellers found their way into propelling ships. Later, iron ships came into being and by the end of the nineteenth century, steel had replaced wood almost completely in ship manufacturing. At the same time, oil was replacing coal as the fuel for steam reciprocating engines. In the beginning of the twentieth century, welding was used for steel construction, which revolutionized shipbuilding. Diesel engines and steam turbines appeared around the same time and started replacing steam reciprocating engines as ship-powering machines. Though the first working submarine was reported to have been built in England in 1623 and was subsequently built and used in the American Civil War, modern submarines appeared during the World Wars in the twentieth century.

The two World Wars saw tremendous advances in ship design and manufacturing technologies: new ship types such as hydrofoils and hovercrafts appeared, submarines appeared as weapons of war and nuclear-powered steam turbine–driven ships (*USS Nautilus*, the world's first nuclear-powered submarine) appeared in the shipping scene. After the Second World War, the demand for a large number of ships grew for moving materials across the globe for facilitating reconstruction. This gave rise to the construction of standard vessels such as Freedom, SD-14 and Fortune class of vessels. Ships also started being classified based on the type of cargo they carried, such as bulk carriers, tankers and general cargo ships. The 1960s was known as the golden era of shipping when ship size increased due to the economy of scale and the world saw big ships such as very large crude carriers (VLCC) and ultra large crude carriers (ULCC). Starting from the 1970s, two developments happened: first, sharp increase in crude oil prices led to a drop in oil cargo movement and second, tanker disasters, starting with the *Tory Canyon* disaster, gave rise to the development of major tanker design modifications leading to the double hull tankers in the twenty-first century. Fossil fuel in the form of natural gas was required to be carried by sea leading to cryogenic vessel design for LNG carriage. Simultaneously, exploration and production of oil from the seabed started, giving rise to structures such as floating and fixed platforms.

1.2 Types of Marine Vehicles, Structures and Facilities

Marine vehicles, structures and facilities or systems are designed and produced to serve some specific purpose such as transporting cargo from one place to another, fighting wars, exploiting living and nonliving resources, harnessing energy from the sea, tourism and sports, support services for all such activities and facilities for such activities at the land and sea interface. Based on the purpose intended, vessels, structures and facilities at sea and on the coast vary in their functionality and design. Accordingly, various items can be classified into different types.

1.2.1 Transportation

The oldest and the most common form of vehicles and structures are ships used for transporting goods and passengers. Such ships can be classified as follows:

- General cargo ships carrying packed cargoes of various sizes, commonly known as break bulk cargo, of various types and sizes: these could be tramps moving from port to port based on demand with no fixed schedule or cargo liners with multi-decks and having their own cargo-handling gear and moving between pre-defined ports with fixed schedule.
- Unitized cargo carriers such as container ships, barge carriers and RORO (roll-on roll-off) vessels.
- Passenger vessels such as passenger ferries, cruise ships, passenger ships of various types and sizes and fast transport vessels.
- Ocean research vessels, coastal research vessels or fisheries research vessels, though not transport vessels, are designed like passenger vessels except that these must have a large deck area for scientific sample collection and analysis. These vessels have large accommodation for scientists and laboratory space.
- Dry bulk carriers carrying various types of bulk cargo such as ore, fertilizer, grain, etc. and combination carriers such as OBO (oil, bulk, ore) ships.
- Liquid bulk cargo carriers such as crude oil carriers, product carriers, chemical tankers and liquefied gas carriers to carry LPG, LNG, ammonia, etc.

Ocean-going cargo vessels can be of various sizes starting from a few thousand tonnes deadweight (dwt) capacity to a few hundred thousand tonnes deadweight. For easy identification with regard to size and seaway negotiation capability, some nomenclatures are commonly used, particularly for bulk carriers and tankers. Some examples are as follows, which is not an exhaustive list and the deadweights indicated are only approximate, which change with change in waterway dimensions and trade convenience. With restrictions of passages through various seaways increasing or easing, the deadweights indicated may change.

- Mini bulk carriers or coastal tankers: less than 10,000 tonnes dwt
- Handy size bulk carriers or tankers: 10,000–35,000 tonnes dwt
- Handymax carriers: 35,000–50,000 tonnes dwt
- Supramax bulk carriers or tankers: 50,000–60,000 tonnes dwt
- Medium range tanker: 25,000–55,000 tonnes dwt
- Long Range I Tanker: 55,000–80,000 tonnes dwt
- Long Range II Tanker: 80,000–150,000 tonnes dwt
- Seawaymax vessels: up to about 28,900 tonnes dwt
- Aframax or cape size vessels: 75,000–120,000 tonnes dwt
- Suezmax vessels: 120,000–240,000 tonnes dwt

- Panamax vessels: 50,000–90,000 tonnes dwt
- Malaccamax vessels: 200,000–315,000 tonnes dwt
- VLCC: super tankers of 160,000–320,000 tonnes dwt
- ULCC: super tankers above 320,000 tonnes dwt
- Very large ore carriers: bulk carriers of more than 200,000 tonnes dwt

The size of a vessel depends on the market demand, whereas design conditions depend on operational requirements. A vessel needs to be ice strengthened if it is to operate in polar waters where it may encounter ice conditions. A large ocean-going vessel may encounter severe sea conditions during its voyage across the Atlantic, Pacific or Indian Ocean, whereas a short sea or coastal vessel may encounter less severe sea conditions. A river–sea vessel, i.e. a vessel going in river as well as coastal waters in a multi-modal transportation system, may encounter still less severe sea conditions, whereas a river vessel need not encounter waves at all though it has to negotiate bends and shallow water unlike an ocean-going vessel. River vessels as well as some sea-going vessels may be dumb not having their own propulsion systems.

1.2.2 Defence

Naval vessels and crafts have been a major part of the navies of various nations over the ages, which have been used for both defensive and offensive warfare. The development of naval military technologies has led to the building of modern advanced naval vessels. The technologies, thus developed, have found their way into merchant ship design and construction.

Naval vessels are characterized by the carriage of offensive/defensive weapons and accessories such as guns of various sizes and ranges, deployment or removal of mines/depth charges, guided missiles (anti-aircraft warfare), torpedoes (anti-submarine torpedoes), surface-to-air and surface-to-surface missiles, including nuclear missiles. These vessels, based on requirement, are capable of different oceanic operations and having different attributes such as

- Surveillance and detection (radar, sonar, etc.)
- Reconnaissance and rescue
- Stealth
- Range—short/medium/long
- Carriage of helicopters/aircraft with platform at sea for loading and take off
- Support services such as troop, tanks carriage, fuel oil replenishment at sea
- Attack and defence power supported by high speed and manoeuvrability

To cater to such needs, naval vessels are generally designed for carriage of personnel with large living accommodation; sonar domes at the bottom; stealth against electronic detection as well as sonar detection (noise); carriage of weapons and their operations, including shock or impact loading; multispeed operation for both high-speed and cruising operation necessitating the need to have multiple engines and screws, including alternative systems such as a combination of gas turbines, steam turbines and/or diesel engines.

As per size and operational requirements, warships may be as follows:

- Patrol crafts patrol the coast equipped with some gunfire power. Corvettes operate in littoral waters to protect a country's assets even far away from the mainland. A corvette can accommodate sophisticated air/surface defence systems, surveillance equipment, even a small anti-submarine warfare helicopter generally having a single propulsion power plant.

- Frigates carry variable depth sonar, towed array, and/or torpedoes, anti-submarine torpedoes, surface-to-air/surface-to-surface missiles, landing deck and hanger to operate helicopters (equipped with sono buoys, magnetic detectors, etc.) and are also used for search-and-rescue operations. Frigates use advanced stealth technology with minimum radar cross section, can be used for high-speed deployment and generally work in a fleet/convoy.

- Destroyers are bigger than frigates with more weapons. Aircraft carriers are the biggest defence vessels having a single normal operating speed. Such a vessel must move in a convoy to protect itself. It carries a large number of aircraft, including landing/take-off facilities as well as storage and maintenances facilities. So deck space required is large, including large runways and hangers below. Modern aircraft carriers have been conceived as multihull vessels such as trimarans or pentamarans.

- Submarines are vehicles designed for going down to a depth below the surface under water and also come up as and when required. The capability of a submarine is known by the depth it can submerge to and for how long it can stay under water. These vessels are characterized by the ballasting and de-ballasting arrangements, multiple propulsion arrangements with diesel engines and electrical (battery) power or nuclear power, living arrangements in confined environment and torpedo carrying and firing capability.

1.2.3 Resource Exploitation

The sea is a storehouse of various living and nonliving resources. Though sea has been providing fish to people for a long time, exploration and production of fossil fuels and minerals are of comparatively recent origin. Still more recent is the extraction of energy from the sea. The demand for extraction of resources from the sea is likely to continue for a long time.

1.2.3.1 *Living Resources*

Initially, the ocean was the primary source of life on Earth. Life force on land evolved much later. Even today oceans are full of life, including microscopic plants and animals. The sea life with which human beings are closely associated include fishes of various types, sea animals such as the blue whale, which is the largest living creature, crustaceans such as prawns, lobsters, crabs and the common barnacles, sea plants and seaweed. Living sea resources also provide useful medicinal extracts and could provide many other useful items in the future.

Fishing is the most frequent commercial activity, which has evolved over many years. Based on commercial viability, fishing vessels can be small, primarily operating in coastal and brackish waters, with the entire fish-catching cycle being not more than 1 or 2 days. If the fishing vessel has to venture deeper and operate in the exclusive economic zones (EEZ), then it has to be bigger with fish catch being more than that of the previous type of vessel. (At present EEZ is defined as a region 200 nautical miles from the coastline as per the United Nations Convention on the Law of the Sea over which a nation has sovereign rights of economic exploitation.) Still larger vessels can operate and catch fish in the deep ocean beyond the EEZ. Fishing vessel design varies based on the fishing gear used. If it uses trawl net to catch fish, the vessel is called trawler, which can deploy the trawl net either from the stern (stern trawler) or from the side (side trawler). A seiner is a vessel with a surrounding gear and a seine net. A dredger is a fishing vessel which collects molluscs from ocean bottom by dragging nets on the ocean bottom. A lift netter uses large lift nets using outriggers. Similarly, a trap setter vessel sets traps to capture crabs, etc.

Based on size and mode of fish handling, trawlers may be classified into different groups. A wet fish trawler is one in which fish is kept in wet and fresh condition and is used for short trips only. A freezer trawler is a medium-sized trawler used for medium duration trips, which may carry ice or has a refrigerating plant to preserve fish. A fish factory ship is a large trawler used for a long duration and has facilities for handling and processing fish on board, including gutting, filleting, freezing, storage and even canning. A mother ship provides services to fishing vessels at sea, such as fuel, provisions, crew, etc. Such a ship can also be a factory ship and can be used to transport fish to the shore from other trawlers.

The design of a fishing vessel, therefore, requires proper selection and installation of fishing gear, arrangements for handling the fishing gear (including 'A' brackets or cranes) and hauling trawl net, appropriate propeller design to provide enough pull for trawling, arrangements for processing and freezing the catch, as well as adequate deck area and space for fish handling and storage.

1.2.3.2 Mineral Resources

The most common nonliving resource from the sea is the common salt used in every food prepared by human beings. Salt is extracted from seawater by collecting and evaporating the water so that only salt remains. But transportation of large quantities of salt to different places by the sea is quite complex. Being hygroscopic in nature, salt requires protection from moisture.

Placer materials are generally found on beach sand around the world. Such materials are normally in the form of rutile, monazite or ilmenite sand. Sometimes placers are found away from the beach in the shelf region. Placers contain rare earth materials such as monazite and rare materials such as zircon, chromite, wolframite, cassiterite, etc. Placer materials to be processed on land are extracted on or near the beach by the method of sweeping or shallow water dredging. This requires shallow water dredgers and support vessels.

Natural gas (mostly methane molecules) combines with water molecules to form gas hydrates, which remain stable either due to low temperature or due to pressure, being submerged under sand or similar shelf material. Hydrates contain a large amount of hydrocarbons, and extraction of these items on a commercial basis is a matter of current study.

Polymetallic nodules, found in ocean bottom at a depth of 5000 m or more, have been formed due to millions of years of geological activity combining metallic compounds with ocean mud and water. These nodules are ellipsoidal in shape and vary from a few millimetres to about 10 cm or more. Apart from ocean mud, these nodules contain iron and manganese, which are found abundantly in land mines. Nodules also contain rare metals such as copper, nickel, cobalt and molybdenum, which are rare in land mines. These nodules lie on the ocean bottom without being buried under sand. Ocean mining involves exploration and production of these nodules. In such mining, a sweeping mechanism collects nodules from the ocean bottom with the help of a bottom crawler moving on its own on a predetermined path and collecting nodules at particular points. In the next stage, a pumping and piping system lifts the nodules to the surface, perhaps in a slurry form (mixed with seawater). Storing, transportation and extraction of metals follow through surface vehicles and factories. Unless the economics justify, this technology is unlikely to be implemented in the near future.

1.2.3.3 Renewable Energy

The sea absorbs most of the radiated heat from the sun and stores this energy in various forms such as waves, currents, tides and temperature difference across the bottom. The source of this energy is the sun and, therefore, inexhaustible. If this energy can be harnessed for human use, this will not only be renewable but also nonpolluting. Apart from technical complexities of having engineering devices at sea and economic difficulty of finding appropriate return on investment, one of the main problems is the transfer of energy to land for human use. The complexity and cost of this part of the entire exercise increase as the device moves further away from land.

Tidal energy requires the construction of a barrage or dam to let the high tide cross the barrage, which could then be released to the sea after passing through a turbine. The turbine, in turn, runs a rotor to generate electricity. Such tidal energy systems are suitable where there is a large tidal variation twice a day. Tides also generate large currents in the water in coastal regions, in river mouths or upstream of rivers. In large rivers, this is a large source of energy. A combination of boat-shaped bodies such as Cockerell crafts can be used so that the differential movement of boats can be captured through the hinges to move a rotor to generate electricity. Water turbines moving in a single direction irrespective of the current direction can be similarly used.

A visible form of energy in the sea is the surface waves. Various devices have been invented to convert the kinetic energy of the waves into rotary motion, which can then run a generator to convert mechanical energy into electrical energy. Such devices include Salter Ducks, Cockrell Craft and Oscillating Water Columns (OWCs). An OWC is the most promising device, which is a floating, half-immersed inverted J tube, open at the bottom and closed at the top, with an escape air valve. As waves pass across the J tube, the air column inside the tube escapes or gets pulled in. This dynamics of air is used to move a unidirectional air turbine, which moves a generator to get electricity. Though this system seems very promising, the prohibitive cost restricts its construction beyond research laboratories.

The temperature difference between the top surface and 1000 m below the surface may be about 18°C to 20°C. This indicates a large amount of heat energy variation across the ocean depth. If this energy can be captured and converted to usable energy, it will be renewable and nonpolluting. Such a system is called the Ocean Thermal Energy Conversion system. There can be many ways of doing this. A possible method is to have a floating platform at a site where the depth of water is more than 1000 m. The surface (hot) water can be drawn to a ship to vaporize a suitable liquid at that temperature at normal pressure. This energized gas can run a turbine, which can in turn run a generator to produce electricity. After this energy is spent, the gas can be condensed and liquefied through a heat exchanger using the seawater pumped to the ship from the depth of 1000 m. Such systems are generally of low efficiency for power generation, but considering the fact that the energy source is unlimited, one can obtain nonpolluting energy at a reasonable price.

Wind energy is a clean source of energy, and it is well known that wind moves at a higher speed over coastal and sea surfaces than on land due to the lack of any obstruction. Therefore, extraction of wind energy at identified locations on coastal and sea surfaces in EEZs is a very attractive proposition. The wind turbine technology, including conversion to electricity, is fairly well established today. The same technology can be used for a wind energy device at sea. However, it becomes more complicated when it comes to designing and housing a platform to hold the wind turbine. The platform technology used in offshore oil industry can be burrowed and converted suitably for this purpose. If it is a near-shore device at low depth, one can use a fixed platform to house a single or multiple wind turbines. If the water depth increases, one can go for a floating platform like a spar platform for a single turbine or a ship-like structure to house a number of turbines. This is known as an offshore wind farm. The other technical complexity for a floating platform is mooring and position keeping.

1.2.3.4 Fossil Fuels

Till date people's energy needs have been met by oil and natural gas (and coal, to a limited extent), which are known as fossil fuels, a large source of which is below the sea surface at a depth of 10–15 km. At such locations, the water depth could be a few metres in near-shore areas, a few hundred metres in the continental shelf or up to a few kilometres in the deep ocean. An oil/gas production process involves various activities such as exploration drilling to locate a production site, well testing, pre-production and production activities, including drilling, receipt of oil and gas from drill head, separation of water, hydrocarbon and gases, water injection into the well head, flaring of excess gas, storage and export of oil and gas. As can be observed, the offshore oil platform has to house many systems and subsystems, which must work in a cohesive integrated manner and the platform must behave in a manner that all the systems work effectively and efficiently.

Around the world, until 2013, there are more than 6500 offshore oil and gas installations distributed in about 53 countries. The numerical distribution of offshore oil platforms around the world is about 4000 units in U.S. Gulf of Mexico; about 950 in Asia; 700 in the Middle East; 490 in Europe, North Sea and North East Atlantic; 380 in West Africa and 340 in South America. An offshore platform is a large structure at sea used to house crew and machinery for exploration and/or production of natural resources such as fossil fuels from under the ocean bed. It is normally located in the continental shelf, but could be in deeper waters, could be fixed to the seabed or floating and could be dumb or mobile. Accordingly, offshore platforms can be classified as fixed platforms, compliant towers, jack-up platforms, semi-submersible platforms, drill ships, tension-leg platforms (TLPs), SPAR platforms and normally unmanned installations.

A fixed offshore platform is normally installed for long-term use as a production platform in shallow depth between 5 and 150 m. It has a substructure, which is fixed to the seabed at the bottom, which stands across the depth of water and comes out of the surface to hold the superstructure or the platform. The substructure can be of steel tubular jacket construction or could be of pre-stressed concrete. The superstructure has modules which house drilling equipment, production equipment, power-generating sets, pumps, compressors, a gas flare stack, revolving cranes, survival craft, helicopter pad and living quarters with hotel and catering facilities. The load on the jacket depends on jacket length, and it can be very severe at some weather conditions. The weight of such a platform could go up to 40,000 tonnes.

Compliant towers consist of narrow, flexible legs or towers attached to a piled foundation supporting a conventional deck or platform for drilling and production operations. With the use of flex elements such as flex legs or axial tubes, resonance with waves is reduced and wave forces are de-amplified. Due to its flexibility, the compliant tower system is strong enough to withstand wind and sea, even hurricane conditions. Therefore, these can be used for conventional oil production from much greater depth up to 900 m.

A jack-up platform is a self-contained drilling rig on a floating barge fitted with long support legs which can be raised or lowered independently with respect to each other. Before installation at site, the barge holds the legs upright above water and it is towed to the desired location floating in water with the help of a supply vessel or tug. Upon arrival at site, all the three legs are jacked down to the seabed adjusting the legs such that the platform floats in normal position irrespective of the bottom bathymetry. Using the lowering/raising mechanism of the legs, the platform is slowly raised above water to a predetermined height. Then the platform is used for exploration or production. The jack-up rig is useful when the platform has to move from site to site for exploration or limited production. A jack-up platform can be used up to a depth of 100 m or even 130 m.

A semi-submersible platform consists of two or more cylindrical submerged hulls and a rectangular top deck or platform above water. The submerged hulls and the top platform are joined by four or six cylindrical pillars or legs. The entire semi-submersible platform remains afloat by weight and buoyancy balance in such a manner that the bottom hulls are fully submerged and the platform stays about 30–40 m above free surface. The draught of the semi-submersible platform can be adjusted by ballasting and de-ballasting the tanks. The semi-submersible platform can be moved from place to place by towing. It stays in position for drilling and production by a multi-point deep-water anchoring/mooring system and/or a dynamic positioning (DP) system such that yaw, sway and surge are limited to a minimum. The platform houses all the systems required for oil and gas production and accommodation for the crew.

A drill ship houses all equipment for drilling and production of oil and gas. Compared to a semi-submersible platform, a drill ship has large volume and, therefore, can carry large payload and also large amount of drill equipment. Being mobile, a drill ship can move from site to site and can work in remote areas. Being a floating platform, it must have station-keeping ability, which may include deep-water anchoring and/or DP system. It can operate in shallow as well as deep water of depth up to 1500 m.

A TLP is similar to a semi-submersible platform in construction, and it is held to the ocean bottom by means of vertical tethers fixed to the ocean bottom. The tethers are slightly less in length than what should have allowed weight and buoyancy equilibrium. The structure is pulled down slightly by the tethers, and the buoyancy force generated by the immersion of the semi-submersible structure is more than its weight. Thus, tethers are always in tension, which is equal to the difference between buoyancy and weight. This restricts the horizontal movement of the TLP in yaw, sway or surge mode. Also, it restricts heave motion of the platform in heavy weather. A TLP can be used for production in very deep water of up to 2 km depth.

A spar platform is a single cylindrical pillar-like structure standing vertically in water by being moored to the seabed with conventional deep-water mooring lines. It holds a production platform above water. A spar can be a conventional spar consisting of a single cylinder, or a truss spar, where the mid-section of the pillar is replaced by a truss structure, or a cell spar, where the pillar does not consist of a single cylinder, but a number of vertical cylinders of varying length grouped together to form the pillar. A spar is more economical to build as small- and medium-sized rigs than TLP and has more inherent stability than a TLP.

An offshore facility designed to operate in remote areas without constant presence of personnel is called an unmanned installation (UI). Normally, these are small, have a helipad on top and are easy to construct and maintain. These are cheap to operate as well. These are suitable for use in shallow water. It may be economical and convenient to use a number of UIs in an oil field rather than use a single expensive production platform such as a drill ship.

Along with production, if storage and offloading to other vessels and pipelines are also required, the system has to be bigger having adequate storage facility. Such a system is called a floating production, storage and offloading (FPSO) system. If the system is not used for production but only used for storage and offloading, it is called a floating storage and offloading (FSO) unit, and if it is only used for storage, it is called a floating storage unit. An FPSO system consists of floating tanks designed to take and store all the oil or gas produced from itself and/or nearby platforms or processes. From the FPSO system, oil or gas is offloaded onto waiting tankers, which shuttle between the shore and the FPSO unit, or sent through a system of undersea pipelines. FPSO systems, along with shuttle tankers, are effective in remote or deep-water locations where seabed pipelines are not cost effective. FPSO systems are economical in smaller oil fields which can be exhausted in a few years and do not justify the expense of installing a fixed oil platform. Once the field is depleted, the FPSO unit can be moved to a new location. Large tankers have been converted to work as FPSO or FSO systems. A new FPSO system similarly resembles a large tanker with all equipment required for drilling, production and processing.

1.2.4 Tourism, Recreation and Sports

We have already discussed passenger vessels such as tourist ferries and cruise vessels for tourism. Small and large boats used for tourism and recreation, normally available at tourist facilities, include paddle boats, motor boats, solar and sailing boats, small catamaran boats, etc. Single and catamaran rowing boat racing is a common recreation at many places where the boats are made hydrodynamically smooth to ply in water with least resistance. At seaside tourism facilities, water sport equipment are available mainly for individual sports, such as surfboards, water skis, wind surfing equipment, planing boats, water scooters and hydrofoils. The main attraction for recreation is yachting. Sailing yachts are long and slim finely made boats with facility for hoisting sails and are powered by wind. Sailing boat enthusiasts have yacht races across the seas regularly. Yachts fitted with motor propulsion are called motor yachts, which can be aesthetically very beautiful with luxurious living facilities inside. Based on facilities, these motor yachts can belong to commoners as well as to royalty and to the rich and famous.

Many high-speed vehicles are used for tourism and sports and also for patrolling by the navy, coastguard and coastal police. The speed of a water-borne vehicle is related to its length or size. Therefore, a high-speed vehicle does not represent high speed in absolute terms but is related to its Froude number F_n based on its speed V, length L or its volume of displacement ∇, which is defined as

$$F_n = \frac{V}{\sqrt{g \cdot L}} \quad \text{or} \quad \frac{V}{\sqrt{g \cdot \nabla^{1/3}}}$$

where g is the acceleration due to gravity, making F_n dimensionless. Normally, for a vessel with a steady forward speed, vessel weight is fully supported by buoyancy or the weight of the water displaced. Such a vessel is called a displacement craft. This happens with a vessel having an F_n up to 0.4. If a vessel is moved at a speed higher than this, the flow around boat bottom generates a vertical lift force thereby lifting a portion of the boat above water. Thus, the weight of the boat is supported by reduced buoyancy plus the hydrodynamic lift. Such a boat is called a semi-planing or semi-displacement craft. As the speed increases further, more lift is generated, thus lifting the boat higher and reducing buoyancy further. When almost the entire weight is supported by vertical lift, the buoyancy support being negligible, the boat is said to be fully planing and such a boat is called a planing craft. At planing or semi-planing condition, the drag to forward motion is proportionately less than that for a displacement craft since only a portion of the boat is in contact with water.

In a hydrofoil craft, the hydrodynamic lift supporting almost the entire weight of the craft at high speed is generated by hydrofoils, which are fully or partially submerged extensions of appendages fixed to the main hull, which is raised clear above water. It is well known that a wing with aerofoil sections can generate large lift force. If a hydrodynamically supported vessel can be fitted with wings above water, these generate lift (in air) and if these wings move close to water, the lift generated is higher than that generated if the wing was high above. If a high-speed vessel is fitted with such wings and if enough lift is generated so that the entire craft can come out of water completely, it can fly like an aircraft close to ground. Such a vehicle is known as wing-in-ground craft of WIG craft.

Another type of dynamically supported vehicle is the hovercraft where the entire weight is supported by a self-generated air cushion. If the air cushion is enclosed in a chamber, then the air-cushion vehicle (ACV) has no contact with the surface (land or water) over which it operates and, therefore, an ACV is amphibious. Another type of hovercraft, the surface effect ship, has air cushion below the craft surrounded by sidewalls and the water surface at the bottom. This is a water-borne vehicle and not amphibious.

When two hulls are rigidly connected so that a large above-water deck area can be obtained, it is a catamaran. Compared to a mono hull vessel with the same carrying capacity, a catamaran has much higher stability and, if properly designed, can have reduced power consumption also. However, the main advantage lies in the large deck area and higher over-deck loading because of higher stability. If in a catamaran, the underwater volume is pushed down, making the water plane small and narrow, we get a small waterplane area twin hull (SWATH) vessel. This does not have the advantages of resistance reduction, and so SWATH is not a high-speed vehicle. But it has the distinct advantage of less motion in a seaway due to the reduced waterplane. Similarly, if three or five hulls are joined together, the vessel becomes a trimaran or pentamaran, respectively.

Apart from snorkeling and scuba diving, underwater activities for tourism and recreation include movement in submarines carrying large number of tourists for underwater viewing and exploration. Submersibles are also used for carrying one or two persons under water. But the depth to which these can submerge is limited by the hydrostatic pressure the craft is designed to withstand. If the vessel can be unmanned and open, then its depth of submergence increases manifold. A remotely operated vehicle operates under water by getting commands from a mother ship on the surface connected by an umbilical cord. An autonomous underwater vehicle (AUV) is designed to work independently with robotic operations based on embedded commands. A mother ship may be required to launch and retrieve an AUV. Various underwater operations such as exploration and mapping, inspection and repair of underwater structures, search and retrieval of lost items and many such activities have given rise to the development of underwater robotics and communications.

1.2.5 Land–Sea Interface

The coast and the river mouths form the interface between sea and land. The coast provides beautiful beaches and also habitation where people depend on the sea for a living. Control of beach erosion, land reclamation for human activity, preservation of pristine as well as tourist beaches, provision of shelter against storm surges and tsunamis and development of profitable fishing activity are some of the coastal zone problems which must be tackled on a continuous basis.

One of the main interfaces providing integration of hinterland transportation with sea transportation is the seaport. On one side, seaports have road, rail or river way providing movement to hinterland and, on the other side, the sea route to distant land. Thus, ports also have to provide adequate handling arrangement for cargo, storage arrangement and logistic support for receiving and discharging cargo at both ends. Ports, therefore, must have adequate land infrastructure and wet infrastructure to handle vessels. This includes the approach channel, identification with buoys, adequate night navigation facility including lighthouses, the turning basin and the quays, berths, jetties, etc. Ports must have adequate logistic and management activity to handle a large number of ships and be profitable from the revenue earned. Ports also have to maintain a fleet of support ships for the maintenance of draught and services to ships. Based on the technology and economics of ship operation, ports could be built for specific cargo. Thus, we have container ports, oil terminals and bulk cargo-handling facilities.

Ports are also required to maintain the air and water environment in the port areas and nearby seas. It is well known that most of the oil spills from ships occur near the coast, particularly ports. So pollution prevention measures must be available quickly if any major pollution occurs. These include pollution-clearing vessels, oil booms and similar devices.

Shipbuilding and repair facilities, including wet basins and drydocks, are land and sea interfaces for ships and have similar facilities as ports but catering to the needs of ship and offshore construction and repair. Marinas are boat-parking places where boats of small and large sizes can be parked, maintained and repaired, allowing their owners to move from land to boat to sail out and come back at their convenience.

Large floating tourism facilities, floating airports and habitats at sea may come up in the near future in the near-shore areas having quick connectivity to land.

1.2.6 Support Services

The large number of engineering devices operating in the ocean require support from a number of smaller vessels. Tugs are required to pull or push a vessel into or out of the port, to bring a damaged vessel to port, for search-and-rescue operation, as well as firefighting and providing general services to ships at port. A tug must provide pull at low speed (identified as bollard pull at zero speed) and also good free running speed. Tug propulsion can be of different types suiting the operational requirement, and it must have good towing arrangement. Further, it must have large working area on deck and should be fit for river/harbour/escort or ocean-going operation. It can push a vessel at sea in a tug barge system or a flotilla of a number of barges in a river.

A supply vessel supports the offshore activities of a platform at sea near shore. It is designed like a tug since it must have pulling/pushing capability. It must also have a large deck area and large storage space for the carriage of supply to the platform and return goods from the platform.

Dredgers are a type of vessel required for capital dredging (new harbour, deepening of waterways), maintenance dredging (maintaining draught in ports and harbours, rivers and canals), land reclamation, harvesting material from seabed, deep-sea mining and removal of trash and debris. Dredgers can be dumb, standing at one location on legs or spuds so that the dredger hull is above water or floating and powered or submersible for mining operation. Dredgers can be purely mechanical for dredging and lifting dredged materials such as the backhoe, grab or bucket dredger, bed leveller dredger or water injection dredger. Sometimes during capital dredging, there is rocky surface, which is required to be cut and removed. This is done by a cutter suction dredger where the cutter is operated in the seabed by mechanical power and the dredged material is lifted by hydraulic suction. One of the efficient methods of dredging large areas with non-rocky surfaces is using a trailing suction hopper dredger where the vessel moves at low speed and the dredge and suction pipes trail behind and lift the dredged materials to the hoppers in the dredger by hydraulic power. The dredged materials can be disposed at a distance.

1.3 Design Definition and Marine Environment

Over the years, the subject of designing a product, process or system has grown into a discipline by itself. Marine vehicles and products, being large and complex, require following such a discipline. Kuo (1991) has compiled a number of definitions of engineering design from various sources. One of the definitions is from Rawson (1979) who appropriately describes the design process succinctly, 'Design is a creative iterative process serving a bounded objective'.

Engineering design can be executed for a product, process or system to satisfy a given need or objective. It should perform certain functions in which case the specified technical and other functional requirements are the objective. The performance criteria, efficiency and safety may be included in the objective. An example could be the design of a new car model or a new generation of aircraft. Sometimes a customer wishing to buy a product may not be able to define the objectives clearly. For example, the local authorities wishing to have a bridge over a river may not be able to define all objectives clearly such as loadings due to direct regular load and occasional environmental load, movement of traffic below the bridge, support system, funds available, maintenance requirement, etc. Similarly, a ship owner may desire to acquire a bulk carrier in a particular route, but he or she may not be aware of the exact size and speed of the vessel, which may depend on a number of techno-economic factors and market demand. Any product, process or system is for the benefit of a number of users. It is necessary to identify the market of such users and find out the buyers' needs to formulate the objectives. Similarly, design must also satisfy the sellers' needs, i.e. it must also consider the manufacturing process itself.

A number of design solutions can be found to satisfy the design objectives, but one of the solutions has to be selected. The selection criterion is the design which satisfies the objective best. Generally, the product, process or system under consideration is a complex system and the design objectives may be numerous. Thus, it becomes a multi-objective, multi-variate optimization process. The solution to such a problem may be mathematical optimization, which may not always be feasible or acceptable. In that case, the choice of a solution is based on decision making to satisfy conflicting requirements of different objectives.

Boundaries on design variables introduce discipline in the design process. The design is directed by the objective but is limited in scope by boundaries. Creative endeavour cannot be allowed to proceed in any direction if it is to be acceptable. It must be contained in areas which are judged to be permissible, feasible or sensible. Boundaries need to be prescribed within which the parameters of the problem can be manipulated to designer's advantage.

Design variables can be free or dependent variables. The bounds or constraints for the free variables can be fixed from the study of the physical world in which the product is to operate. The performance criterion in any one aspect of the product can be defined as a function of a number of variables. A number of such performance criteria may be conflicting in nature. Defining the lower and upper limits on such criteria is necessary so that a full design solution satisfying all design requirements may be found. Further, the boundaries on a design may be of varied nature such as technical, economic, environmental, political, geographical and industrial.

Creativity is the undisciplined element in the process of designing, which introduces uniqueness to a design. This activity is basically new and not a copy of past creations though it may draw upon the past knowledge gained. In the past, when design was based on intuition rather than scientific facts, it was perhaps more creative, having its own unique features and appeal. Today, when design activity is based on scientific analysis and facts, creations as distinctly different from previous products are rare. It must be appreciated that creations which, by definition, are different from previous ideas or products can be in any field of design such as a component, its operation, its maintenance, its economic viability or even the management of design activity itself. For creative ideas to emanate, it is necessary to have adequate knowledge of past products, processes or systems and also the knowledge to analyse the effects of any creative idea. The creative idea must be discussed with affected groups so that it can have appeal to the buyer.

When the design process starts, knowledge about the product, process or system is non-existent. It is necessary to start with some values assigned to the design variables based on past or statistical data. This generates further information based on scientific calculations. The results need not confirm to predefined performance criteria. Then it is necessary to change the design variables and start the design process again. As the design progresses and more information is generated, the same process of analysis and decision making is carried on during the entire design process. This process is an iterative process shown diagrammatically in Figure 1.1.

A marine vehicle or structure is a complex technology-intensive system operating in marine environment and satisfying various laws of physics and mechanics. The design of such a complex product would require understanding of the marine environment and loads imposed on the structure, including extreme loads, mechanics of floatation and stability both in intact and damaged conditions, structural integrity considering normal and extreme loads, corrosion and fatigue. Reliability and risk assessment, fluid structure interaction, motion response and control, station keeping, etc. are some other considerations. Also, design aspects must look at the comfort and safety of the personnel working in the harsh marine environment as well as reduction (if not elimination) in environmental pollution. In a competitive environment, it is necessary for the designer to ensure high design efficiency in terms of cost of production, cost of operation and ease of dismantling. Based on these facts, a more elaborate definition of design has been given by the

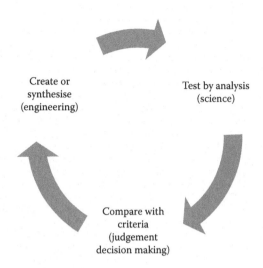

Create or
synthesise
(engineering)

Test by analysis
(science)

Compare with
criteria
(judgement
decision making)

FIGURE 1.1
Iterative design process.

Design Council (1991), Kuo (1991): 'Engineering design refers to that activity necessary to provide assurance of the efficiency, performance and safety of a product, project, process or system to satisfy a market need. It includes the management of that activity, and the necessary instructions for realization, maintenance and use'. Perhaps this definition can be marginally modified (additions shown within round brackets) to suit the present-day requirement: 'Engineering design refers to that activity necessary to provide assurance of the efficiency, performance and safety (and sustainability) of a product, project, process or system to satisfy a market need. It includes the management of that activity, and the necessary instructions for realization, maintenance and use (and dismantling)'. Thus, a good engineering design must efficiently satisfy the technical and economic (market need) requirements. It must also address the issues of production, operation, maintenance and dismantling. The sphere of design has become very large and encompasses different and varied skills and knowledge sets. Therefore, the management of this activity (knowledge, human resource and facility and equipment) also forms a part of design.

2

Marine Environment

Any design process must start with identifying and understanding the environment in which the vehicle, structure or engineering device/facility is to function. Such products operating in the marine environment include ships and marine transportation vehicles required for both trade and military purposes, underwater vehicles for exploration and war, structures for exploration and exploitation of living and nonliving resources, tourism vehicles and facilities, ports as interface between land and sea and coastal structures.

A coastal nation has the right to protect 12 nautical miles of its territorial waters. This gives coastal and island nations the right to develop defence and coast guard services along its coast and also the right to have port state control activities. The Law of the Sea Convention under the aegis of the United Nations has conferred the rights of economic exploitation of the coastal waters up to a distance of 200 nautical miles; this coastal area is known as the exclusive economic zone (EEZ). This may undergo change again from 200 nautical miles to the end of the continental shelf. In any case, the area for the economic exploitation of living and nonliving resources as well as energy available to a coastal country or island is very large, in many cases even larger than the land area. The area beyond the EEZ, the deep ocean, is considered the 'common heritage of humankind'. However, certain countries and a consortium of institutions have invested large funds in exploring resources in these regions, which have been given the pioneer investor status with an identified ocean area over which these investors will have exclusive right of exploration. Resources from such and other areas of the oceans will have to be shared by all nations through a mechanism being developed by the United Nations. This vast scope leads to the development of large engineering devices to be deployed in the marine environment. The first part of this chapter discusses the distribution of water mass around the world in the form of oceans, rivers and lakes. Then some important straits and waterways around the world which play a major role in ship movement and determination of trade routes are discussed. Also discussed are the different physical properties of water which affect the design of small and large products.

2.1 Oceans

Oceans cover a large portion of the earth. The subject of oceanography is devoted to understanding the science of oceans. It is neither possible nor necessary to describe ocean science in detail to a marine designer. However, the designer must have an overall idea of the marine environment so that the ocean properties can be utilized to the advantage of functioning of the designed product. The reader can get enough information on oceanography from various sources from Internet, lecture notes from the Internet and NASA websites. Some particular references McLellan (1965), Morgan (1990), Barltrop (1998), Lamb (2003a) and Talley et al. (2011) deal with oceanographic information applicable to marine vehicles and structures.

2.1.1 Ocean Bottom

Earth is a water planet, perhaps the only one of its kind in the universe known to human-kind. Water is essential for the existence of human beings and life in general as we know it. Though Earth looks like a sphere from space, its crust is far from a smooth spherical surface, going down to nearly 10 km below the mean sea level (MSL) and at places going to a height of 8 km or more above MSL. Figure 2.1 shows schematically the undulations on the sea floor on the surface of the earth.

Continents are formed on continental plates, which are formed and moved across the world over the earth's crust by a process known as tectonic activity. Today the continents on which we live have thus been formed over millions of years. The earth's crust is broadly divided into two parts, namely continental crust and oceanic crust. The continental crust extends up to the continental margin, which is a region comprising the continental shelf, continental slope and continental rise. Hence the continental margin is the extension of the continental crust into the oceans. Towards the edges of the landmass, the continental plate is relatively flat and gently slopes into the oceans. This ocean-covered extension of a continent is called the continental shelf. Seaward, the shelf ends abruptly at the shelf break, the boundary which separates the shelf from the continental slope. The shelf occupies an area of approximately 7% of the total ocean area. The average slope of the continental shelf is 1.9 m/km. The average depth of the continental shelf is about 135 m with a maximum depth of 350 m near Antarctica. The distance of the end of the continental shelf from the shoreline is, on an average, 70 km. But the continental shelf can also extend to a large distance, sometimes up to 1300 km. At the end of the continental shelf, the ground suddenly goes down, which is known as the continental slope, and then the continental rise to the oceanic crust at large depth, which is known as the abyssal plain. The abyssal plain is not

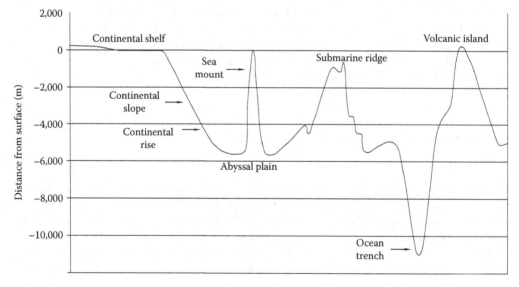

FIGURE 2.1
Undulations of the sea floor.

a plane surface as the name indicates but has various sharp as well as gentle rises known as sea mounts, submarine ridges and dormant or active volcanic rises. Some of these rise above the sea surface forming islands. Sometimes islands are also formed when the surface near the continental landmass rises due to erosion and geological activity. Thus, the oceanic region can be broadly divided into three zones: interface with the shoreline (coastal zone), continental shelf to abyssal plain (shallow) and deep ocean and abyssal plain (deep ocean).

2.1.2 World Water Resources

The North and South Americas extend from near the North Pole to the South Pole. Similarly, Asia and Europe in the northern hemisphere join Africa near the equator, and the African landmass extends down to the north of the Antarctic landmass. These landmasses also demarcate the three major oceans: the Pacific, the Atlantic and the Indian Oceans. In early days, to sail from the East African coast to the western side, one had to circumnavigate around the southern tip of Africa. Though Cape of Good Hope is conventionally known as the southern tip of Africa, Cape of Agulhas is the actual southern tip of Africa, which is about 150 km east of Cape of Good Hope. Cape Agulhas is also the point where the Atlantic and the Indian Oceans meet. To the south of Cape Agulhas is the Southern Ocean, which runs around the southern tip of Cape Horn, the southernmost point of the South American landmass. In early days, sailing from the east coast of North America to the west coast of America meant circumnavigating around Cape Horn where the Pacific and the Atlantic oceans meet.

The total oceanic water mass, which is saline, is divided into five oceans: the Pacific, the Atlantic, the Indian, the Arctic and the Southern Oceans. Apart from the major oceans, there are some saltwater sources such as lakes connected to the sea, bays and seas outside the demarcated ocean areas such as the Mediterranean Sea, Persian Gulf, Baltic Sea, Red Sea, Hudson Bay, Bering Sea, etc.

The landmass of the world has an undulating surface with depressions formed during the continent formation and due to geological processes such as water run-offs and erosion and elevations forming hills and mountains and mountain ranges. The freshwater sources of the world include freshwater lakes formed at the depressions of the landmass. The five Great Lakes of North America – Lakes Superior, Michigan, Huron, Erie and Ontario – situated on the borders of the United States and Canada form the major transportation route from the north central regions to the east coast of the North Americas. There are many small and big freshwater lakes distributed in all continents, including the Caspian Sea in Asia. Rivers are the other source of fresh water, which carry large amounts of sediments to the sea. Large and small rivers distributed all around the world landmass are used for transporting goods and passengers. Rivers have a winding passage from a height to the sea level where they meet the sea. The other major source of fresh water is the ice caps and glaciers in the North Sea near the North Pole and in Antarctica near the South Pole. If the water locked up in the polar ice and terrestrial glaciers were to completely melt, the oceans would rise about 240 ft above its present level. Fresh water is also stored in underground aquifers and present in the atmosphere as water vapour and clouds. Table 2.1 gives the areas of the world covered with sea water, fresh water and land. Figure 2.2 gives the same information in a comparative polar diagram.

Most of the landmass on the earth is distributed in the northern hemisphere, and so most of the water mass is distributed in the southern hemisphere. Figure 2.3 (Lamb 2003a) shows the distribution of ocean surface as a function of northern and southern latitudes. It can be seen that at the southernmost point, the earth is covered by the continent of

TABLE 2.1

Areas of Land and Oceans

Item	Area (km²)	Area (%)
Pacific Ocean	155,557,000	30.5
Atlantic Ocean	76,762,000	15.0
Indian Ocean	68,556,000	13.4
Arctic Ocean	14,056,000	2.8
Southern Ocean	20,327,000	4.0
Other saltwater areas	15,318,000	3.0
Freshwater area	11,042,700	2.2
Total ocean area	335,258,000	65.7
Total land area	148,647,000	29.1
Total surface area	510,265,700	100.0

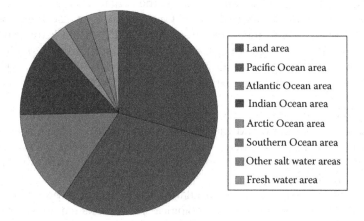

FIGURE 2.2
Distribution of land and water on the surface of the earth.

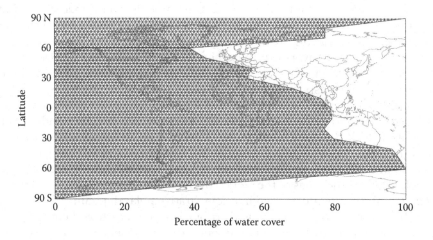

FIGURE 2.3
Distribution of water cover as function of latitude.

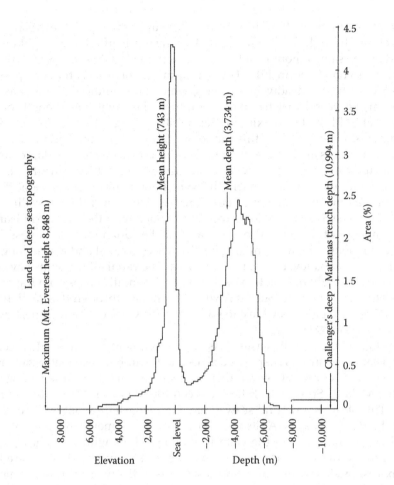

FIGURE 2.4
Distribution of the earth's surface above and below the mean sea level.

Antarctica with no water cover, whereas in the northernmost point, there is no landmass and the water cover is 100%.

As has been already discussed, the earth's crust does not make a perfect sphere. The deepest point in the oceans is Challenger's Deep in Marianas Trench in the eastern Pacific Ocean with a depth between 10,898 and 10,916 m, whereas the average depth of the ocean is about 3734 m. Similarly, the highest point on land is Mount Everest, which is 8850 m above MSL, whereas the average height of land is 743 m. The distribution of land above MSL and the distribution of ocean depth below MSL is shown in Figure 2.4 taken from Talley et al. (2011). If the earth, including its water, were made into a perfect sphere, the landmass would be 2440 m below the sea level.

2.1.3 Straits and Waterways

To reduce the distance of travel from the east to the west coast of the Americas, the Panama Canal was built in Panama, which separates the Caribbean Seas of the Pacific and the Atlantic Oceans. The Panama Canal is an engineering marvel and is known as one the seven modern wonders. It is 77.1 km long, consists of three sets of lock systems, Lakes

Miraflores and Gatun as well as rivers and waterways which join the Atlantic with the Pacific. Travelling through the Panama Canal, each ship is lifted to a height of 26 m and also lowered by the same amount. Until 2014 more than 900,000 ships passed through the canal since it was completed in 1914. The optimum size of vessels that can pass through the canal is 50,000–90,000 deadweight tonnes, which are commonly known as Panamax vessels. The largest vessel that has passed through the canal has a length of 296.57 m and a breadth of 32.31 m. The maximum permissible draught through the canal is 13 m. Vessels of dimensions larger than this cannot go through the canal and must go around Cape Horn. Such vessels are bulk carriers, tankers and containerships of large dimensions. There is a plan to have new, bigger and separate lock systems which will allow large 12,000 TEU carrying container vessels with length 366 m, breadth 49 m and draught 15 m.

A similar human made canal is the Suez Canal, which allows ships travelling between Asia and Europe. The Suez Canal joins the Gulf of Suez with the Mediterranean Sea and between landmasses of the Sinai Peninsula and the Egyptian main landmass. The main canal is 162.25 km long with a 22 km long northern access canal and a 9 km long southern access canal. This is a sea-level canal with no locks. The canal allows ships having a beam up to 77.5 m and a draught of 20 m to pass through it. Normally, ships carrying deadweight up to 240,000 tonnes, known as Suezmax vessels, can pass through the canal. If the canal is closed or ships are bigger, they have to go around the Cape of Good Hope, travelling an extra distance of about 4850 km.

The sea distance between the eastern Asian subcontinent and Australian continent is strewn with islands of various shapes and sizes and the in-between waterways are treacherous with regard to depth. However, there is a safe waterway for the passage of ships, known as the Malacca Strait, which lies between the Malayan Peninsula and the island of Sumatra. This strait is a natural seaway having a length of about 805 km, joining the Indian and the Pacific Oceans. At its narrowest location, somewhere south of Singapore, it is 2.5 km wide. The depth of the waterway allows ships with draught up to 25 m to pass safely through the strait. The vessels that can pass through this strait are known as Malaccamax vessels. This is one of the busiest straits of the world through which around a quarter of the international oil cargo passes, going from the Middle East to the Far East.

Though not a main waterway, the Bering Strait, which is the waterway between the extreme east of Russia and the extreme west of Alaska, has a width of 85 km with a varying depth between 30 and 49 m. It joins the Chukchi Sea in the Arctic Ocean and the Bering Sea on the south, which is a part of the Pacific Ocean. This is not a main shipping route. However, if a ship negotiates the Bering Strait, it has to be of adequate strength to negotiate ice on its path. Similarly, there are a number of straits and canals across the world which have been used as passages for large ships and small boats, reducing the distances of travel and cost. One such strait is the Palk Strait, which is a strait between Tamil Nadu in India and the Mannar District of the Northern Province of the island nation of Sri Lanka. It connects the Bay of Bengal in the north-east with the Palk Bay and thence with the Gulf of Mannar in the south-west. The strait is 53–80 km wide. It is studded at its southern end with a chain of low islands and reef shoals, which are collectively called Adam's Bridge. This chain extends between Dhanushkodi on Pamban Island in Tamil Nadu and Mannar Island in Sri Lanka. Because of this, ocean-going vessels do not pass through this strait. Only small passenger and fishing boats negotiate this strait.

The St. Lawrence Seaway, the river way system consisting of a system of locks, canals, channels and the St. Lawrence River, permits ocean-going vessels to travel from the Atlantic Ocean on the east coast of North America to the Great Lakes, as far inland as the western end of Lake Superior on the borders of the United States and Canada. The riverine

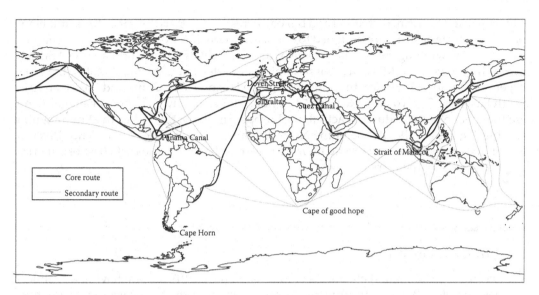

FIGURE 2.5
Major sea trade routes of the world.

section consists of several stretches of navigable channels, a number of locks, as well as canals along the banks of the St. Lawrence River to bypass several rapids and dams along the way. There are 15 locks on the river, lifting the vessels to a height of more than 180 m over the entire stretch of the seaway. The size of vessels which can traverse the seaway is limited by the size of locks. The maximum allowed vessel size is 225.6 m long, 23.8 m wide and 8.1 m draught. Vessels designed for the Great Lakes following the opening of the seaway are informally known as Seawaymax vessels. This is one of the main trade routes of the United States connecting the hinterland to the Atlantic Ocean and carrying bulk, general and grain cargo and further along the American east coast as well to other countries. Figure 2.5 shows the major trade routes of the world. Note the importance of canals discussed earlier (Rodrigue 2013).

2.1.4 Freshwater Resources

Apart from the seas, there are a number of freshwater lakes in the world. These lakes are not only freshwater reservoirs but also a medium of transportation, tourism and recreation. The Caspian Sea is the largest lake in the world, having a total area of 371,000 km² and located on the borders of Asia and Europe. The five Great Lakes in North America are Lake Superior (82,414 km²), Lake Huron (59,600 km²), Lake Erie (25,719 km²), and Lake Ontario (19,477 km²) bordering the United States and Canada and Lake Michigan (58,000 km²) in the United States. The large lakes of Africa include Lake Victoria (69,485 km²), Lake Tanganayika (32,893 km²) and Lake Malawi (30,044 km²). Lake Baikal of Russia has a surface area of 31,500 km² and Lake Great Bear of Canada has a surface area of 31,080 km².

The countries of the world have many rivers flowing through them carrying a large amount of fresh water and sediments regularly to the sea. This creates large delta regions where these rivers meet the sea giving rise to heritage mud flats, mangroves and a variety of flora and fauna. Though rivers carry fresh water, the river mouths have a mixture of sea water and fresh water fluctuating with river currents and tidal variations. Apart from irrigation, all the rivers of the world have been used for cargo transportation, passenger movement and tourism, recreation and biological resource generation since time immemorial. Any vehicle or structure designed to operate in a river has to work primarily in shallow water and the winding river routes. Some of the major rivers in the world are the Amazon in South America, the Missouri–Mississippi river system in the United States, the Nile, Niger and Zambezi in Africa, the Yangtze and the Huang Ho in China and the Mekong in South East Asia. The Volga in Russia and Danube and Rhine rivers in European mainland are the lifeline of the industrial activity of Europe. Similarly, the Indus, the Brahmaputra and the Ganges are the lifeline of the North Indian peninsula. The St. Lawrence River in the United States is the trade route to bring hinterland cargo to the east coast of America and then to rest of the world.

2.2 Properties of Water

Water is the most abundant chemical compound on the earth's surface having a chemical composition of H_2O in its purest form. At room temperature, it is a tasteless and odourless liquid, nearly colourless with a hint of blue. Many substances dissolve in water. Water is miscible with many liquids forming a single homogeneous liquid. On the other hand, water and most oils are immiscible, usually forming layers according to increasing density from the top. Water is commonly referred as the universal solvent. Because of this, water in nature and in use is rarely pure and some of its properties vary slightly from those of pure water.

2.2.1 Physical Properties

The heat retention capacity of water is very high. Table 2.2 gives the specific heat values of water at atmospheric pressure or at 101.325 kPa. The specific heat capacity of ice at −10°C is about 2.05 J/g·K and that of steam at 100°C is about 2.080 J/g·K. Both these values are nearly half of the heat retention capacity of water. Water also has a high latent heat of vaporization (257 J/g) at the normal boiling point. The water mass covering the earth's surface, with

TABLE 2.2

Some Physical Properties of Water

Temperature (°C)	Specific Heat Capacity of Water (J/g·°C at 100 kPa)	Vapour Pressure of Water (kPa)	Density of Fresh Water (kg/m³)	Density of Sea Water (3.5% Salt) (kg/m³)	Kinematic Coefficient of Viscosity-FW (m²/s * 10⁶)	Kinematic Coefficient of Viscosity-SW (m²/s * 10⁶)
0	4.22	0.646	999.83	1028	1.78667	1.82844
4	4.21	0.813	999.97	1027.7	1.56557	1.6094
5	4.2	0.872	999.93	1027.6	1.51698	1.56142
10	4.19	1.252	999.70	1026.9	1.30641	1.35383
15	4.19	1.710	999.10	1025.9	1.13902	1.18831
20	4.19	2.306	998.21	1024.7	1.00374	1.05372
22	4.18	2.911	997.77	1024.1	1.95682	1.00678
25	4.18	3.005	997.05	1023.2	1.89292	1.94252
30	4.18	4.186	995.65	1021.7	1.80091	1.84931

such properties of heat retention, allows moderation of the earth's climate by buffering large fluctuations in temperature.

Water has a high surface tension of 72.8 mN/m at room temperature, the highest of the common nonionic, nonmetallic liquids. This can be seen when small quantities of water are placed on a Teflon-coated surface; water stays together as drops. Similarly, due to disturbance on water surface, air is trapped to form air bubbles in water. Another surface tension effect is capillary waves, which are the surface ripples that form around the impacts of drops on water surfaces and sometimes occur with strong subsurface currents flowing to the water surface. The apparent elasticity caused by surface tension drives the waves. Due to the interplay of the forces of adhesion and surface tension, water exhibits capillary action whereby water rises into a narrow tube against the force of gravity. Water adheres to the inside wall of the tube and surface tension tends to straighten the surface causing a surface rise and more water is pulled up through cohesion. The process continues as the water flows up the tube until there is enough water such that gravity balances the adhesive force. Some major properties of sea water, namely density, temperature, salinity, electromagnetic and acoustic propagation characteristics affect the operation of marine vehicles and structures at sea.

2.2.2 Density

The density of water is approximately 1 g/cm^3 at room temperature, and it varies with temperature. When cooled from room temperature, liquid water becomes increasingly dense, as with other substances, but at approximately 4°C (39°F), pure water reaches its maximum density. As it is cooled further, it expands to become less dense. The solid form of most substances is denser than the liquid phase; thus, a block of most solids will sink in its liquid. However, a block of ice floats in liquid water because ice is less dense. Upon freezing, the density of water decreases by about 9% or the volume of ice is about 9% more than the corresponding volume of water. Thus, a block of ice floats in water with about 9% of its volume being above water. The density of water also changes, i.e. increases with salinity. Normal sea water has a salt content of about 35 parts/1000 by weight or 3.5% (=35‰) and its density at room temperature is approximately 1.025 g/cm^3. Table 2.2 gives the density of fresh water and sea water between 0°C and 30°C. Thus, when a ship or boat moves from fresh water to sea water, the ship gradually moves from liquid of lower density to higher density and, following the laws of floatation, it experiences a slight reduction in draught. Similar to density, the viscosity of water also changes with temperature and salinity. Table 2.2 also gives the kinematic coefficient of viscosity of fresh water and sea water at various temperatures.

Water is a poor conductor of heat. Therefore, the solar heat received by the surface of water does not percolate down to large depth. In freshwater lakes, water becomes very cold as depth increases. But since fresh water is densest at 4°C, bottom water can reach that temperature. If water becomes cooler, say 3°C or 2°C or 1°C or 0°C, being lighter, it does not stay at the bottom, but rises up. If the surface temperature falls below the freezing point, ice is formed and it floats in water on the surface. Since water is a poor conductor of heat, the lower layers of water retain the heat and the entire water column does not freeze and the temperature at the bottom does not fall below 4°C. With an increase in the salinity of water, the temperature of maximum density reduces further below 4°C and also the freezing point below 0°C. At a salinity of 24.7‰, the temperature of maximum density and the freezing temperature coincide. At a salinity of 35‰, the freezing point is at approximately –1.9°C, which is higher than the temperature of maximum density. This would mean that if sea water was to freeze, it would become denser than sea water itself and sink. This does not happen because of a process known as 'brine rejection'. Thus, as temperature reaches the freezing point of fresh water, salt precipitates out of sea water and ice is formed out of fresh water, which floats to the top. Thus, the ice formed is that of fresh water with densely saline water just below the ice. Being heavier, this water sinks to the bottom. Thus, density increases rapidly with depth for about 1 km from the surface, and this is known as pycnocline. Sea water density remains nearly constant beyond this depth, as shown in Figure 2.7c.

2.2.3 Temperature Distribution in the Oceans

The temperature distribution on the ocean surface varies based on the amount of heat received due to solar radiation. Accordingly, the climatic zones have been defined based on the temperature distribution on the ocean surface. The polar region gets the least heat and is known as the polar or frigid zone, which is normally to the north of latitude 66.33°N and to the south of 66.33°S. Here the temperature goes below 10°C up to −2°C or less. The temperate climatic zone lies roughly between 33° and 66.33° latitude in both northern and southern hemispheres. This climatic zone is biologically the most productive zone where temperature varies between 10°C and 18°C. In this zone, there are warmer regions known as warm temperate zones such as the Mediterranean Sea where temperature is somewhat warmer. The equatorial zone between the Tropic of Cancer (23.27°N) and the Tropic of Capricorn (23.27°S) is known as the tropical zone, where the ocean surface temperature is above 18°C, sometimes going up to as high as 35°C. The area between 23.27° and about 33°, north and south, is known as subtropical region. Figure 2.6 gives the National Oceanic

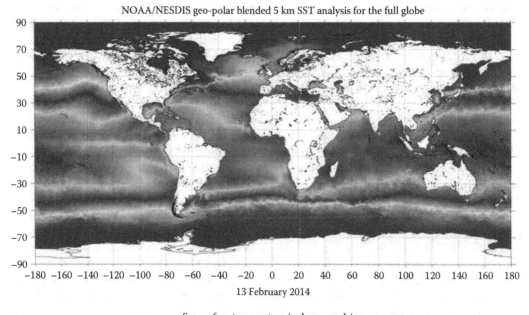

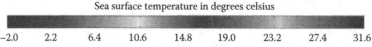

Sea surface temperature in degrees celsius

FIGURE 2.6
Temperature distribution over the ocean surface in February 2014.

and Atmospheric Administration (NOAA)/National Environmental Satellite, Data, and Information Service geo-polar picture showing the ocean temperature distribution in February 2014 over the entire earth. A vessel moving from one climatic zone to another experiences change in draught, and its equipment, machinery and structure must be suitable for the different temperatures in the waters, particularly if it moves to polar regions.

Oceans are a great source of biodiversity. Ecological development, both geological and biological, varies from place to place depending on local conditions. If the ecological balance is disturbed, the damage caused may be irreversible. The International Maritime Organization (IMO) issues suggestions and recommendations for ocean structures and vehicles to reduce environmental degradation due to oceanic engineering activities, particularly with respect to ships. There are Particularly Sensitive Sea Areas (PSSAs) where ecological damage may cause irreparable harm. Such areas have been identified by IMO, and very strict and elaborate rules regarding environmental protection are adhered to in these areas. With regard to shipping, some important PSSAs are the Western European waters, the Wadden Sea and the Baltic Sea.

Solar heat gets absorbed by the first few centimetres of water in the vast oceans during the day. During night, this water cools by losing some heat to the atmosphere. Due to the wave action of surface currents, the top layer mixes with slightly deeper layer and thus heat gets uniformly distributed over this mixed layer, which may be about 100 m deep. Then the temperature of water drops sharply with depth until about 1000 m, and this drop is known as thermocline. Below this depth, the temperature remains nearly constant, dropping only slowly until about 0°C to 4°C at the bottom, as shown in Figure 2.7a. Saline water does not freeze until this temperature. The existence of thermocline and its depth depend on the surface temperature, wave and current action and the depth of the sea. Sometimes in polar regions, thermocline is nearly absent since the surface temperature is very low and ice sheet works as an insulating layer.

2.2.4 Transmission of Electromagnetic Radiation in Water

Water absorbs electromagnetic radiation of almost the entire range of frequencies, right from infrared to ultraviolet rays, including visible range of light rays, radio waves and very short wavelength gamma and other cosmic rays. Therefore, although communication through electromagnetic waves is most effective outside water, the same is not possible under water. Light energy is rapidly absorbed by water but not as rapidly as other higher frequency electromagnetic waves. Light waves penetrate up to a few metres of water depending on intensity. Thus, there is some visibility in clear water up to a limited depth. Suspended particles in water (such as turbid water and river or coastal waters) reflect light waves in multiple directions, a phenomenon known as scattering. Light also gets refracted while entering water from air, and the refraction changes with salinity. Thus, light penetration may be drastically reduced in water with a large amount of suspended particles. Beyond this depth, the water world is extremely dark with no visibility.

2.2.5 Salinity

Over millions of years, rains, rivers and streams have washed rocks containing various types of salts into the sea. Undersea volcanoes and hydrothermal vents have added to this salt content. Also solar radiation evaporates water from the ocean surface, leaving salt behind. Thus, over millions of years, the seas have developed a rather noticeable salty taste.

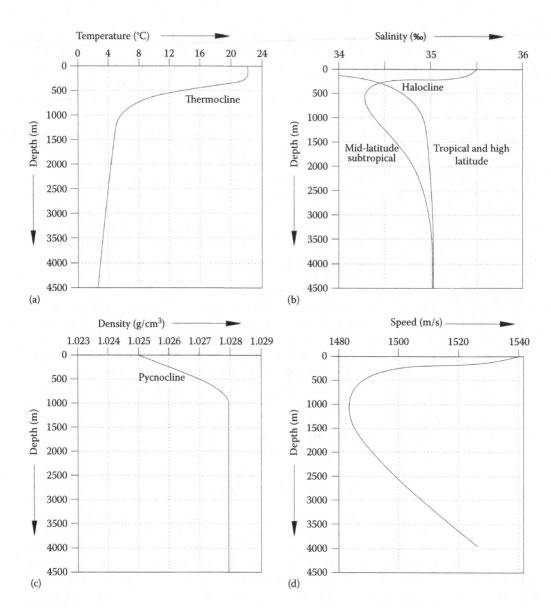

FIGURE 2.7
Distribution of (a) temperature, (b) salinity, (c) density and (d) speed of sound in water across depth.

Salinity is measured as grams of salt in 1000 g of water or ppt expressed as ‰. The salinity of sea water varies between 32 and 37, the average being 35. Freshwater run-offs into the sea may affect salt content sometimes. Thus, Black Sea is diluted with fresh water to such an extent that salinity is only 16‰. Water with salinity less that 1.5 ppt is considered fresh water. Brackish water, typically at river mouths and estuaries, has a salinity varying between 1.5 and 17 ppt.

Surface salinity varies from location to location and is affected by river run-offs, water evaporation and absorption and rain in the tropics as well as freshwater injection due to melting of ice in the polar regions. Salinity also gets affected by ocean currents, particularly thermocline circulation. Thus, salinity variation in the top layer of depth 500–1000 m could show an increasing or decreasing trend. This variation of salinity in the top layer is known as halocline. Beyond this depth, salinity rises very slowly with increasing depths, remaining more or less constant at about 35‰. Figure 2.7b shows two typical salinity profiles across the ocean depth.

2.2.6 Sound Properties in Water

Sound wave propagation and, therefore, transmission of acoustic energy are fundamentally different from electromagnetic waves. Sound waves are longitudinal waves and get transmitted by the vibration of adjacent molecules. Therefore, sound travels through a medium and cannot travel in vacuum. The denser the medium, the faster the sound travels in the medium. Sound travels nearly four times faster in water compared with air. The speed of sound in dry air at 15°C is 343 m/s or 1234 km/h, whereas in sea water near the surface, it is about 1500 m/s. The speed of sound decreases with a decrease in temperature and salinity and increases with an increase in pressure. There is a linear increase in hydrostatic pressure as water depth increases. Noting the temperature and salinity changes shown in Figure 2.7a and b, one can conclude that sound speed will decrease up to a depth of about 1 km and then increase as hydrostatic pressure goes on increasing. The variation of sound speed with depth is shown in Figure 2.7d.

Sound waves are transmitted through water with little loss of intensity, but they are refracted as they move through water of varying density and they are reflected at interfaces between layers of different densities. Because of this play of sound in water, there could be a shadow zone at mid-depth where there is no further refraction of sound to higher depth such that no sound sent from a surface location can penetrate this zone. But if the sound source is located at this depth where sound speed is minimum, say about 1 km under water, the emitted sound waves are refracted back and forth within a narrow channel known as 'sound fixing and ranging (SOFAR)'. Sound can travel great distances within this sound channel with very little loss in intensity. Many marine species utilize these properties for communication by acoustic means.

2.3 Atmosphere

The earth is covered by a thick layer of atmosphere, the lowest layer being commonly known as air. Air stays in contact with the solid surface as well as the fluid surface of the oceans. It rotates and revolves around the sun along with the earth. It does not escape to space due to the gravitational force of the earth. Theoretically, the atmosphere extends to about 10,000 km above the earth's surface and the entire atmosphere is divided into five layers based on the density and temperature of the air. The lowest layer, where our interest lies in the present case, is known as the troposphere, which is up to about 10 km high and is covered with a layer known as stratosphere, then mesosphere, thermosphere and finally exosphere. The atmospheric temperature and density normally reduce as height increases. The lower layer of the atmospheric air, which is in contact with the earth's surface, has a composition of 78.09% nitrogen, 20.95% oxygen, 0.93% argon, 0.03% carbon dioxide and minute quantities of other gases. Apart from these, air also contains dust, pollen, sea spray, volcanic ash and industrial pollutants. Moisture content varies in air. Generally, cold air is dry and warm air can retain moisture to a large extent. The composition of atmosphere is such that it maintains a temperature range between about −10°C at the poles and 45°C in equatorial region in summer, the average temperature being 15°C.

The atmosphere contains, on an average, about 0.3 ppm of ozone (O_3), which is beneficial to life form. In the lower layer of the stratosphere about 20–30 km above the earth, there is an ozone layer, which contains about 10 ppm of ozone. This layer absorbs the ultra-violet rays of the sun and protects life from their harmful effects such as occurrence of skin cancer, skin-related diseases, cataract and effects on immune suppression systems. Excessive emission of chlorofluorocarbons, halons and other ozone-depleting substances has reduced the concentration of ozone in the stratosphere, creating an ozone hole. The hole is particularly visible in the stratosphere above the South Pole. Industrial pollutants can cause irreparable harm to the atmosphere, causing immense harm to the life on earth. Increase in carbon dioxide increases the earth's temperature, causing what is commonly known as the greenhouse effect. This can melt the ice caps and increase the water level. Similarly, harmful gases such as sulphur dioxide cause acid rains.

The density of dry air at 15°C is 1.217 kg/m³. The density reduces as height increases. The atmospheric pressure on the surface of the earth is normally denoted by 1 atm where

$$1 \text{ atm} = 101.3 \text{ kPa} = 1.013 \text{ bar} = 760 \text{ torr}$$

The atmospheric air is not static but blows over the surface at a certain speed. Due to the interplay of temperature, pressure and many other factors, air may blow at a slow speed, when it is known as wind or light breeze, or at a high speed, when it is known as storms or cyclones. To characterize the wind, an internationally accepted wind scale known as the Beaufort scale has been devised. Table 2.3 gives the wind speed against the corresponding Beaufort scale. Near the surface of water, wind speed reduces due to the friction between air and water. The relative velocity of air at the water surface is nearly zero, and this speed increases rapidly as height increases, soon to reach its full value. The wind velocity V at height h is given as $V/V_0 = (h/h_0)^{0.2}$, where h_0 and V_0 refer to any standard measured values.

TABLE 2.3

Beaufort Scale for Wind Speed

Beaufort Number	Description	Mean Wind Speed (knots)	Mean Wind Speed (m/s)	Probable Mean Wave Height (m)
0	Calm	<1	0.5	1.0
1	Light air	1–3	0.5–1.7	1.1
2	Light breeze	4–6	1.8–3.3	1.2
3	Gentle breeze	7–10	3.4–5.4	1.6
4	Moderate breeze	11–16	5.5–8.4	1.0
5	Fresh breeze	17–21	8.5–11	2.0
6	Strong breeze	22–27	11.1–14.1	3.0
7	Near gale	28–33	14.2–17.2	4.0
8	Gale	34–40	17.3–21.8	5.5
9	Strong gale	41–47	21.9–24.4	7.0
10	Storm	48–55	24.5–28.5	9.0
11	Violent storm	56–63	28.6–32.6	11.5
12	Hurricane	>63	>32.6	14.0

2.3.1 Coriolis Effect

Due to the earth's rotation, each particle forming the earth system – the land, sea and atmosphere – move at the rate of one rotation in 24 h, which works out to an angular velocity (ω) of 7.29 × 10^{-5}/s. The tangential velocity of a particle on the surface of the earth is, therefore, $2\pi\omega r$, where r is the distance of the point from the axis of rotation at that latitude. Thus, the tangential or transverse velocity or the velocity of a particle along the latitude will be the highest at the equator and will reduce to zero at the poles. Thus, the transverse velocity at the equator is approximately 1600 km/h, whereas it is 1400 km/h at 30° latitude and 800 km/h at 60° latitude. A body moving on the surface of the earth from the equator towards the North Pole starts with an inertial transverse velocity of 1600 km/h. As it moves north, the transverse velocity should reduce; therefore, it will have a tendency to move towards left or westwards. Similarly, in the south also, it will have a tendency to move westwards. This is known as the Coriolis effect (Lamb 2003). The Coriolis force is zero at the equator and increases as one move towards the poles.

2.3.2 Atmosphere Circulation

Of all the heat directed towards the earth by solar radiation, about 30% is lost to space by reflection and scattering. About 50% is absorbed by the surface and 20% by the atmosphere. This heat absorbed by the earth must be given back to space through radiation from the earth to maintain the heat budget. The equatorial region gets maximum solar heat, and the polar regions get the least heat due to its inclination from the direct path of radiation. To maintain heat balance, it is necessary that this uneven receipt of heat must be distributed across the earth's surface so that heat can be radiated back from the earth more uniformly. The amount of heat received equals the heat radiated back more or less at latitudes 38° north and south. The distribution of heat over the earth's surface is achieved through atmospheric and ocean circulation.

Due to the intense radiation from the earth's surface, the atmosphere over the equatorial region heats up and rises upwards. The ocean surface also heats up and water evaporates mixing with hot air, which can retain a large amount of water vapour. As the air rises it cools, and the water vapour condenses forming water droplets in the form of clouds. The heat released due to the latent heat of condensation causes secondary heating of atmosphere, rising up to cool at a height. The cooled dry air then moves polewards, to the North Pole in the northern hemisphere and to the South Pole in the southern hemisphere. As it moves away from the equator, it comes down and at about 30° latitude, it starts moving along the surface of the ocean, partially back towards the equator and partially polewards. Thus, the cell is completed so that the air rising up at the equator returns back to the equator to rise again. This is called the Hadley cell. Due to Coriolis effect, the atmosphere moving towards the equator near the ocean surface has a tendency to move from east to west generating the so-called easterlies in the zone near the equator in both the northern and southern hemispheres. Similar situation occurs again at about 60° latitude in both the northern and southern hemispheres. The atmosphere rises up due to heat dissipation from the earth and the oceans taking water vapour with itself. As it rises up, it cools, clouds are formed, secondary heating of atmosphere takes place and the air moves polewards, and from the poles it moves back towards the 60° latitude along the ocean surface. Similar to Hadley cell, the air at the surface level moves from east to west causing the polar easterlies. Between 30° and 60° latitudes, the atmosphere forms the so-called Ferrel cell where the atmosphere at the surface level moves polewards, rises up at about 60° latitude and moves

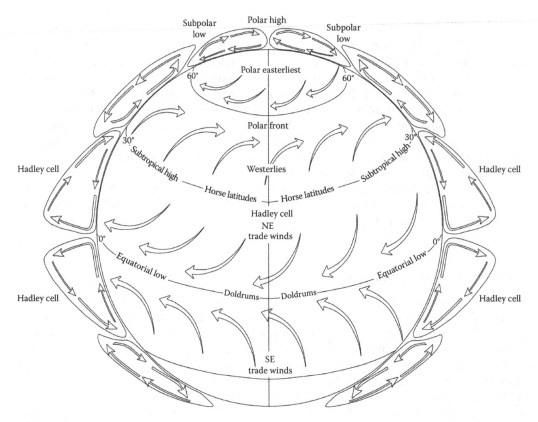

FIGURE 2.8
Atmospheric movement of trade winds.

towards the equator to come down at 30° latitude. Due to the same Coriolis effect, this surface air moves from west to east as it goes polewards, thus causing the westerly trade winds. Figure 2.8 shows the atmospheric movement diagrammatically.

2.4 Ocean Circulation

Wind blowing across the ocean surface moves the surface waters due to a frictional drag on the surface where ripples or waves create the surface roughness necessary for the wind to couple with the surface waters. A rule of thumb is that the wind blowing steadily over deep water for 12 h at an average speed of about 100 cm/s would produce a 2 cm/s current (about 2% of the wind speed). In an imaginary stationary earth, frictional coupling between moving air and the ocean surface would push a thin layer of water in the same direction as the wind. This surface layer, in turn, would drag the layer beneath it, putting it into motion. This interaction would propagate downwards through successive ocean layers. However, because the earth rotates, the shallow layer of surface water set in motion by the wind is deflected due to the Coriolis effect, which is zero near the equator and increases towards the poles.

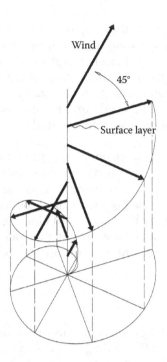

FIGURE 2.9
Ekman spiral.

2.4.1 Ekman Spiral

In an ideal case, a steady wind blowing across an ocean of unlimited depth and extent causes surface waters to move at an angle of 45° to the right of the wind in the northern hemisphere (45° to the left in the southern hemisphere). Each successive lower layer moves more towards the right and at a slower speed. If the magnitude of current is depicted by the length of the arrow and the direction by the arrow itself, the current varies in the form of a spiral known as the Ekman spiral, as shown in Figure 2.9 (McLellan 1965). At a depth of about 100–200 m (330–660 ft), the Ekman spiral goes through less than half a turn. Yet the water moves so slowly (about 4% of the surface current) in a direction opposite to that of the wind (i.e. 235° to the right of the wind) that this depth is considered to be the lower limit of the wind's influence on ocean surface movement. Approximately, this depth is the start of the thermocline. If one adds all the vectors in Figure 2.9, the resulting flow is at 90° to the right of the wind direction in the northern hemisphere and to the left of the wind direction in the southern hemisphere. This mass flow of water is known as Ekman transport.

Ekman transport piles up surface water in some areas of the ocean and removes water from other areas, producing variations in the height of the sea surface, causing it to slope gradually. One consequence of a sloping ocean surface is the generation of horizontal differences (gradients) in water pressure. These pressure gradients, in turn, give rise to geostrophic flow.

2.4.2 Geostrophic Flow

Geostrophic flow is the steady time-invariant horizontal flow in the interiors of the ocean, which is predominant below the surface layer where the flow is predominantly as per

Ekman spiral and above the bottom flow region. This steady current accounts for about 98% of the ocean volume. The geostrophic flow is due to the balance of the pressure gradient and the Coriolis force. In the northern hemisphere, the Coriolis force is to the right of velocity or eastwards and the pressure gradient is directly opposite to the Coriolis force, pressure being higher to the right. The pressure gradient may further be coupled with the ocean surface gradient due to the pile up of water caused by Ekman spiral. In the southern hemisphere, the same process occurs with the Coriolis effect being to the left of velocity or eastwards.

Near the ocean bottom, there is a frictional effect due to the movement of water over the bottom surface. This is further coupled with the Coriolis effect like the surface layer. The bottom flow is thus modified in the same way as the Ekman spiral on the surface. The top of this bottom region is the bottom of the geostrophic flow region of the ocean interior. Thus, in the deep ocean with steady wind blowing, there are three regions along the depth: the top layer having the surface current, the middle layer (98% of ocean volume) having the deep current due to geostrophic flow and the bottom layer having the bottom current.

2.4.3 Gyres

A gyre is a large system of rotating ocean currents due to large wind movements coupled with the Coriolis effect. There are five main gyres, one in each ocean basin – North Pacific, South Pacific, North Atlantic, South Atlantic and the Indian Ocean basin – as shown in Figure 2.10 (Talley et al. 2011). Apart from the gyres, there could be formation eddies with warm or cold core formed at the edges of permanent current lines. An eddy could have a diameter between 150 and 300 km, whereas a gyre is much larger being of the order of up to 2500 km.

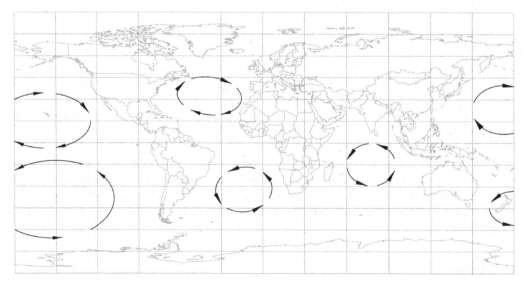

FIGURE 2.10
Rotating ocean circulations: gyre.

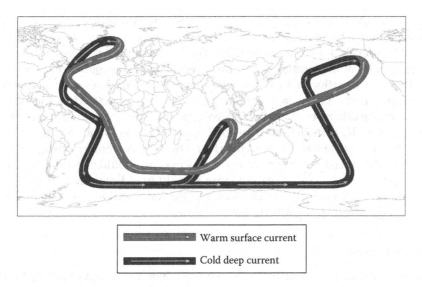

FIGURE 2.11
Thermohaline ocean circulation.

2.4.4 Thermohaline Circulation

Apart from the circulations mentioned earlier, there is the so-called thermohaline circulation due to the density variation caused by temperature and salinity differences along the depth as well as between the poles and the equator, as has been discussed in Section 2.2.2. This current causes surface and deep-water interaction as well as mass movement through an ocean conveyor belt, as shown in Figure 2.11 (Lamb 2003).

2.4.5 Circulation in Basins other than Deep Ocean

So far we have seen the ocean circulation patterns in deep oceans in steady condition. In oceans and seas other than deep oceans, the current pattern changes based on depth, unsteady wind patterns and other local effects. Marginal ocean basins are basins which are transition between deep ocean basins and coastal continents where depth keeps changing from deep to shallow with slopes varying based on undersea surface topographic details. Coastal waters are waters above the continental shelf in the coastal sea region. The extent of coastal seas up to the end of the continental shelf varies from location to location with no average value. In shallow water regions, the Ekman spiral may not be completely developed. Because of the lack of depth, the frictional effect may extend from the surface to the bottom of the basin. This gets aggravated when there is stormy weather. Thus, the current pattern in shallow areas, particularly near land, is affected by local factors. There are also tides, flow from rivers and tidal currents which affect the coastal current patterns.

2.4.6 Tides

Tides are caused due to the attraction between the earth and the moon and to a lesser extent between the sun and the earth. The semidiurnal tides are due to the attraction between the earth and the moon occurring at an interval of half a lunar day or 12 h 25.235 min. Apart from this, twice in a month the earth–moon attraction is complemented by the earth–sun attraction due to the linear orientation, and a large tide is resulted and is known as the spring tide. Similarly, twice a lunar month, the two attractions oppose each other causing reduction in tide levels, known as neap tides. The height of tidal rise depends on the location, the geography of the landmass and the orientation of the continental shelf. At some locations, the tidal variation is as high as 15 m. When the tide rises near a coastal region, there is a tidal current landwards, the strength of which is related to the tidal height. Similarly, when the tide recedes, there is a strong tidal current seawards. If a river joins the sea in the region under consideration, the total current is a combination of the tidal current and the current due to the water flow in the river. Thus, the current in coastal regions, estuaries, river mouths and bays is a very complex process and should be studied by modelling all the aforementioned processes. Also a large amount of sediment is transported mainly from the land to the sea near the river mouths. Also some sediment is transported based on permanent current lines parallel to the coastline.

2.4.7 Ocean Currents

Major ocean currents are primarily due to the five major gyres influenced by local eddies and other currents described in the preceding discussion. On the western boundary of the North Atlantic Gyre, moving in a clockwise direction along the east coast of North America is the Gulf Stream, which is a warm current. This gyre, continuing across the North Atlantic basin, moving from the North American coast to Europe is the North Atlantic Current, which is a cool current due to the influence of the northern winds. This gyre then turns south along the west coast of Africa and is known as the Canary Current. Turning westwards and flowing towards the Caribbean Sea across the Atlantic Ocean, this gyre influences the North Equatorial Current. The South Atlantic Gyre similarly influences four currents: the Brazil Current on the east coast of Brazil moving south or anticlockwise, the West Wind Drift or Antarctic Circumpolar Current moving east near the South Pole, the Benguela Current moving north along the west coast of Africa and finally the Falkland Current near the Falkland Islands moving westwards. Similarly, the North Pacific Gyre influences four major ocean currents on its four sides. The North Pacific Current is on its northern side moving across the Pacific from the east coast of Asia towards the west coast of the United States. The California Current moves south on the west coast of the United States; the North Equatorial Current moves across the Pacific towards the eastern Asian coast and then moves northwards as the Kuroshio Current. The South Pacific Gyre, bound by the equator on the north, the eastern side of the South American peninsula on the east and the Australian continent on the west, has the Atlantic Circumpolar Current on its south side. Thus, both the South Atlantic and South Pacific Gyres influence the Atlantic Circumpolar Current near the South Pole. The direction of the current due to the influence of the Indian Ocean Gyre is very complex and is highly influenced by the monsoon. The major current influenced by this gyre is the Indian Monsoon Current, which is warm and moving eastwards in summer. The current changes its direction westward in winter and becomes a cold current. The North Equatorial Current merges with the Monsoon Current and changes direction accordingly. Figure 2.12 shows the major currents in the ocean around the world, including the ones mentioned earlier.

A massive amount of water is moved due to currents. Surface currents make up about 8% of all the water in the ocean and are generally restricted to the upper 400 m. Current speed varies from location to location, based on global as well as local factors, between 0.5 and 1.5 m/s or 1–3 knots occasionally going up to 2.5 m/s or 5 knots. The deep-water currents, owing to the mass of water involved, move at a much lower speed less than 0.5 m/s.

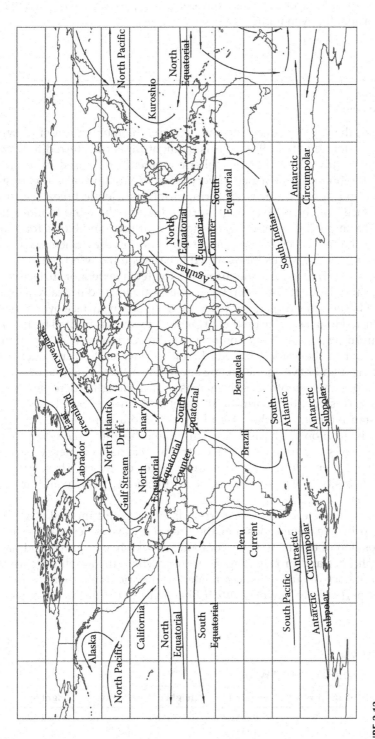

FIGURE 2.12
Ocean currents.

Currents at river mouths vary due to freshwater flow into the sea as well as tidal effect. The direction of such currents reverses twice a day depending on tides. This current can go up to 4 knots or occasionally, even higher.

2.5 Ocean Waves

At the interface of the ocean surface and atmosphere, the momentum of blowing wind is transferred to water and about 97% of this momentum is transferred to water to create ocean currents. Only 3% of the atmospheric momentum is used in the generation of ocean waves, the quantum of which is very large in human scale, affecting production, installation and operation of engineering artefacts. Gentle breeze causes ripples on the water surface, which create a rough surface helping the wind to move the water. The ripples die down quickly as wind ceases to blow. If the wind blows for a long time, ripples grow into larger waves forming short and choppy seas. Such waves are steep and chaotic. The size of the waves thus formed is based on the wind speed, the time for which the wind is blowing in the same direction, the fetch (extent of surface over which wind is blowing) and the depth of water. Large waves are formed if steady wind blows at a constant speed in the same direction over large deep ocean surface over a long period. The energy imparted to the surface on the formation of waves is limited by the formation of steep waves until waves start breaking or when they cannot absorb any more energy. The waves break when their slope exceeds 1:7 in deep water. In a fully developed sea, waves become longer and have an increasing speed, even exceeding that of wind. In this condition, waves become smoother with increasing wavelength, increasing period and reduced steepness or slope, transforming themselves to swells. Swells, thus generated, travel a long distance far beyond the fetch of wind without losing any energy. Sea swell characterization is done on the basis of the length and height of the wave, as shown in Table 2.4. Ocean waves are transverse waves and are carriers of energy over long distances without loss of energy if they do not encounter obstacles. These waves are not affected by the viscosity of water; they are only affected by gravity and are, therefore, called gravity waves.

Ocean waves are characterized by their irregularity in both time and space. However, according to the principle of superposition, the irregular wave system can be described as a linear superposition of the many wave components which are regular and have different lengths, amplitudes and propagating directions, where the amplitudes are assumed to be small. The theory of surface waves related to marine applications is discussed in many references, particularly Newman (1977), Bhattacharya (1978), Price and Bishop (1974),

TABLE 2.4

Characteristics of Swell Waves

Length of Swell Waves	Length (m)	Height of Swell Waves	Height (m)
Short	0–100	Low	0–2
Average	100–200	Moderate	2–4
Long	>200	Heavy	>4

Comstock (1967), Lewis, vol III (1989c). A brief summary is as follows, which may be useful in the present context.

2.5.1 Potential Theory of Water Waves

A velocity potential $\phi(x,y,z,t)$ is a function – a mathematical expression with space and time variables – which is valid in the whole fluid domain when the fluid (in this case, water) is assumed to be incompressible, irrotational and nonviscous. This potential function has been defined in such a way that for any point in the fluid, the derivative of this function in a certain direction provides the velocity component of a fluid particle at that point in that direction. The velocity potential of a harmonic oscillating fluid in the x-direction is given by $\phi = Ux \cdot \cos \omega t$. If the magnitude of the potential function ϕ is doubled, then the velocity component is doubled too.

Since $u = d\phi/dx = U \cdot \cos \omega t$. Thus, the velocity potential is a linear function. For this linearity to be valid, it is necessary to assume that the slope of the water surface or wave steepness is small or the wave elevation itself is small compared with the length of wave so that higher order terms of amplitude or slope can be ignored. This means that all potential flow elements (pulsating uniform flows, sources, sinks, etc.) may be linearly superposed.

Assuming that the linear theory holds, harmonics such as displacements, velocities and accelerations of water particles and also pressures have a linear relation with the wave surface elevation. The profile of a simple wave with a small steepness looks like a sine or a cosine function, and the motion of a water particle in a wave depends on the distance below the still water level (origin of the coordinate system). The general equation of wave potential in water of any depth at any point at a depth of h below the surface can be written as

$$\phi = \frac{\zeta_a g}{\omega} \cdot \frac{\cosh k(h+z)}{\cosh kh} \cdot \sin(kx - \omega t)$$

Waves are dispersive, i.e. they disperse outwards. They run at speeds which depend on their length (and water depth). The relationship between c and λ in deep water, or equivalently between ω and k, can be established from the condition that fluid particles in the surface of the fluid remain there (the free surface condition is that there is no flow across it) as

$$\omega^2 = k \cdot g \cdot \tanh kh$$

These relations are valid for all water depths, but the fact that they contain hyperbolic functions makes them cumbersome to use. For deep water, $h \to \infty$, and the expressions for the velocity potential and the dispersion relation reduce to

$$\phi = \frac{\zeta_a g}{\omega} \cdot e^{kz} \cdot \sin(kx - \omega t)$$

and

$$\omega^2 = k \cdot g$$

where

$$k = \frac{2 \cdot \pi}{\lambda} \text{ and } \omega = \frac{2 \cdot \pi}{T}$$

where
  λ is the wavelength
  T is the time period

This dispersion relation provides other definitions for the phase velocity of waves in deep water too:

$$c = \frac{\omega}{k} = \frac{g}{\omega} = \sqrt{\frac{g}{k}}$$

Simple and very practical relations between the wavelength (m) and frequency (rad/s) or period (s) follow from this:

$$\lambda = \frac{2\pi g}{\omega^2} \approx \frac{61.6}{\omega^2}$$

$$= \frac{gT^2}{2\pi} = 1.56T^2$$

$$\omega = \frac{7.83}{\sqrt{\lambda}} \quad \text{and} \quad T \approx 0.8\sqrt{\lambda}$$

These relations between the wavelength and the wave frequency or the wave period are valid for regular deep-water waves only. The resulting velocity components – in their most general form – can be expressed as

$$u = \frac{\partial \phi}{\partial x} = \frac{dx}{dt} = \zeta_a \omega \cdot e^{kz} \cdot \cos(kx - \omega t)$$

$$w = \frac{\partial \phi}{\partial z} = \frac{dx}{dt} = \zeta_a \omega \cdot e^{kz} \cdot \sin(kx - \omega t)$$

At the crest of a wave, the water movement is with the wave. At the trough, it is against the wave. This can be seen easily by watching a small object, such as a bottle, floating low in the water. It will move more or less with the water particles. The combined motions in the x- and z-directions become circles. The circular outline velocity or orbital velocity, V, follows as

$$V = \sqrt{u^2 + w^2}$$

$$= \zeta_a \omega \cdot e^{kz}$$

This velocity, which consists of two harmonic contributions, is not harmonic; in this case it is constant and the particles move in a circular path with a constant velocity at a particular

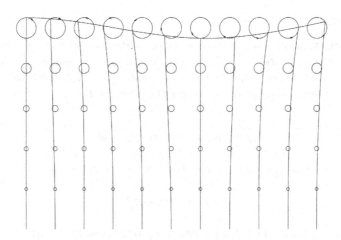

FIGURE 2.13
Trajectories of water particles in a deep-water wave.

depth z and are tangential to the trajectory circle. The trajectory radius r decreases exponentially with increasing distance below the free surface as a result of the negative value for z in the exponential term e^{kz}. Figure 2.13 shows the exponentially decreasing wave effect below the surface in deep water.

The pressure p in the first-order wave theory follows from the Bernoulli equation:

$$\frac{\partial \phi}{dt} = \frac{1}{2}(u^2 + w^2) + \frac{p}{\rho} + g \cdot z = 0$$

$$p = -\rho g z - \rho \frac{\partial \phi}{\partial t} - \frac{1}{2}\rho(u^2 + w^2)$$

It can be shown that

$$p = -\rho g z + \rho g \zeta_a e^{kz} \cdot \cos(kx - \omega t) - \frac{1}{2}\rho \zeta_a^2 \omega^2 e^{2kz}$$

Three parts can be distinguished in this expression for the pressure:

1. The first part $p^{(0)} = -\rho g z$ is the (zeroth order) hydrostatic pressure as used in Archimedes' law.
2. The second part $p^{(1)} = +\rho g \zeta_a e^{kz} \cdot \cos(kx - \omega t)$ is the (first order) dynamic pressure due to the wave form and is used when treating 'first-order wave loads' on floating bodies.
3. The third part $p^{(2)} = -(1/2)\rho g \zeta_a^2 \omega^2 e^{2kz}$ is the (second order) pressure due to the local kinetic energy in the waves – also called radiation pressure – and is used when treating 'second-order wave drift forces' on floating bodies.

2.5.2 Regular Waves

A sinusoidal surface gravity wave is an isobaric surface or surface of constant pressure equal to the atmospheric pressure. There are similar isobaric surfaces below the surface also. If ζ_a is the surface wave amplitude in deep water, k the wave number, ω the circular frequency of the wave and h the depth of point below the flat free surface, then it can be shown that

$$\zeta(x,t) = \zeta_a e^{-(2\pi/\lambda)h} \cos(kx - \omega t)$$

giving the regular surface gravity wave or free surface elevation as

$$\zeta(x,t) = \zeta_a \cos(kx - \omega t)$$

This cosine wave is a harmonic wave propagating in the positive x-direction in a Cartesian coordinate system where x and y are the perpendicular axes in the horizontal plane on the water surface and z is vertically upwards. Figure 2.14a shows the wave contour at a certain time t_i. If we observe the free surface elevation in deep water at a certain position x_j, we can draw the time history of the free surface elevation λ as shown in Figure 2.14b.

It has been shown that the wave celerity c is given as

$$\frac{\omega}{k} = \frac{\lambda}{T} = c$$

where c is the wave celerity or the phase velocity. It should be pointed out that only the wave form moves with this phase velocity, not the water particles. The wave contour described in Figure 2.14a and b is a contour of constant pressure equal to the atmospheric pressure. The contour of equal pressure at any depth h is also cosine curves with an amplitude $\zeta_a e^{-kh} = \zeta_a e^{-(2\pi/\lambda)h}$, as shown in Figure 2.14c, and the pressure at any point is given as

$$p = -\rho g(z - \zeta) = \rho g h + \zeta_a e^{-(2\pi/\lambda)h} \cos(kx - \omega t)$$

The last term of the right-hand side of the aforementioned equation denotes the difference in pressure due to the wave elevation ζ at $z = -h$, which decays proportionally to e^{-kh}. This difference is due to the so-called Smith effect.

2.5.3 Irregular Waves

An irregular sea surface is a random process. The probabilistic analysis of random waves at sea has been discussed by St. Denis and Pierson (1953), Price and Bishop (1974), Longuet-Higgins (1984) and others. An irregular sea surface can be usefully represented as the sum of a large number of regular waves, each component having a particular frequency, amplitude, direction and randomly distributed phase angles. Figure 2.15 shows a part of a simple time history of an irregular wave. When such a time history is available, a simple analysis can be carried out to obtain statistical data from this record.

The average wave period T can be found from the average zero upcrossing period or from the average period of the wave crests or troughs. The simplest way to do this is to divide the record duration by one less than the number of upward (or downward) zero crossings found. The significant wave height $H_{1/3}$ is defined as the average of the highest one-third of the waves in the record. The significant wave height $H_{1/3}$ plays an important role in many

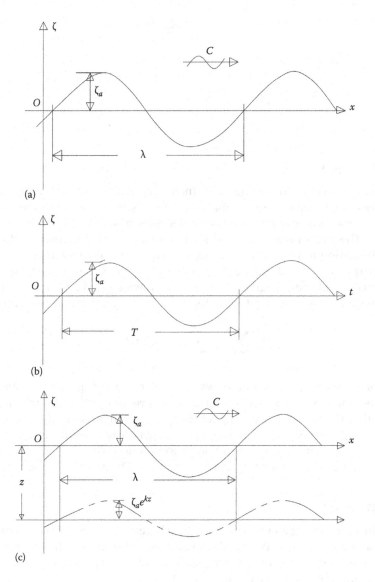

FIGURE 2.14
Typical sinusoidal water wave: (a) wave contour at time t_i, (b) surface elevation at position x_j and (c) isobaric contours across depth.

practical applications of wave statistics. Often there is a fair correlation between the significant wave height and a visually estimated wave height. This comes, perhaps, because higher waves make more impression on an observer than do the smaller ones.

Statistical information can be obtained from the probability density function $f(x)$. For example, the probability that the wave height H_ω exceeds a certain threshold value a in this record is given by

$$P\{H_\omega > a\} = \int_a^\infty f(x) \cdot dx$$

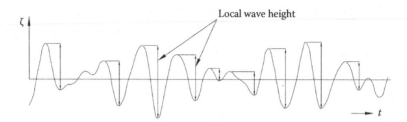

FIGURE 2.15
Time history of sea surface.

Although ocean waves are characterized by their irregularity in both time and space, mathematically, they can be described as random or stochastic process with statistically steady, or stationary characteristic appearance under short-term statistically stationary conditions. Therefore, the theory of probability and statistics can be used to analyse the wave characteristics. According to the principle of superposition, irregular waves can be described as linear superposition of infinite number of simple, regular harmonic wave components having various amplitudes, lengths, periods or frequencies and directions of propagation. Hence, an irregular wave propagating in the positive x-direction can be written as

$$\zeta(x,t) = \sum_{i=1}^{\infty} \zeta_i = \sum_{i=1}^{\infty} \zeta_{ai} \cos(k_i x - \omega_i t + \varepsilon_i)$$

where $\zeta_{ai}, k_i, \omega_i, \varepsilon_i$ are the wave amplitude, wave number, wave frequency and stochastic phase of the ith wave component. Since ε_i is stochastic, the wave contour ζ_i of the ith wave component is stochastic. If all the regular wave components propagate in the same direction, long-crested waves, which are 2 dimensional irregular waves, are generated. On the other hand, if the wave components propagate in different directions, short-crested waves, which are 3 dimensional irregular waves, are obtained leading to a confused sea. The latter is the more general case.

2.5.4 Energy Spectrum

The energy in a train of regular waves consists of kinetic energy associated with the orbital motion of water particles and potential energy resulted from the change of water level in wave hollows and crests. Under a wavelength λ, the kinetic energy E_k and the potential energy E_p per unit breadth of a wave are given by

$$E_k = \frac{1}{4}\rho g \zeta_a^2 \lambda, \quad E_p = \frac{1}{4}\rho g \zeta_a^2 \lambda$$

The total average energy per unit area of free surface is $E = E_k + E_p = (1/2)\rho g \zeta_a^2 \lambda$

Since ocean waves can be regarded as the linear superposition of simple, regular harmonic wave components, the energy of ocean waves can be obtained from the summation of the energies of the wave components. The total average energy per unit area of free surface for wave components of frequencies ($\omega_i, \omega_i + \Delta\omega$) is given by

$$E = \sum_{\omega_i}^{\omega_i + \Delta\omega_i} \left(\frac{1}{2}\rho g \zeta_{ai}^2 \right) = \frac{1}{2}\rho g \sum_{\omega_i}^{\omega_i + \Delta\omega_i} \zeta_{ai}^2$$

Denoting this energy by $\rho g S_\zeta(\omega_i)\Delta\omega$, where S_ζ is the so-called energy spectrum, we have

$$S_\zeta(\omega_i) = \frac{(1/2)\sum_{\omega_i}^{\omega_i+\Delta\omega}\zeta_{ai}^2}{\Delta\omega}$$

where

$S_\zeta(\omega_i)$ denotes the wave energy density at ω_i

S_ζ shows the distribution of energy of the irregular waves among the different regular components having different frequencies

As $\Delta\omega \to 0$, $(\omega_i + \Delta\omega) \to \omega_i$; therefore, as $\Delta\omega \to 0$, $S_\zeta(\omega_i)d\omega = (1/2)\zeta_{ai}^2$. Figure 2.16 shows a typical energy spectrum which clearly indicates which wave components are important in the energy distribution on the sea surface.

Figure 2.17 shows diagrammatically how statistical analysis can lead to obtaining the energy spectrum of a seaway by obtaining the wave height record over a period of time.

In general, we can define the nth order moment of the area under the curve $S_\zeta(\omega)d\omega$ with respect to the vertical axis at $\omega = 0$:

$$m_{n\zeta} = \int_0^\infty \omega^n S_\zeta(\omega)d\omega$$

$$\text{For } n = 0, \quad \text{we have } m_{\zeta 0} = \int_0^\infty \omega^0 S_\zeta(\omega)d\omega = \sigma_\zeta^2 (\text{variance of } \zeta)$$

$$\text{For } n = 2, \quad \text{we have } m_{\zeta 2} = \int_0^\infty \omega^2 S_\zeta(\omega)d\omega = \sigma_{\dot\zeta}^2 (\text{variance of } \dot\zeta)$$

$$\text{For } n = 4, \quad \text{we have } m_{\zeta 4} = \int_0^\infty \omega^4 S_\zeta(\omega)d\omega = \sigma_{\ddot\zeta}^2 (\text{variance of } \ddot\zeta)$$

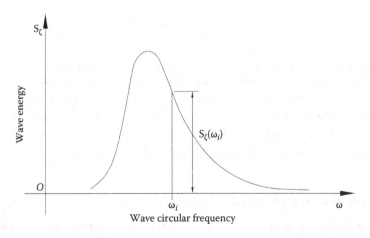

FIGURE 2.16
Energy spectrum.

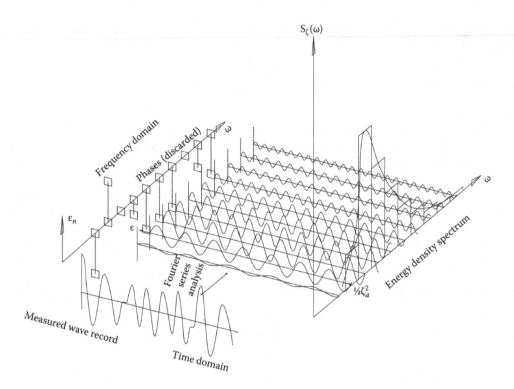

FIGURE 2.17
Wave record analysis.

The variance σ_ζ^2 can be used to estimate the statistical characteristics of the free surface elevation, for example the average wave amplitude $\zeta_a = 1.25\,\sigma_\zeta$; the significant wave amplitude $\zeta_{a(1/3)} = 2.00\sigma_\zeta$ and the average amplitude of highest 1/10th waves is $2.55\sigma_\zeta$

The bandwidth of the spectrum is given by ε where

$$\varepsilon^2 = 1 - \frac{m_2^2}{m_0 m_4}$$

$\varepsilon \to 0$ indicates a narrowband spectrum with a dominant frequency, and the wave height follows a Rayleigh probability density function. On the other hand, $\varepsilon \to 1$ indicates a wideband spectrum with Gaussian probability density function. The average zero-crossing period for a wideband spectrum is given by $T_C = 2\pi\sqrt{(m_0/m_4)}$ and the crest-to-crest period is $T_C = 2\pi\sqrt{m_2/m_4}$. Correspondingly, the average zero-crossing and average crest-to-crest wavelengths are $\lambda_C = 2\pi g \cdot \sqrt{m_0/m_4}$ and $2\pi g \cdot \sqrt{m_2/m_4}$, respectively.

2.5.5 Representation of an Irregular Seaway

A sea is represented by its energy spectrum, and numerous formulae have been suggested (Pierson–Moskowitz spectrum, Bretschneider spectrum, Derbyshine spectrum, British Towing Tank spectrum, Newman spectrum, JONSWAP spectrum, etc.) to describe the various sea states as given by Lewis, vol III (1989c), Price and Bishop (1974) and Bhattacharya (1978). The International Towing Tank Conference (ITTC) has suggested a standard spectrum to be used in normal work as follows:

$$S_\zeta(\omega) = \frac{A}{\omega^5} \cdot e^{-B/\omega^4}$$

where

$$A = \frac{173 h_{1/3}^2}{T_1^4}$$

$$B = \frac{691}{T_1^4}$$

T_1 is the characteristic period and $h_{1/3}$ is the significant wave height, which are given by

$$T_1 = 2\pi \cdot \sqrt{\frac{m_0}{m_1}} \quad \text{and} \quad h_{1/3} = 4.0\sqrt{m_0}$$

Further, the ITTC has suggested the following relationship between wind speed (V) and significant wave height ($h_{1/3}$) from which the energy spectrum can be derived.

V (knots)	20	30	40	50	60
$h_{1/3}$ (m)	3.1	5.1	8.1	11.0	14.6

The physical description of the sea state has been codified by the IMO and is given in Table 2.5.

2.5.6 Shallow Water Waves

As has been already described, the regular wave equation changes to include the depth term h and the shallow water properties change. This happens in shallow water basins,

TABLE 2.5

IMO Sea State Code

Sea State Code	Significant Wave Height (m)	Characteristics
0	0	Calm (glassy)
1	0–0.1	Calm (rippled)
2	0.1–0.5	Smooth (wavelets)
3	0.5–1.25	Slight
4	1.25–2.5	Moderate
5	2.5–4	Rough
6	4–6	Very rough
7	6–9	High
8	9–14	Very high
9	>14	Phenomenal

estuaries, rivers near the mouth to the sea and near the beaches or surf zone. Shallow water waves are generally steeper, the steepness increasing with decreasing depth. When a deep-water wave enters the shallow region near the beach, the wave height and consequently the wave steepness increase. When wave height reaches about 1.8 of the water depth and wave steepness reaches 1.7, waves break generating surf. As the wave enters the shallow region, its speed reduces, while the deep-water wave moves faster. Therefore, the waves entering shallow water change direction, becoming parallel to the beaches. The surf zone is a belt of nearly continuously breaking waves parallel to the shoreline, submerged bank or bar even though the origin of the wind-driven waves might have been in a different direction.

Waves near the shore generate turbulence, which in turn generates localized currents and sediment transport. Both these phenomena depend on the bottom topography of the shoreline, generation of wind-driven waves and long-period swells that might have travelled from the distant sea.

2.5.7 Seiches

Seiches are standing or stationary waves generally found in restricted basins with steep sides (e.g. the Red Sea) or in water tanks (e.g. wave basin for ship model testing). Here, waves do not move or are of zero velocity. Certain portions of the surface do not go up and down at all and remain stationary. These are called nodes. Certain other portions go through maximum oscillation from crest to trough and back. These are called antinodes. Seiches occur mainly due to the interference of the original waves with the reflected waves.

2.5.8 Storm Surges

Storm surges are long-period surface waves generated by strong winds which drive water along a coast causing the sea level to rise near the shore. This causes flooding of low-lying areas near the shore. The oscillation of the sea surface set in motion by the generation of waves continues even after the wind stops. The inundation of coastal areas continues until these oscillations subside.

2.5.9 Tsunamis

If the ocean bottom is disturbed due to earthquake or volcanic eruption, the sea level immediately above it rises, generating a very long wave known as tsunami. A tsunami moves at a very high speed, and in deep ocean, the wave height may not be noticeable by shipping activity. But when the waves approach the shore or shallow water region, the water piles up to a height and moves like a wall inland from the shore. This can sometimes be very devastating. Similar to storm surges, the oscillation of water does not stop immediately but continues for some time causing flooding again and again.

2.5.10 Internal Waves

Generally, one would like to believe that waves form only at the interface of air and water. In fact, gravity waves form wherever there is a stratification of fluid based on density difference. In the ocean, due to sharp continuous or discontinuous change in density in the pycnocline caused by temperature difference in the thermocline, an internal wave can be generated, which moves horizontally. Similar waves can form in the continental shelf region where low-density brackish water overlays saline sea water at river mouths. Such waves have very little effect on ocean surface. The speed of propagation of the waves depends on the density gradient across the surface. Internal waves move at a much slower speed and have much lower frequency and higher amplitude than surface waves since the density gradient is much smaller than that between water and air at the surface. The internal wave frequency can vary from seconds to hours and the wavelength from centimetres to kilometres.

Sometimes a curious phenomenon occurs when a ship moving in visibly calm water experiences high resistance, reducing its speed drastically, even not moving forward with full power. Such an incidence can occur when a ship moving in fresh water experiences a large drag due to an internal wave caused by a stratified layer of salt water below. This can occur at the polar region where fresh water on top has salt water of high density below because of salt precipitation due to ice formation. Such water is called dead water, which was first reported by Norwegian explorer Fridtj of Nansen.

3

Design Process

A maritime product or system design requires the understanding of the construction, maintenance, operation and dismantling of the product (discussed in Chapter 1). Therefore, understanding the entire life cycle of a maritime product, from concept to disposal, is necessary. Figure 3.1 shows the life cycle of such a system. Though the time spent on designing the product, sometimes only a few months, is relatively less compared to manufacturing and operating time, the design determines the efficiency and cost-effectiveness of the product or venture over the entire lifespan of 15–25 years.

3.1 Mission Requirement

Identifying the mission requirement is the first step in the design process, and this depends on the client. The broad mission requirement or the venture objective for the owner of a maritime product, typically a merchant vessel, may be profitability of the entire venture, i.e. acquisition, operation and disposal. On the other hand, a builder's mission requirement may be profit in product construction. An offshore platform may have least acquisition cost or low operational cost as the mission requirement. Similarly, a naval vessel may have offensive or defensive effectiveness as the main objective. This type of design objective, known as 'soft' information, is very broad and does not give enough information of the product requirement. It is necessary to have well-defined or 'hard' product requirement so that the design process can start. For this purpose, market research and market study are necessary.

3.2 Market Study

The definition of design given in Section 1.3 highlights that design cannot be sustained or be successful without a proper market study. Marine vehicles and structures have characteristics different from that of other products. Unlike a consumer item or a mass-produced item based on a standard design, marine products are custom built with limited repeat orders and clientele. Manufacturing such items is generally labour intensive, capital intensive and technology intensive. These are built in harsh environment, in the open and near the waterfront such as sea or river. Ventures with such products have long-term return on investment. Very often governmental support such as tax concessions, soft loan terms, order support or direct subsidy is extended to the marine construction industry so that it remains competitive in the international market and ensures good product quality and effective operational performance. The marketing and design of

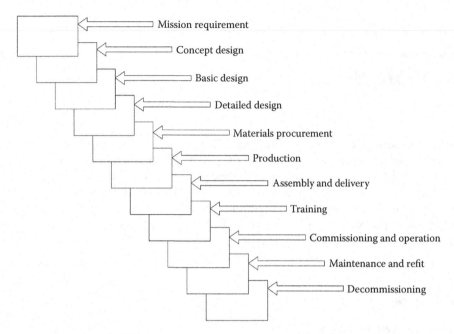

FIGURE 3.1
Maritime product life cycle.

such a product are concurrent activities and are complementary to each other. Marketing consists of the following steps:

- Identifying customer needs
- Design product to meet these needs
- Relate product to enterprise
- Promotion

3.2.1 Identifying Customer Needs

Customer requirements can be identified by predicting customer needs through a market forecast. Normally, short-term forecast is not preferred in the marine industry since the business environment in the lifetime of the product is long term. Therefore, the development of new products depends on long- and medium-term market forecast. This is done by studying trends in transportation, identifying transport/marine activity areas with the fastest growth rate, identifying geographical areas having potential increase in transport/ marine activity and studying intergovernmental agreements. Market study also requires estimating the present availability of the product and decommissioning of existing products, thus identifying future requirements. As an example on merchant ship requirement, Figure 3.2a shows new building deliveries in 2013 and Figure 3.2b gives ships sold for demolition in 2012, taken from maritime transport review 2014 and 2013, respectively. Figure 3.3 shows ships on order in 2013 and 2014, which indicates a drop in ship construction in 2012 and 2013, but a slight increase in 2014. However, further market study should indicate demands of the future by identifying customer needs.

Societal awareness and demands as well as technology development lead to changes in customer requirement of future products. In recent years, the range of marine products

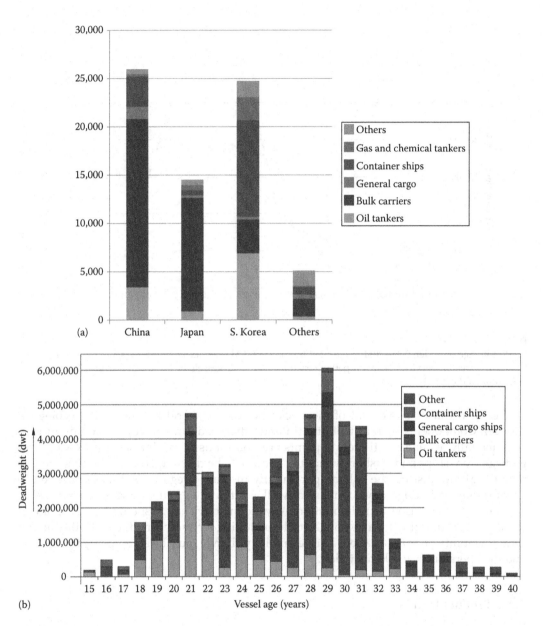

FIGURE 3.2
New vessel deliveries in 2013 and scrapings in 2012. (a) Deliveries of new buildings, major vessel types and coun-tries where built, 2013, and (b) tonnage sold for demolition by age, 2012. (From UNCTAD secretariat, on the basis of data supplied by Clarkson Research Services; *Note:* Propelled seagoing merchant vessels of 100 GT and above.)

has expanded tremendously leading to demands for safety and environmental protection. Whereas burning of fossil fuel at sea causes air pollution adding to the threat of global warming, acid rain and ozone-layer depletion; the ocean gets polluted by the discharge of oil from ships and offshore structures and accidental discharge of oil at sea, particularly at coastal region with devastating effects. Ballast water moved across the seas through ships is of the order of 10 billion tonnes per annum, and this is causing the movement of inva-sive species across the oceans leading to ecological disaster. Fouling on ship's underwater

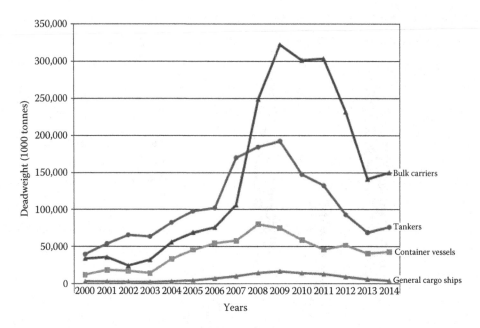

FIGURE 3.3
World tonnage on order, 2000–2014. (Compiled by the UNCTAD secretariat, on the basis of data supplied by Clarkson Research Services; *Note:* Propelled seagoing merchant vessels of 100 GT and above; figures at the beginning of the year.)

surface also causes movement of biospecies across the world adding to the ecological disaster. Components such as tri-butyl tin and other lead, tin and copper compounds in commercially available anti-fouling paints are poisonous to marine species. Sewage and garbage from ships and offshore structures require to be discharged in sea so that they do not biologically disturb the habitat, particularly in coastal regions. All these have led to new product and design developments as well as alternative and renewable energy usage at sea to restrict environmental pollution. Any technology change should be incorporated in the product design after a thorough study, including techno-economic feasibility analysis and viability of its application. This technology change must be accepted by the builder and the operator for implementation.

3.2.2 Product Design

Preliminary market survey techniques are used to define market requirements. The preliminary market survey includes forecasting demand and customers as well as classifying customers, competitor's products, deliveries, price and price constraints, financials, guarantees and penalties. This leads to customer needs, which are used to define the mission requirement firmly. A concept exploration model of the product is then developed as the first step of the design exercise to study the feasibility of the product design and manufacture.

3.2.3 Relate Product to Enterprise

At this stage, it is necessary to identify enterprises that would be interested in manufacturing the product and relate the product to the enterprises. This includes the following activities: checking the total capacity of the enterprise(s) with regard to plant, building, machinery and equipment; checking labour resources; checking output volume with regard to demand and required deliveries; preparation of cost estimate and financial arrangements; preparation of feasible production schedule. This information is used for a feasibility study and concept exploration of the product so that it can be manufactured easily and in a cost-effective manner within the contractual delivery period.

3.2.4 Promotion

Promotion of the product normally means dissemination of product information to the identified market segment through presentations in seminars, limited client conclaves and technical literature. A number of designs of a particular product by different design groups are all expected to satisfy the performance requirements. To promote a particular product through its design, it is necessary to project to the client the benefits of the particular design against the alternative designs of the competitors with regard to cost benefit during manufacture and operation, ease of operation and maintenance, technological benefits and aesthetics.

3.3 System Design

The last half century has seen a tremendous change in the normal design concept. Any major engineering product or system consists of a number of subsystems, each of which should be designed to serve its purpose. Further, all subsystems must be integrated in such a manner that the whole system works efficiently and effectively. Today, no system works in isolation. It must necessarily interact with the world around it on issues involving technology, business, environment, society, law and such other items which may be technical, semi-technical or non-technical. The design must address these issues if the product or system has to succeed in the world of its operation. This is the systems engineering approach to design necessary for the product to succeed in today's competitive environment.

An individual engineering discipline concentrates on the problems of the real world related to that area, and the solution is found following the laws governing that area. Systems engineering, on the other hand, focuses on the analysis and synthesis of problems involving multiple aspects and disciplines of the real world. The International Council on Systems Engineering has given an explanation of systems engineering as 'In some real sense, systems engineering means always looking toward how things fit together at a level above where one's day-to-day challenges reside'. Quoting Checkland (1981), Mistree (1990) gives a definition of systems engineering as 'An epistemology which, when applied to human activity is based upon the four basic ideas: emergence, hierarchy, communication, and control as characteristics of system'. Mistree (1990) adds that system thinking then emphasizes both the emergent properties of the system as a single entity and the separate and collective properties of the subsystems and their components in their intrinsic environment. A most visible area of application of systems engineering is the design of large complex marine vehicles and structures.

For example, ship design requires the knowledge and application of various science and engineering disciplines such as hydrostatics, hydrodynamics, mechanical engineering, electrical and electronics engineering, structural and materials engineering, physics, mathematics, etc. The existence of knowledge in these individual disciplines may ensure individual component design involving a particular component or subsystem. But this does not ensure an efficient design of the ship as an overall system. For this it is necessary that all subsystems must be integrated to work as an efficient overall system using systems engineering approach, which is cross-disciplinary in nature. System thinking also includes non-technical disciplines at a higher level. A design, though technically suitable, may not be implementable due to cost, time and other non-technical issues. Such non-technical issues may include market identification, cost of production, environmental standards, profitability during operation, maintainability, dismantling or recycling, etc. Then this becomes a system problem where paramount importance is given to economic and other issues at a higher level while integrating the technical issues simultaneously.

The systems approach to marine design requires that the whole system must be broken down to individual systems, subsystems and components and establishment of their cross connectivity. Methods must be evolved to solve the problem at the individual discipline level or at the lowest level. The next step is to establish the hierarchy of system integration. Then mechanisms for integrating the individual solutions to each subsystem and system must be evolved. The methodology for final system integration at the highest level must also be established. Thus, the mechanism of systems engineering involves a structured and hierarchical approach.

The subsystems/components at the lowest level must be designed correctly at the beginning. Therefore, it is necessary to implement sophisticated scientific design and analysis tools such as computer-aided design (CAD) software for surface generation, computational fluid dynamics (CFD), finite element method (FEM) for structural analysis and other numerical computation software at the concept level itself so that errors at a higher level of system integration do not occur.

At a higher level of system integration, the solutions to all subsystems must be integrated to give the overall system success. The objectives of each subsystem may be different; for example, the hydrodynamics subsystem may have power minimization as an objective, whereas structural design may require weight minimization and so on, the overall system objective being minimum cost of construction (say). Thus, at a higher level, the design problem becomes a multivariate, multiobjective optimization problem subject to linear and/or nonlinear constraints. Defining such a complex design problem mathematically is very difficult and, even if it is defined, a mathematical solution is difficult to find. Often such a solution may not be acceptable from practical considerations and uncertainties involved in the data provided for optimization. Therefore, the selection of design variables involves decision making, which may be heuristic in nature.

A marine product is generally a large product and a complicated large system consisting of a number of subsystems, which must work in an integrated manner so that the product works efficiently and effectively. The product must also satisfy the economic or such other requirements. For appreciating the various subsystems and their connectivity, it is necessary to understand the features of a marine product.

3.3.1 Features of a Marine Product

Marine products, including ships, are generally custom built unlike other large products such as cars and aeroplanes. Even when a repeat order is placed, a number of changes are incorporated into the design. Also repeat orders are limited to a few orders only.

So the design of such a product has to be done on an individual basis, and the advantage of repeatability or standardization is limited. Marine products are capital intensive and require financing from financial institutions. The period of return on investment of a marine product venture is large, generally 10 years or more, sometimes going up to 25 years with a high level of maintenance. Construction of a marine product, for whatever operation it is utilized, is highly technology intensive, involving both production work and assembly work (approximately 50% each). Construction of such a product is also highly labour intensive and involves multiple skills, and it is built in a harsh environment. Labour rate being low in the eastern countries, maritime construction has moved from Europe to Far East.

Government support to marine construction can be of various forms. Such support also varies from country to country. Marine construction industry is an internationally competitive industry, and government support plays an important role in the pricing of ships. Construction of such a product is complex, and the contract-to-delivery period is large.

3.3.2 Sustainability

Business related to the maritime product must be sustainable, i.e. it must be profitable and operationally successful without harming the business and the environment in the future. Product design must ensure long-term survival of the business, may it be profits in shipbuilding, marine operations or naval effectiveness. Also the product design must ensure safe operation involving life and property to ensure credibility and survival. Operational sustainability should ensure that the product must work as a successful component of a bigger system, e.g. ship in the transportation chain, a production platform in the overall oil production system or a high-speed vessel in coastal security. Technological sustainability means reduction in the emissions of pollutants to air, water and land, the measures for which must be incorporated during the design stage. This includes reduction in fossil fuel consumption, use of renewable energy, reduction in the deposit of pollutants to the sea and such other measures necessary during the process of construction, operation and disposal. The design should go beyond the statutory requirements to set technology trends in future products and should be recognised for technology leadership.

3.3.3 Subsystems and System Components

At a lower level, each subsystem consists of a few components with defined component specifications. Combining these components into a subsystem, a particular task of the overall system is performed. Generally at this stage, the subsystems are fairly independent and can be designed on a stand-alone basis. Similarly, combining a few of these subsystems forms a bigger subsystem identified for a bigger task or well-defined function. Keeping this task in mind, the smaller subsystems are integrated to form a compatible unit.

The higher-level subsystems can now be identified. These are economy or business, sustainability and performance. These are not stand-alone functions, rather cross-functional. Figure 3.4 shows the various tasks or subsystems in a typical ship design project. Figure 3.5 shows functional and cross-functional objectives. At a higher level, the major subsystems, starting from the mission requirement, are technical, economic, productional, sustainable and operational, as shown in Figure 3.6, which must be integrated to give one or more feasible solutions.

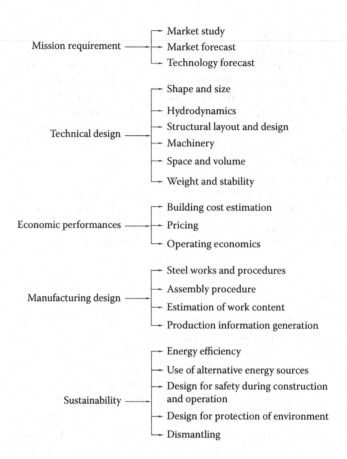

FIGURE 3.4
Typical tasks and subsystems in a maritime design project.

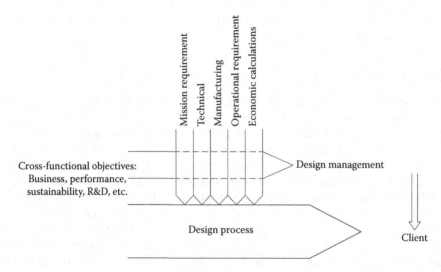

FIGURE 3.5
Functional and cross-functional objectives of the design process.

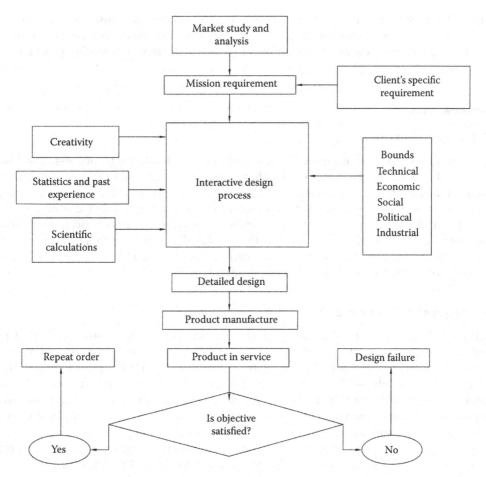

FIGURE 3.6
Design process.

3.3.4 System Integration

At any level, system design involves optimizing the objectives defined as a function of a set of free variables subject to a number of constraints. At the lowest level, a component or subsystem may be well defined mathematically. If such is the case, the design objectives can be written down mathematically as functions of identified free variables. If the constraints can be mathematically defined, a mathematical solution to the design problem can be found. An example can be the design of a gear for the transmission of a predefined power at a given rotation rate. But marine design problems are seldom so well defined. So a mathematical solution may not be possible. Then the solution can be arrived at by doing a parametric study and comparing the output of each set of parameters. Sometimes sensitivity analysis can lead to identifying the variable which must be defined accurately. But when one moves up the system ladder, one has to find an optimum solution to two or more systems at the same time. Many a time the solutions to the various objectives of different systems may oppose one another. A solution satisfying all objectives should be found. Normally, in complex design problems such as ships or other marine products, finding a mathematical solution is difficult, and even if such a solution is possible, its acceptability by the client is suspect. Therefore, the methodology to find a feasible solution is to find solutions to individual subsystems and integrate them into the larger system through a decision-making process, which is made easier and faster by adopting concurrent engineering practices.

3.4 Design Process

Figure 3.6 gives a brief diagrammatic description of the total design process. The tasks in each subsystem can be performed either sequentially or concurrently. Designing a complex marine structure or vehicle is a time-consuming process involving a large number of people with different skill sets. There could be a number of feasible solutions. It is imperative that the internal design groups must agree on a few feasible design solutions. Subsequently, at a higher system level, different stakeholders have to agree on a single feasible solution for which a detailed design can be worked out. The design process can be based on point-based or set-based design paradigm.

3.4.1 Sequential Design Process

Starting with a design solution, one moves from one activity to another such that the requirements of the activity or subsystem are satisfied. One moves in a sequential manner, which means that the activities of a later subsystem depend on the outcomes of one or more previous subsystems. If the performance requirement of the subsystem is not satisfied, one has to go back to the beginning for design modification. Multiple iterations are necessary to arrive at a feasible design space. The technical subsystem, as shown in Figure 3.4, can be shown as a sequential design process as per Figure 3.7a.

The basic tenets of a widely accepted sequential approach to ship design have been captured successfully by the so-called design spiral given by Taggart (1980), which is a point-based design system. Figure 3.8 shows the traditional design spiral where the spokes of the spiral indicate various subsystems whose requirements must be met. The spiral converges to a product at the centre. Mistree (1990) makes two important observations regarding this spiral:

1. The spiral converges towards a product, but the process is divergent with regard to information, i.e. increasing detail of definition. Perhaps this divergence aspect is represented in the spiral given by Buxton (1971) in which the direction of movement along the spiral is outwards. At any stage the design information is a combination of 'hard' or exact information and 'soft' or qualitative information. As the design progresses, the ratio of hard to soft information increases.

2. The spiral approach is sequential and iterative, and it has truly represented the state of the art (state of research) and the state of the industry (state of practice) of any product design in the past years. The computer, in this definition, has been used more as an efficient calculator where certain performance calculations can be done by scientific algorithms as and when required. This method has been improved over the years where the interaction between the designer and the computer has been more direct (CAD).

Such a solution necessarily leads to a satisfactory solution though not an optimum solution. Being slow and laborious, this method may be effective when the marine industry

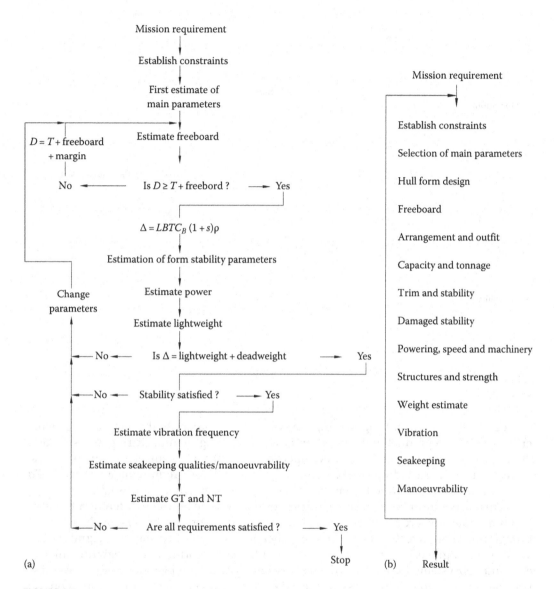

FIGURE 3.7
Sequential and concurrent design processes: (a) sequential design and (b) concurrent design.

is doing well and the manufacturer's order book position is good. Since the design is based on a large amount of previous data, such a design is almost always built and small improvements can be incorporated from product to product.

3.4.2 Concurrent Engineering in Design

A useful technique of system design is doing various individual tasks at a time, and through a communication process between such tasks, an integrated solution can be achieved. This process is known as concurrent engineering. It is a common-sense approach to product development in which all elements of a product's life cycle from conception through disposal are integrated into a single continuous feedback-driven design process. According to

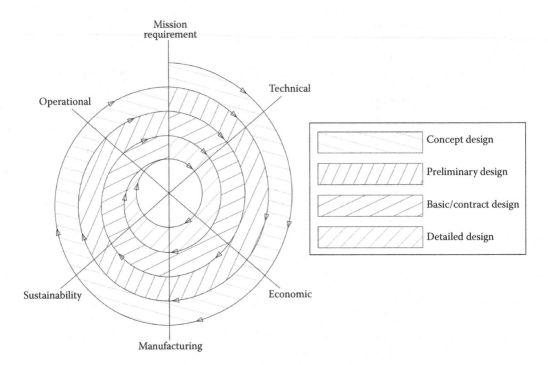

FIGURE 3.8
Design spiral.

Mistree (1990), 'Concurrent engineering is a systematic approach to the integrated, concurrent design of the product and related processes including manufacturing and support'. This approach intends to make the developers consider from the outset all elements of the product life cycle, including quality, cost, schedule and user requirements. Figure 3.7b shows this approach modified from the earlier approach shown in Figure 3.7a.

Concurrent engineering or design requires communication between different subsystem design modules for proper integration. This communication can be effectively achieved through information technology (IT) tools. Therefore, IT tools become an integral part of any systems engineering application. Without adequate information, integrating all systems and subsystems into a unified whole is impossible. Complex data and knowledge bases have to be evolved to access and use data and generate further information on the product in a concurrent manner. The concurrent engineering approach is embedded in the iterative design process, which also includes the decision-making process. Though a unidirectional flow of information is shown, it can be observed that the design must have the life cycle information of the product while designing a new product.

This design methodology by Mistree (1990) is represented by the frustum of a cone (Figure 3.9). Let the surface generator of this frustum represent the locus of one of the design activities (as depicted in the traditional design spiral) such as stability, shape generation, etc. As the design process advances in the traditional sense, one works outside this frustum, i.e. the design activity is sequential and iterative. Alternatively, if the design process is viewed as taking place inside this frustum, all design activities can be sighted at any inside point. Thus, the notion of concurrence can be accommodated by working inside. Each interplay between design considerations can be modelled effectively with information flowing through a 'ring of interactions'. Concurrence should, therefore, replace iterative processes where a balance

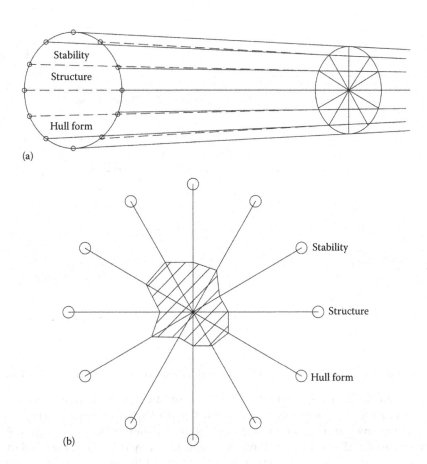

FIGURE 3.9
Frustum of a cone representation of ship design, spokes indicates various design activities, a few being shown as illustrative.

between conflicting requirements and objectives is sought. In this sense, at any phase of design, the information available/generated is disk-like but of irregular shape. In the limit, at the end of the design process, the information disk is geometrically circular and complete. Thus, the design time is reduced and perhaps a more efficient design is obtained.

3.4.3 Point-Based Design

Point-based design is a sequential design process. To reduce rework time due to iterations, concurrent engineering principles can be adopted at this level; when all lower-level sub-systems can be analysed simultaneously, all results can be compared with respective criteria. Expectedly, this process reduces the design time. This gives a feasible design, which can be modified to start the iteration process from the beginning to get another feasible solution. One can generate a design space consisting of a number of feasible solutions and select the one that satisfies the design objectives the best.

3.4.4 Set-Based Design

In the set-based design process (Singer et al. 2009), developed and used by the Toyota Motor Corporation, defining a design space where all feasible and possible designs will

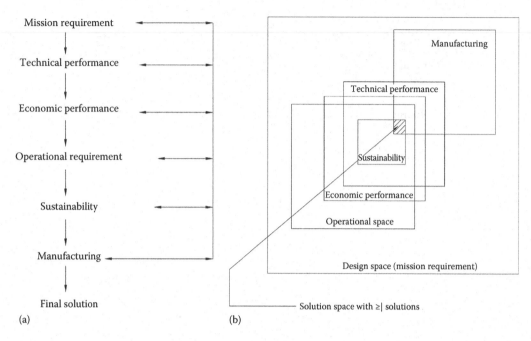

FIGURE 3.10
Point- and set-based design processes: (a) point-based design with concurrence and (b) set-based design.

reside is necessary. The design space for each major subsystem is defined such that all design requirements of the subsystem are met within this space. It is necessary to consider all design alternatives and explore trade-offs between different sets of designs. After all the design spaces for all subsystems have been defined, they must be integrated to get the overall design solution. This is done by the intersection of the design spaces such that the space common to all subsystems is identified. This common space may contain a number of solutions. A solution is then found from among these solutions such that system performance is robust in all respects, the solution is less sensitive to minor changes in design and it can absorb a good bit of uncertainty in data used without compromising with performance. To reduce the design time, concurrent engineering concepts can be adopted at different stages in an overall manner to obtain one feasible solution. Figure 3.10a and b shows the conceptual difference between point-based and set-based design processes.

3.5 Design Stages

A marine product design is a long and, sometimes, tedious process. Various activities such as tendering and contract signing take place during the design activity. Generally, marine design follows the following stages as described by Taggart (1980):

1. *Definition of objectives:* Once the mission requirement is established, design objectives specifying the design space can be defined. The space is bounded by the business objectives and operational requirements with regard to performance. This space is limited by statutory requirements regarding safety and sustainability as

well as quality requirements related to construction and operation. If the client or owner requires any new or advanced technology, that must also limit the design space accordingly.

2. *Concept exploration:* In this phase, a feasible design space is identified. Alternative sets of design solutions are synthesized and a feasibility study is carried out to see if each set satisfies the objectives. At this stage, product information is very little, but performance has to be estimated to see the feasibility, and this estimate, in the absence of a detailed scientific analysis, must be accurate enough so that rework can be avoided at a later stage. In the case of a merchant ship, at this stage a number of sets of main parameters such as length, beam, depth, draught and block coefficient are determined. Also considered are the different hull forms such as mono- or multi-hull vessels, with and without bulbous bow; different construction materials such as steel, aluminium or fibreglass-reinforced polymer; different machinery arrangements such as single or multiple screws and podded propeller. For each set, economic performance or building cost estimation is carried out with this scanty information based on past experience and statistical data. This stage provides a basis for discussion with the client and other stakeholders for moving towards a single solution. If it is found that such a design solution is not feasible, further design effort should be abandoned so that the minimum expenditure of effort is spent up to this stage.

3. *Preliminary design:* At this stage further detailed calculations are carried out for determining performance characteristics subject to constraints which should have been defined earlier and a single design solution is arrived at by freezing the design variables. In the case of a ship, it would be the main parameters and the main machinery and equipment. A preliminary general arrangement drawing and a brief product specification are prepared. Due to increasing competitiveness and client demand on performance excellence, it has become necessary to study the hydrodynamic and structural behaviour of the design solution by using computer software such as CFD and FEM techniques. The building cost estimates are then carried out. Normally, a builder quotes for a vessel or platform at this stage, where the tender documents include a preliminary general arrangement drawing, a brief specification, the shape of hull form, performance estimates and the quoted price. If the tenderer fails to win the contract, all design efforts must stop and the expenditure on this activity should be absorbed by the organization.

4. *Contract design:* Once the tender is successful, work on contract design starts, which includes all major drawings, material and equipment list, detailed specifications and any model test and analysis results. At this stage, schematic diagrams or line diagrams of all systems are prepared. In the case of a ship, this includes the main propulsion system, all piping systems, electrical system, cargo handling, air conditioning, ventilation, refrigeration, manoeuvring and control system, navigation, life-saving and firefighting, environmental control systems, etc. Also prepared are the structural and machinery drawings to be submitted to the Classification Society for approval and preliminary stability calculation to be submitted to regulatory authorities. All these documents are parts of the contract document. On the completion of these design stages, the contract is signed for a price. So it is necessary that the cost estimate, which is more detailed and accurate, should be within acceptable limits of variation from the earlier quoted price.

5. *Detailed design:* After contract signing, detailed designing starts, which is the preparation of production drawings to be transferred to planning or production departments for production and installation. All information provided in the design
   drawings for production must provide detailed information up to the workman level.
   The information should be detailed enough that rework can be avoided. With the
   use of computers in manufacturing such as using computer-aided manufacturing
   or computer-instructed manufacturing, information must be provided in numerical form for machines to read them and operate accordingly. Such manufacturing
   information must also emanate from the detailed design group, and this activity is
   sometimes known as manufacturing design. This includes development of hull shell
   plates, information for 2D and 3D plate forming, information for double- and single-
   flanged plates, nesting of plates, disintegration of 3D structural models for stiffener
   and frame parts and NC information generation, development of curved plates and
   stiffeners, zig information generation for support of curved panels, drawing information for building and assembly sequences and similar production information.
   It is generally said that 'beauty is in details'. The better the design information at the
   detailed design stage, the better and accurate the execution of work.

3.6 Information Generation and Management

Any design activity is an information-generation process in which with very scanty information such as mission requirement, product information is generated in stages such that
at the end of the design process, complete information about the product is available. For
effective information generation, it is necessary that various subsystems communicate
between themselves for coordination and updating of information. Thus, an information
system in marine design has to have two main functions:

1. Collection of information so as to
   a. Define needs which can be understood and translated into guidance for mak-
      ing the desired product
   b. Acquire knowledge so as to grasp the theories and principles, standard prac-
      tices, performance criteria so that correct inferences can be made
   c. Acquire data (numerical, textual, graphical, etc.) regarding past records, facts,
      experiences, relationships, decisions, etc.
   d. Acquire information regarding the organization so as to take necessary action
      by the right people at the right time
2. Processing of information so as to
   a. Create or generate ideas and concepts, both innovative and traditional, to sat-
      isfy the identified needs
   b. Carry out necessary technical and economic calculations for performance
      prediction
   c. Take decisions at the end of any small or large design activity to select one
      among many alternatives feasible
   d. Perform the right task at the right time

To carry out the aforementioned functions and the design process effectively, an efficient computer-based information management system is necessary, which can collect, store and update information, selectively retrieve information as and when necessary and utilize this information to generate further information. The efficiency of such a system is determined by the ease with which information is entered into the system, the flow of information in the computer network, the bank of sophisticated software to generate correct and precise additional information, storage of generated information for future use and retrieval as and when required through a query system. Such an information system can be developed as a life cycle modelling, generally known as a 'product model'. A product model can be defined as 'A data structure of a product in a computer system which covers all scope of related design and manufacturing'. Another definition of product model is 'A digital representation of a real or abstract object described by a collection of graphical and non-graphical attributes as well as relationships with other objects such that the collection spans the full life cycle of the product and conceptually appears to reside in a single repository'. Both the definitions support the design definitions given earlier. To build a product model, it is necessary to use not only high-speed and high-capacity computers, but also high level of information communication. Modern tools for such a job include object-oriented computing, sophisticated numerical computing software in the domain area, parallel processing, cloud computing, etc.

3.7 Communication

It is now evident that the design of a marine product is a complex and multi-skill activity involving a number of specialist groups. It is necessary that effective communication exists between people and groups involved in design. It is also necessary to have continuous communication with production/manufacturing groups as well as operating groups and statutory authorities. The various groups between whom communication must be good and effective can be listed as follows:

- *Designers:* Internal communication between groups within
- *Manufacturers:* Production information confirming to design details
- *Owners:* Owners, charterers and ship management groups
- *Users:* Engineers and officers on ships and offshore structures
- *Statutory bodies:* Local, port state control and flag state control authorities
- *Classification societies:* Technical quality assurance and inspection
- *Vendors:* For technical specification and cost
- *Researchers:* Future technologies

3.8 Design Tools

Design activity is a complex activity involving scientific analysis, decision making and synthesizing various requirements through system integration. To do this job properly and efficiently, it is necessary to use modern tools and practices.

3.8.1 Data Collection and Statistical Analysis

Various types of data and information are required to be collected and analysed for supporting different design activities at different stages in both technical and economic areas. A few examples are as follows:

1. Approximate relationships to arrive at the initial values of variables are an essential requirement of the engineering design of a product, which is new. In the case of ship design, one has to start with a set of approximate dimensions until dimensions are finalized through scientific analysis. This initial dimension can be arrived at by empirical relationships arrived at through data collection and analysis. For ships this has been discussed further in Section 5.1.

2. Collection of information can provide guidelines for designing items for preliminary evaluation so that a design cycle at concept design stage can be completed to go ahead. An example can be hull form design; if the basic ship or similar ship data of shape (hull form) are available, a new ship shape can be generated easily though it may be changed later after CFD analysis and model testing.

3. Collection of data and proper analysis can be used for future predictions and forecasting, though this has to be done with care.

4. In many data items, there could be a large amount of uncertainty. For example, in economic data there could be large variations in variables such as fuel oil price or freight rate. Similarly, in structural analysis, there could be a huge amount of uncertainty with regard to load on ship or even resistance to load by the structure. If a large amount of previous data is available, it is possible to model the uncertainty by plotting data in the form of histograms and obtaining probability density. One has to be careful if data samples are limited, which may lead to erroneous modelling of uncertainty.

3.8.2 Scientific Knowledge Base and Computer Software

The design team must have adequate knowledge base to study all design alternatives and select the best. Sometimes it is necessary to take help of specialized knowledge, which may not be available within the design team. It may also be necessary to do model test or prototype test for the prediction of performance or for selecting the best alternative. For this it is necessary to have adequate knowledge to select the specialist or the experimental facility where required simulation study and analysis can be conducted and prediction done.

Shape-generation and simulation software are now available to generate 3D forms of ships and offshore structures. CAD software is now available for space and equipment layout and design drawings, including stability studies of floating and submerged bodies. Hydrodynamic simulation studies in water are very complicated, and numerical simulation around ships and platforms, propellers, around tow lines and fixed pipelines is being done using CFD techniques. Structural design and analysis of various kinds of stresses are similarly very complicated mainly because of the 3D nature of the structure and uncertainty in loads on such structures and vehicle. There are sophisticated computational software commercially available for complete structural analysis using FEM techniques. Similarly, software for modelling and analysis of machinery and different kinds of equipment are also commercially available. Software for the design and development of each system are now available commercially, and these are no more limited to research institutions only. It is possible to analyse hydrodynamic and structural behaviour of the vehicle or structure at the concept design stage itself so that uncertainty with regard to performance at a later stage of design is minimized if not eliminated. Decision making becomes easier once the analytical tools predict the performance of various systems properly. Product lifetime modelling may be used to handle all design activity easily, avoiding multiple data entry and data errors, and quick transmission of information to all concerned.

4

Engineering Economics

Before a large investment is made in an engineering product, an investor considers the economic viability of the product. The investor compares different alternatives to choose the best one for optimum economic performance. The necessary economic calculations have to be made on projected costs, investments and income over an assumed life of the product. These calculations are for a future period for a product which is still to be manufactured. The effectiveness of any judgement based on such calculations depends on the understanding of the identified economic criterion and expressing this criterion in terms of design variables. In this chapter, we discuss the time value of money, different economic criteria and their evaluation and economic complexities which affect economic evaluation.

4.1 Interest Relationships

Money has an appreciable time value. A certain amount of cash can buy more things today than, say, after 1 year. Money kept in a bank grows as per the contracted interest rate. A rent, interest or reward must be paid if money is to be lent to compensate the lender for postponing spending it. This reward or interest is fundamental to all economic calculations while considering cash handled over a considerable time.

A nominal rate of interest can be negotiated or contracted on a loan taken from a financial institution or a time deposit in a bank or in government securities. This is the predetermined rate of return on the money spent or invested. On the other hand, if money is invested in an engineering project or process, the effective rate of return or equivalent interest rate is not known a priori and is calculated based on the excess of income over expenditure over a period. It is necessary for an investor to know if this equivalent interest is higher than the interest accrued by depositing the money in a bank.

Let A be the annual return, i.e. the excess of income over expenditure or the annual repayment consisting of principal and interest, F be the future sum of money, P be the present sum of money or the principal amount of investment or borrowed capital, N be the number of years and i be the interest rate or effective rate of return or discount rate per annum.

For a borrowed sum P with simple interest i per annum, the repayment after N years becomes

$$F = P \, (1 + Ni)$$

In the case of compound interest, the principal is increased by the amount of interest due at the end of the year and further interest is calculated on this amount. This is known as interest being compounded each year. In such a case, the repayment amount after N years is

$$F = P\,(1+i)^N$$

Sometimes, interest is compounded at a lesser period than a year, say T times in a year. Then the repayment after N years is

$$F = P\left(1+\frac{i}{T}\right)^{NT}$$

If interest is compounded on a continuous basis, then

$$F = Pe^{iN}$$

It can be observed that the present sum of money P is equivalent to the future sum of money F, and the conversion of P at present to a value F after N years depends on the type of interest contracted. The factor or multiplier by which a present sum of money is converted to a future sum is called the compound amount factor CA, which is given by

$$CA = \frac{F}{P}$$

Similarly, a factor or multiplier called the present worth factor PW can be utilized to convert a future sum of money F to present worth P, which is the inverse of the compound amount factor. Thus,

$$PW = \frac{P}{F} = \frac{1}{CA}$$

CA and PW are independent of the actual cash transaction and depend only on the interest rate. It is easy to convert any amount P to F or F to P based on these factors:

$$F = CA \cdot P$$

and

$$P = PW \cdot F$$

In retail banking of small investors, single repayments, as described earlier, are used (Figure 4.1). For example, an investor can put an amount P as fixed deposit in a bank to get back an amount F after N years based on a contracted rate of interest i. Today, a simple rate of interest is rarely used in any commercial transaction. In fact, to woo retail investors, financial organizations frequently resort to quarterly or even monthly compounding of interest.

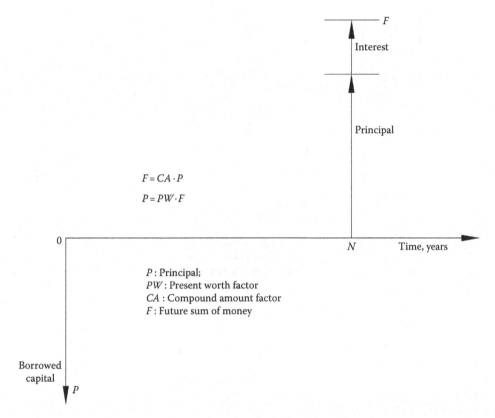

FIGURE 4.1
Single repayment.

In commercial banking, where large sums of money are lent by financial organizations or the government to prospective investors, rarely is the repayment demanded in one single instalment. The repayment is done on a regular basis over a number of years. There are two common methods of repayment prevalent in the market:

1. Principal repaid in equal instalments and interest paid on declining balance. Then, the annual repayment is high at the beginning years and decreases in the later years. At any year $j = 1$ to N, the repayment is given by

$$\frac{P}{N} + \left(P - (j-1)\frac{P}{N}\right)i = \frac{P}{N} + (N - j + 1)i\frac{P}{N}$$

2. Equal instalment repayment where the principal component of total repayment is low at the beginning and rises in later years and interest is high in initial years, which decrease over years (Figure 4.2).

The most prevalent repayment terms are based on the later principle. Examples are the commercial house-building loans and pension plans. Here an annual repayment A made

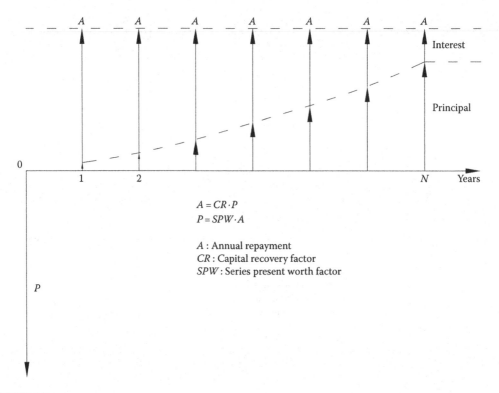

FIGURE 4.2
Equal instalment repayment.

over N years is equivalent to the present sum of money P. Thus, the amount A is calculated based on the capital recovery factor CR based on the contracted interest rate. CR is given by

$$CR = \frac{A}{P} = \frac{\left[i(1+i)^{N} \right]}{\left[(1+i)^{N} - 1 \right]}$$

$$= \frac{i}{[1-(1+i)^{-N}]}$$

A series of annual repayments A can be brought back to a single sum at present worth by a factor called the series present worth factor SPW, sometimes called annuity factor, and is given by the reciprocal of CR:

$$SPW = \frac{P}{A} = \frac{1}{CR}$$

Another rarely used term is the sinking fund factor, which is used to calculate the annual amount A for N years to make a future payment F where the interest component is higher in later years (Figure 4.3).

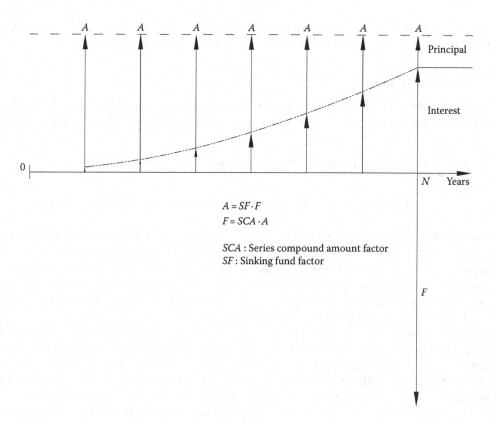

FIGURE 4.3
Future sum of money.

Or $$A = SF \cdot F \text{ where}$$

$$SF = \frac{i}{(1+i)^N - 1}$$

The inverse of SF is the series compound amount factor SCA, which is used to obtain a future sum of money F from a series of annual investments A. This is the reciprocal of SF and is given by

$$SCA = \frac{1}{SF} = \frac{(1+i)^N - 1}{i}$$

4.2 Economic Criteria

To evaluate the economic performance of a product, it is necessary to identify an economic performance indicator or economic criterion which can be expressed as a function of a number of technical and economic variables. Depending on the data available for the variables, one may choose the required economic criterion for the purpose intended. Some economic criteria normally used in engineering design are discussed here.

4.2.1 Net Present Value

The net present value (NPV) or venture worth of a product is essentially the net worth or net profit earned from the product over its life period. The general form of NPV is given by

$$NPV = \sum_{0}^{N} [PW(\text{annual income}) - PW(\text{annual operating cost})$$

$$- PW(\text{first cost or investment})] + PW(\text{scrap value of second-hand sale price})$$

where N is the life of the product in number of years for which investment is planned. N could be the life from concept until scrapping or could be from any time of its life to any later period. This is essentially the accumulated profit over the period under consideration at present worth. The type of income is product dependent. Whereas the income due to a ship is calculated as freight earning per year, that due to a tourist complex may be income due to the number and types of tourists per year.

Similarly, the annual operating cost depends on the type of operation of the product. This could comprise management and administration costs, rates and rents, running costs, maintenance costs, insurance and other statutory costs. Similarly, investment pattern may vary depending on whether the product is new or old and negotiations between the financiers and the owners of the product. Normally, the owners do not invest the total amount from their own funds. They may invest a nominal amount, say 10%–20% of the product price, from their own resources and get the remaining amount as loan from commercial or governmental institutions. In such a case, their actual investment is the loan repayments they make in cash towards capital and interest each year. Thus, the NPV calculated for the same product but with different investment patterns may be different. If the annual cash flow is uniform over the life of the product, it may be possible to use series present worth factor or series present worth of any 1 year's cash flow to calculate the NPV.

The NPV is a measure of the net profit and hence a good item evaluating product performance. For it to be dependable, it implies that the values of economic parameters for both income and expenditure must be well predicted throughout the future period for which NPV is calculated. If the income or expenditure is uncertain, NPV may not be the right criterion to use.

As a function of design variables, the NPV can be a suitable objective function to be optimized (maximized) and accordingly the design variables may be obtained. Alternatively, given different designs of the same product, the design offering the highest NPV is sought. However, for comparing products of different sizes such as a trawler with a very large crude-carrying vessel, the economic scales being different, the NPV is not the suitable criterion. But NPV, as a fraction of total investment, could be a suitable basis for comparison. This is known as the net present value index (*NPVI*), which is defined as

$$NPVI = \frac{NPV}{\text{First cost}}$$

In products related to defence applications, NPV is irrelevant since there is no income. Also investment is generally one-time investment and is made based on criteria other than economic. Mission effectiveness is perhaps the deciding criterion for many decisions related to defence product. However, with competition in supplying defence products increasing and economic considerations becoming significantly important in operating the product, a criterion called average annual cost (*AAC*) can be used to compare different designs having the same mission effectiveness where

$$AAC = \text{Average annual operating expenses} + CR(\text{first cost})$$

Between alternative designs, the design having minimum AAC is preferred.

4.2.2 Required Income

The acquisition of a product generates income, which is normally based on income per some kind of a unit of service or product provided to customers. Thus,

$$\text{Annual income} = \text{number of units sold} * \text{rate per unit}$$

In the case of a ship, the annual income can be freight rate * total cargo carried; in the case of a tourist complex, it is charge per tourist * number of tourists per year; in the case of an aircraft or luxury liner, it is passenger fare * number of passengers per year. If the so-called unit rate fluctuates frequently or cannot be estimated accurately, the NPV calculated is likely to be erroneous. In such an event, it is necessary to calculate the required income which should give zero NPV. The required income can be calculated by estimating the NPV for different unit rates and obtaining the required income to get zero NPV.

Once the required income is obtained, it can be compared with the income based on the prevailing unit rate in the market. If it fluctuates widely, the average value should be higher than the calculated required income to have positive NPV. Therefore, if this item is used as an economic criterion for evaluating different designs, the design giving the lowest required income should be chosen.

In the case of a commercial ship, the required income can be further brought down to the required freight rate (RFR), which gives the required income to cover all expenses so that the NPV is zero. To calculate RFR one must know the ship acquisition cost, the required rate of return or internal rate of return (IRR), the operating expenses and the annual cargo carried. Thus,

$$RFR = \frac{\sum_0^N PW(\text{annual operating cost}) + \sum_0^N PW(\text{ship acquisition cost})}{\sum_0^N PW(\text{annual cargo quantity})}$$

For uniform cash flows, a useful simplification is

$$RFR = \frac{\text{Annual operating cost} + CR(\text{ship acquisition cost})}{\text{Annual cargo quantity}}$$

In comparing different designs, the design with the lowest RFR is sought. Sometimes the RFR is called the shadow price. Then the RFR can be used to calculate the freighting cost and compared with the actual freighting price.

4.2.3 Internal Rate of Return or Yield

IRR or yield is also known as the 'discounted cash flow rate of return' or 'equivalent interest rate', which is that discount rate or interest rate used to bring back the future sum of money to the present worth, that gives zero NPV. This value is obtained in an iterative manner. The NPV is calculated for different discount rates and by interpolation the IRR giving zero NPV is obtained. To evaluate this value, it is necessary to know the income and expenditure reasonably well.

IRR is also called the investor's method since using this the investor can decide between widely different types of investments, say between pipeline laying, refinery set-up and off-shore production facility installation in an oil company without being biased by the amount of investment. This can also be used for comparing investments between widely differing options such as engineering product procurement or government stocks. It is also a useful criterion for making a decision between the alternatives of purchasing a new product or procuring an old product with a few years of service life. Given a number of design alternatives, the one that gives the highest IRR or yield is chosen since the profit (proportional to investment) from this alternative would be higher than the other alternatives.

4.2.4 Permissible Price

Though not used frequently, the permissible price which can be paid for the product to give zero NPV is sometimes required for investment decisions. This assumes that the income and expenditure can be accurately estimated and a predefined rate of return is available. Since the price paid for the product is not a one-time payment but spread over a number of years in terms of loan repayment, the permissible price can be obtained by an iterative process, which gives zero NPV. The design which gives the maximum permissible price is sought so that NPV is maximized between different design alternatives.

The permissible price is a good decision-making tool for purchasing a used product which has got a few years of service life left. The market price of a used product depends on the supply and demand of such old products. If the permissible price is more than the market price, the investment yields positive NPV.

4.2.5 Payback Period

The payback period is the minimum value of the variable N, i.e. the number of years required to get zero NPV if income, expenditure, investment and the required rate of return are known. This is also an iterative calculation method as discussed before. This is sometimes required to estimate the period for which there will be no annual profit. The design giving the lowest payback period is chosen. In this method, the profit is not calculated. It is assumed that the product will only earn profit after the payback period. Therefore, its useful life could be extended.

4.3 Economic Complexities

While negotiating for procuring a large engineering product, a price is arrived at which the future owner of the product must pay the builder. However, the negotiated price is not the owner's investment at present worth or in real terms. This is influenced by many factors, three of which are loans or borrowings, stage payment and subsidy.

4.3.1 Loan

It has been mentioned earlier that the owner's own investment in terms of down payment to the manufacturer is a small percentage of the negotiated price, say up to 20%. The remaining 80% is taken as loan from some financial organization with fixed payment terms. The payment terms are normally similar if borrowing is done from commercial firms in the national region with marginal variations from one organization to another. But if international borrowing could be taken, the loan terms may vary considerably. Sometimes commercial lending terms are eased to get the customer. Frequently, national governments give financial assistance in terms of easy loans to assist a particular industry. Such favourable loans may also be available from foreign governments (may be through private commercial organizations) as per government-to-government agreements. The loan terms may differ in three different ways: (1) interest rate, (2) moratorium or delay in the start of repayment period and (3) period of repayment.

The actual investment is the present worth of money paid as owner's own contribution and that of the repayments made. Generally, if the negotiated interest rate is less than the IRR for the investment, it can be considered favourable.

4.3.2 Stage Payment

A second-hand product, on the other hand, is available 'off the shelf' for which the total payment has to be made immediately. But that is not the case in the procurement of a new product. A large engineering product takes a long time to be manufactured and delivered after contract signing. Manufacturers generally do not have (or does not wish) to invest their own money in material and labour in large quantity. Normally, the contracted agreement includes percentage payment to the manufacturers at pre-defined stages of manufacture. Loan is released as per this agreement. If the time of delivery of the product is taken as the 0th year of the product, the payments are made before this period with only a percentage being paid at the time of delivery. Thus, the value of the investment has already increased as per the interest rate (or as per IRR in the case of own investment) at the start of the 0th year based on which repayments have to be made.

4.3.3 Subsidy

Governments encourage industries as per their priorities in various ways. One of these is subsidy in the form of direct cash support to the manufacturing industry. In such a case, a percentage of the manufacturing cost is provided by the government and hence the negotiated contractual price can be reduced, thus giving a competitive advantage to the manufacturer particularly against international bidders for the same job. There are many other forms of direct and indirect subsidy provided by the government. Some of these are preferential income tax or corporate tax rates, reduction in excise and customs duty, soft loans as discussed earlier and cash support for building up facilities.

Subsidy is the prerogative of the government and is a necessary step particularly for nascent industries or industries which face unfair international competition. But subsidy given to a particular industry for long periods tends to make the industry less competitive and inefficient.

4.3.4 Escalation

In a stable economy, there is small annual inflation, generally ranging between 2% and 5% required to maintain the purchasing power of money. In a growing economy, this may be more, say going up to 10% per annum as may be the case in some developing economies. Inflation shows the average price rise of commodities. But the cost of each item involved in the product manufacture escalates at a different rate from the general inflationary trend. Some items may escalate more than the inflationary rate, whereas others may escalate at a lower rate or may even remain static without any rise in price over a period of time.

The escalation rate effectively reduces the discount rate used to bring a future sum of money to present worth. For example, if one has an income of 100 after 1 year, taking an equivalent interest rate i (=10% say), the present worth factor is $1/[1 + i]$ and the present worth of income is 90.91. Now, if the income of 100 includes an escalation e (=3% say), the actual income without inflation would have been $100/[1 + e]$ or 97.09. Then the effective equivalent interest or discount rate per year is $[1 + i]/[1 + e]$ or, in this case, $(97.09/90.91 - 1) = 0.068\%$ or 6.8%. Thus, due to inflation, the effective discount rate reduces to $[1 + i]/[1 + e]-1 = [i-e]$ if i and e both are small.

In a projected cash flow calculation, there may be a number of items affecting income and expenditure. The projected cash flow for each of these items may escalate at different rates. The escalation may depend on a number of factors such as international supply and demand (e.g. fuel oil price, sea cargo freight rate, etc.), national supply and demand (e.g. indigenously manufactured items), statutory control, taxes and wage negotiations. In a cash flow calculation, it may be desirable to escalate the cash under each head at its own rate and the final net cash flow may be brought back to present worth using a fixed discount rate.

4.3.5 Depreciation

Depreciation is the reduction of the book value of an asset with time. Depreciation is not an actual cost or cash transaction but is only a book transaction used both for tax and accounting purposes. For accounting purposes, the capital can be balanced each year against the depreciated value of fixed assets. On the other hand, depreciation used in tax calculations is used to decide tax to be paid and later to decide on reserve funds and dividend to be paid to shareholders. Often the two calculation methods are different. Depending on taxation laws, depreciation can be accounted for in various ways. But generally depreciation can be calculated on 'straight line' basis or 'declining balance' basis. In the straight line method, the total asset value (P) is reduced to resale or scrap value (S) over a pre-fixed number of years (N) on a straight line basis. Then,

$$\text{Depreciation per year} = \frac{(P-S)}{N}$$

In the declining balance method, the allowance is a fixed percentage of the remaining asset value. Thus, if r is the actual depreciation allowance (i.e. $r * 100$ is the percentage), then in the nth year (n varying between 1 and N),

$$\text{Depreciation allowance (in percentage)} = 100 \cdot r * (1 - r)^{n-1}$$

and

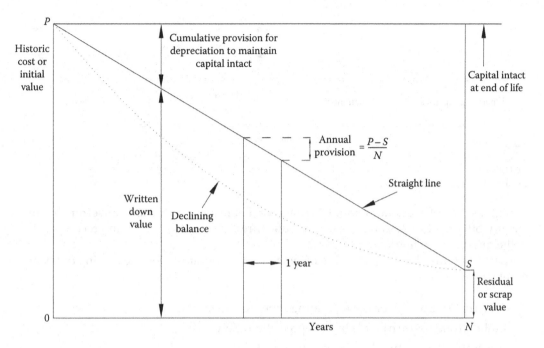

FIGURE 4.4
Depreciation over N years.

Accumulated depreciation up to N years (in percentage) $= 100 * (1 - (1 - r)^N)$
 Then

$$\text{Resale or scrap value } S = P \left[\frac{100 - \text{Accumilated depreciation up to } N \text{ years}}{100} \right]$$

The declining balance rate r is given by

$$r = 1 - \left(\frac{S}{P} \right)^{1/N}$$

Figure 4.4 shows the difference between the two types of depreciation methods discussed earlier.

4.3.6 Taxes

Large industrial products consist of a large number of components. Most of these are manufactured items of other industries procured either within national boundaries or imported from abroad. So the cost of these items includes payment of various taxes (sales tax, excise duty, etc.), customs duty and similar other statutory levies. Upon completion, the item, if sold to another party, local and national duties are further imposed on the product and if it is imported, customs duty has to be paid. Thus, such a product has to bear a multi-tier taxation burden. Sometimes, the total taxes paid at various stages can be

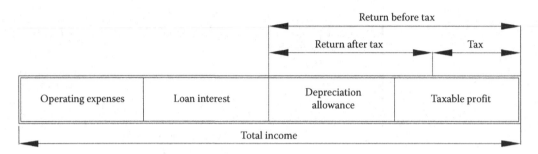

FIGURE 4.5
Taxable profit from total income.

as high as 30% of the total manufacturing cost. Realising this, various governments frequently offer tax concessions to such manufacturers so that they can remain competitive in the international market.

Apart from this, the company pays income tax or corporate tax on the trading or taxable profit (Figure 4.5).

Return before tax = income – (operating expense + loan interest)

Taxable profit = return before tax – depreciation allowance

Return after tax = return before tax – tax on taxable profit

Net profit = taxable profit – tax.

The return after tax is used to allow for capital repayment, if any, developmental work, procurement of new equipment and payment of dividend to shareholders. It can be observed that there is no provision of actual capital repayment in the aforementioned calculations. However, it is assumed that capital invested is converted to an asset of the company and a capital (or depreciation) allowance given to reduce the return before tax.

In design work, it is necessary to understand the implication of investment on design and so the concept of depreciation is not used. Rather, use is made of actual cash flow due to investment based on interest relationships or the actual cash repayment, including capital and interest as per agreed terms.

4.4 Cash Flow Calculation

Annual cash flow calculations are carried out to calculate tax or for accounting purposes. But in design, cash flow calculation is necessary to be carried out over the estimated life of the product so that the identified economic criterion can be evaluated. As has been discussed in the earlier section, the basic economic criterion is the NPV, and cash flow calculation must be carried out to estimate the same. A (non-uniform) cash flow calculation is basically a tabular calculation where the rows indicate the year (n) for which calculation is done with n varying from 0 to N. The columns represent the following items:

- Own investment
- Actual repayments on loan

- Scrap value or resale value at the Nth year
- Total expenses on first cost
- Operating expenses item wise (could be inflated as desired)
- Total operating expenses
- Income item wise (could be inflated as desired)
- Total income
- Gross cash flow = total income − total expense − total expense on first cost
- Present worth factor (based on predetermined discount factor or IRR or yield)
- Discounted cash flow (DCF = Gross cash flow * PW factor)
- Cumulative DCF (=sum of DCF up to the current year)

The contents of the cell of the last row and the last column represent the NPV of the investment. Using the same method in an iterative manner, the other economic criteria can be evaluated.

In products which require heavy investment, the repayment on loan is generally a large amount of the total annual expense. Therefore, it is likely that in initial years, the product demonstrates annual loss and also the cumulative loss increases. Only after the repayment of total loan can the venture earn profit and thus the NPV may be positive. Therefore, it is necessary to do the cash flow calculations for the whole life to evaluate a design. However, there are other uses of cash flow. For example, the financial institution sanctioning the loan would like to know if the loan can be repaid annually.

4.5 Building Cost Estimation

Taggart (1980), Carreyette (1978), Lamb (2003a), Schneekluth (1987) and Watson (2009) have discussed shipbuilding cost estimation, which is done at the stage of tender quotation first time. The total building or manufacturing cost of a marine vehicle or structure can be broken down to material cost, labour cost, direct expenses and indirect or overhead expenses.

4.5.1 Material Cost

Material cost can be further broken down to steel, machinery, outfit and electrical and miscellaneous items.

Total steel cost is based on gross steel procured for the ship where

$$\text{Gross steel} = \text{Nett estimated steel} + \text{scrap}$$

Steel material includes rolled plates and sections used for hull and superstructure construction, main and minor steel bulkheads, double bottom, decks, bulwark and steel seatings for machinery and equipment. The weight of some steel items could be grouped under steel weight or outfit weight based on shipyard practice, which may include steel forgings and castings and deck fittings, steel hatch covers, rudder plating, etc. Scrap depends on effective utilization of steel material. One way to minimize scarp is to use computerized optimal nesting of plates. The scrap percentage varies from 8% to 12% of gross steel.

The outfit items include materials such as timber, internal and external paints, primers and thinners, deck coverings, insulation, non-steel and synthetic bulkheads, ceilings and linings, furnishings and fittings, galley and pantry equipment, ship's inventory, stores and spares, anchoring and mooring equipment, cargo-handling equipment such as cranes, air-conditioning and refrigeration equipment, ventilation fittings and equipment, life saving and firefighting equipment, ship's piping and pumps, steering equipment, navigational equipment, hatch covers, ladders, companion ways and access hatches, windows and portholes, etc.

Machinery items consist of all equipment, piping and valves and auxiliary support system components of main power-generating systems. In the case of a ship, this includes the propulsion machinery and electrical generation system and their auxiliaries. This also includes the CO_2 system for fighting engine room fire, ER workshop and crane and pollution control system.

Electrical equipment includes main and auxiliary switch boards, cables, automatic and control equipment, radar, wireless, DF and VHF equipment and all electrical fittings.

Material cost must include all taxes, freight and insurance during transit. Taxes consist of local and central government taxes. An imported item may be delivered at the origin sea port on free-on-board (FOB) basis where freight for shipping, insurance and transportation cost from destination port to the client's site and customs duty must be added to the quoted price. On the other hand, if the quoted price is on cost, insurance and freight (CIF) basis, the transportation cost from the destination port to the client's site and customs duty must be added to arrive at the total acquisition cost.

4.5.2 Labour Cost

Labour cost is generally taken as the costs of direct labour allocated for the construction of the product. This can be computed as the total wages, including salary and all perquisites of the labour as per job allocation for the product. Then

$$\text{Labour cost} = \text{labour rate} * \text{number of man days for the job in question}$$

$$= \text{Labour rate} * \left(\frac{\text{Total quantum of work}}{\text{Work per labour manday}} \right)$$

There are three components for estimating the labour cost: (1) labour rate or average cost to company per labour man-day, (2) the total quantum of work based on the type, size and complexity of the product and (3) the amount of work done in one man-day or production rate, which also depends on the facilities available in the yard and the yard size, including size of building dock and/or slipways. To some extent this would also depend on the design, planning and management. The number of man-days is the manufacturing yard's norm based on its production efficiency and is estimated from previous data over recent years. For initial estimate of labour cost, it can be broken into four main components: steel, machinery, electrical and outfit labour. For final estimate, each of these components can be broken down further into particular jobs in different shops in the shipyard and labour required for each such job.

4.5.3 Direct Cost

Direct costs can be directly billed to the ship but cannot be attributed to material or labour. The examples of this cost are launching, transportation, installation and delivery expenses, insurance during the period from contract to delivery, expenses due to external supervision by classification societies, statutory bodies, owner's representatives, etc. If steel fabrication and assembly work are subcontracted to a vendor, the subcontractor's charges are direct expenses.

Normally, a negotiated contract is arrived at between the owner and the builder to decide a stage payment agreement such that the builder's expenses are matched by the payments received. Despite this, if the builder has to take assistance from financial institutions, the repayment is a direct expense on the product. Further, if there is a delayed payment due to delayed production, there is a loss in the real value of money in addition to the penalty the builder has to pay. If the owner defaults in stage payment in the right time, there is also a loss of real value of money and the owner must compensate it by paying a penalty.

4.5.4 Indirect Expenses

A large portion of the total cost cannot be billed on the product's account directly but is incurred due to the construction. These indirect expenses include the following:

- Full company establishment expenses such as rates, rents, water, electricity and other services
- Supervision cost of production such as wages of managers up to the chief executive officer
- Design and drawing office expenses
- Expenses of purchase, planning and production control departments
- Expenses due to quality control, administration, sales, marketing, etc.
- Capital and interest repayment on investment made in the yard on borrowed capital and inventory cost

Figure 4.6 gives a break-up of the total ship cost. Very often it is enough to consider the total cost made up of three components – material, labour and overhead – each with its own individual break-up. Thus, all expenses other than material and labour can be termed overhead. The overhead expenses are dependent on production as well as duration of production. Most of the expenses mentioned under indirect expenses are dependent on the duration of production, whereas the expenses mentioned under direct expenses are more or less dependent on total production. Thus, the overhead expenses can be subdivided into two subgroups as (1) fixed, which is dependent on the duration of production and (2) variable, which is dependent on the quantum of production. When the annual accounting of the yard is carried out, all other expenses not billed under material or labour are termed overhead expenses, which are then proportioned to the ships (fully or partially) constructed that year. Doing a statistical analysis of these expenses, one can arrive at some norms regarding the percentage in which the fixed and variable overhead expenses can be apportioned.

Very often labour cost and overhead cost components are difficult to estimate separately and, therefore, these are expressed together under the broad heading of labour cost. Some factors that affect this labour cost are discussed as follows.

4.5.5 Production Quantum

The quantum of work involved in constructing a marine structure or vehicle can be expressed in terms of weight such as deadweight or steel weight. But either of these items is not a proper measure since it does not take into account the dimensions or size of the

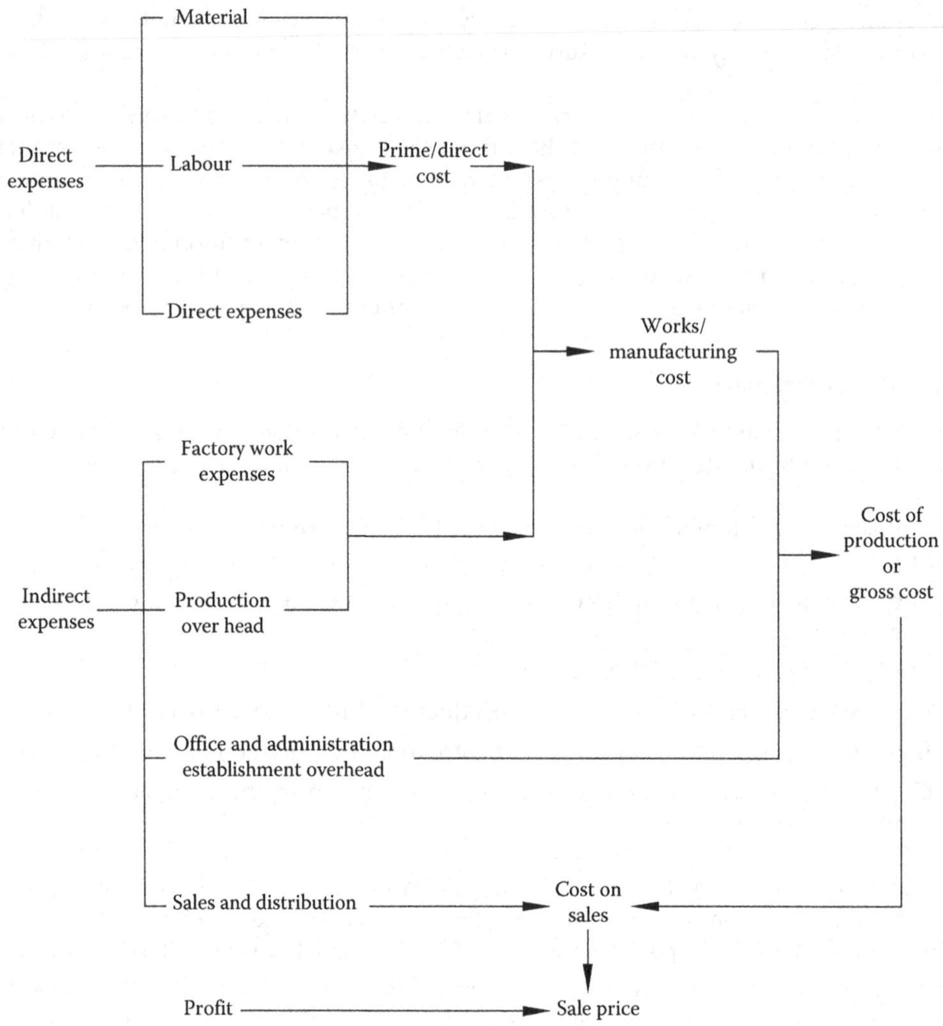

FIGURE 4.6
Shipbuilding cost components.

vessel and the machinery installed. The volumetric measure of a ship's volume is *GT* given by

$$GT = k * V,$$

where

- *k* is a coefficient based on the total volume *V* defined by Tonnage Measurement Rules, 1969, of the International Maritime Organization
- *V* is the total enclosed volume of the ship

Also the complexity of the vessel plays a major role in determining the quantum of work involved. The complexity can be thought of as directly related to the number of components in the unit, and in the case of steel construction, plate preparation required, weld

TABLE 4.1

A and *B* Values for Calculation of CGT

Ship Type	A	B
Oil tankers (double hull)	48	0.57
Chemical tankers	84	0.55
Bulk carriers	29	0.61
Combined carriers	33	0.62
General cargo ships	27	0.64
Reefers	27	0.68
Full container	19	0.68
Ro–ro vessels	32	0.63
Car carriers	15	0.70
LPG carriers	62	0.57
LNG carriers	32	0.68
Ferries	20	0.71
Passenger ships	49	0.67
Fishing vessels	24	0.71
NCCV	46	0.62

length, type and complexity of machinery and equipment, etc. Keeping this in mind, the Organization of Economic Co-operation and Development (OECD) Working Party on Shipbuilding (WP6) developed compensated gross tons (cgt) along with the Community of European Shipyards Association, the Shipbuilders Association of Japan and Korean Shipbuilders Association. The cgt system is a statistical tool developed in order to enable a more accurate macroeconomic evaluation of shipbuilding workload which is possible on a pure deadweight (dwt) or gross tons (GT) basis. Compensated gross tons is a unit of measurement intended to provide a common yardstick to reflect the relative output of merchant shipbuilding activity in large aggregates such as 'world', 'regions' or 'groups of many shipyards'. *Cgt* is defined as

$$cgt = A * GT^B$$

where *A* and *B* are given in Table 4.1 as per the report published in 2007 (NCCV means non-cargo-carrying vessel).

If series construction takes place, later ships in a series can be built with lesser quantum of work because (1) the jigs and fixtures built for the first ship can be used for subsequent ships, (2) learning process improves as more ships in a series are built and (3) not only work can move faster but also rework due to mistakes and errors reduces. OECD data show the total labour effort (*y*) required for subsequent ships reduces from the first ship effort by a ratio given by

$$y = 1 - 0.1483 \ln(i)$$

where *i* is the number of ship in the series varying from 1 to 10. Table 4.2 shows reduction in effort *y*.

Though no data are available, it is assumed that there will be some reduction in the total effort if similar ships are built even if these may not be sister ships.

TABLE 4.2

Reduction of Effort in Series Construction of Ships

Number of the Ship in Series	1	2	3	4	5	6	7	8	9	10
y	1.0	0.8972	0.8371	0.7944	0.7613	0.7343	0.7114	0.6919	0.6742	0.6585

4.5.6 Production Rate

Production rate can now be defined as cgt per man-hour and is a function of a number of variables which are not easy to define. Some of the factors can be stated as follows: (1) sophistication of equipment, lifting capacity and facilities; (2) yard size or total area of building berths and (3) use of modern management practices utilizing modern IT practices and product lifecycle modelling. Based on these technology factors, the National Shipbuilding Research Program of the United States (2001) has defined six levels of shipyards (Lamb 2007). The first level of shipyard has the so-called traditional shipbuilding practice with least sophistication of equipment and using slipways for ship construction. Sophistication of equipment, shipbuilding and management practices improves with the increasing level reaching the highest sophistication and minimum crane lifts at level 6.

The production rate increases with increase in the level of the shipyard. Sometimes production rate is also given as cgt per m^2 of shipyard area or shop floor area or building berth area since all these have a bearing on production as discussed earlier. Further, production rate also varies based on the workload in a particular shop, being maximum with optimum work load. If workload is less, the production rate reduces resulting in idle labour. On the other hand, if the workload is more than the optimum capacity, errors in work result causing increased rework. Thus, production rate varies widely from nation to nation, reflected in the sophistication of the yard, and also in the same country based on yards' order book position. This is reflected in Table 4.3 prepared in 2012.

4.5.7 Financial Complications

Considering financial terms in a small established shipyard which builds small boats and launches or in a well-established medium or a large shipyard, there may not be any financial input to the shipyard which needs to be taken care of in the shipbuilding cost under overhead expenses. On the other hand, in a shipyard which is being upgraded or a new shipyard, there is a large amount of financial input due to infrastructure build-up, which may reflect in the shipbuilding cost under overheads. Of course this expense should be compensated against improved production rate. It is well known that financial inputs play a major role in shipbuilding cost, but it is rather difficult to define it exactly. Many a time,

TABLE 4.3

Shipbuilding Productivity in Various Countries

Area/Country	Year	MH/CGT
Europe	2007	12–15
Japan	2007	9–15
Korea	2007	16–21
China	2007	52–103
India	2007	42–117

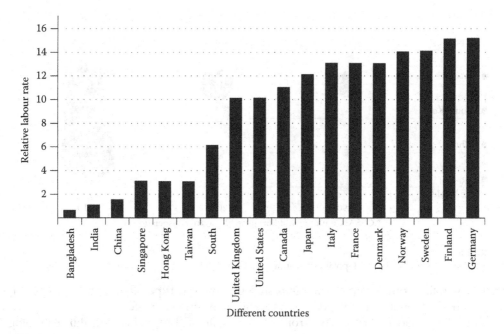

FIGURE 4.7
Relative labour rate in different countries.

the builder may compromise with material and construction quality for better cost control, which also compromises with design and performance.

4.5.8 Labour Rate

Labour rate is related to the living standards of the country of the shipyard. As can be expected, labour rate is low in developing countries and high in developed countries. Figure 4.7 (Zakaria et al. 2010) shows the relative labour rate in various countries around the world. The ratio of the highest to the lowest labour rate can be as high as 20 times. Thus, even if the level of shipbuilding yard in developed countries is higher, the total ship cost may be lower due to low labour rate. This is reflected in the recent order book positions around the world, as shown in Figure 4.8 (Sea-Europe Ships and Marine Equipment Association 2012).

4.5.9 Stages of Building Cost Estimation

Costing is normally done at three different stages: pre-contract cost estimation, contractual cost estimation and final or actual costing.

4.5.9.1 Pre-Contract Cost Estimation

Pre-contract cost estimation is done at the preliminary/concept design stage so that a quick analysis of various feasible alternatives with regard to cost can be made and a suitable decision for an optimum design taken. Also the quoted price is based on the estimated cost at this stage, which must be fairly dependable. At this stage, since costs of detailed items of material and labour are difficult to estimate, previous cost data are analysed with regard

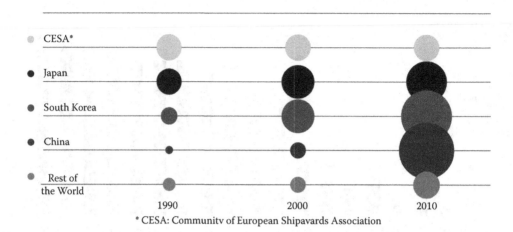

* CESA: Communitv of European Shipavards Association

FIGURE 4.8
Shipbuilding indicator (order book position) around the world.

to a few variables only, such as the main ship parameters, type of ship and major equipment such as propulsion plant, generators, cargo-handling gear, etc.

Taking the statistical data of cost from an aggregate of shipyards of similar technology level, one can estimate cgt, cgt per man-hour and labour rate and thus arrive at the first estimate of building cost after applying corrections for all the factors discussed earlier.

If one has adequate data for detailed costs for a shipyard, the cost of a new ship can be estimated by analysing these data. Several methods are available for this purpose: two of these are by Carreyette (1978) and the Product-Oriented Design and Construction cost model of estimation of the U.S. Navy (Miroyannis 2006). Carreyette has suggested the total building cost C_t can be given by

$$C_t = \frac{A \cdot W_s^{2/3} \cdot L^{1/3}}{C_B} + B \cdot W_s + C \cdot W_o^{2/3} + D \cdot W_o^{0.95} + F \cdot P^{0.82} + G \cdot P^{0.82}$$

$$= \frac{A \cdot W_s^{2/3} \cdot L^{1/3}}{C_B} + B \cdot W_s + C \cdot W_o^{2/3} + D \cdot W_o^{0.95} + E \cdot P^{0.82}$$

where A, B, C, D, E, F and G are constants based on shipyard data where A, C and F are constants for steel, outfit and machinery labour (including overhead) cost and B, D and G are the respective material cost constants; $E = F + G$, W_s and W_o are steel and outfit weights, respectively, and P is the engine power.

The advantage of this formulation is that one can quickly estimate the cost of a new ship based on difference principle. Using Katsouli's formula for C_B (Watson 2009), with ship parameters such as length, breadth, draught and depth constant, as

$$C_B \infty V - 0.6135$$

where V is ship speed and using difference equations for small change in V, the increase in ship cost is

$$\delta C_t = 2.125(E \cdot P^{0.82})P \cdot \frac{\delta V}{V} + 0.2045 \left(\frac{A \cdot W_s^{2/3} \cdot L^{1/3}}{C_B} \right) \cdot \frac{\delta V}{V} - 0.6135(B \cdot W_s)\frac{\delta V}{V}$$

The change in building cost due to small changes in main parameters such as L, B, T, D, C_B or size can be estimated in a similar manner.

The data from various sources regarding shipbuilding cost indicate that steel material cost is between 11% and 15%, propelling machinery cost is between 14% and 20%, other machinery and equipment cost is between 17% and 25%, total labour cost (including sub-contractor's labour) is between 17% and 25% and overhead costs are between 24% and 32% of the total building cost.

4.5.9.2 Pre-Contract Cost Estimation of Value-Added Structures and Vehicles

The data given earlier are based on ships which are built in large numbers across the globe, such as bulk carriers, tankers and container ships. The cost components of such ships normally fall around the middle of the range suggested. But in value-added ships (a ship with high value of equipment and facilities for specific mission purpose) and structures such as oil rig platforms, ocean research vessels, pollution control vessels, etc., there will be a larger component of equipment, and cost in that category would move towards the higher end of the range. Similarly, steel material cost may go up if additional structural requirements, such as ice strengthening and use of alternative materials such as high speed or high-strength low-alloy steel, are contemplated. Therefore, in such vessels, cost components are to be suitably modified at the preliminary stage itself.

Naval platforms are considered large value-added structures. The combat equipment system in a naval vessel is a very large value addition, which increases the work content in terms of weapon and ancillary system design, system integration with the floating and mobile platform, installation, tests and trials. This combat system cost could be as high as 55%–60% of the total cost, leaving 40%–45% only as the platform (ship) cost.

4.5.9.3 Contractual Cost Estimation

The contract for building a marine structure or vehicle is obtained based on the quoted price, which is a function of the pre-contract stage cost estimation. At the contract-signing stage, it is necessary to estimate the cost again to ensure that costs are within control. At this stage the contractual drawings, detailed specifications and material list are prepared, and so it is possible to estimate the detailed material cost, labour cost based on the estimated labour content and the yard's production rate and the overhead cost.

4.5.9.4 Actual Costing

Actual costing is carried out during the construction of the ship by noting the materials and labour booked against the ship. This detailed costing is necessary to generate the required formulation for shipbuilding cost based on the enterprise.

4.6 Determination of Price

Maritime structures and vehicles operate at the international environment following international economic norms. Therefore, the price of such a product is determined based on the internationally competitive environment. The demand for sea-borne trade determines

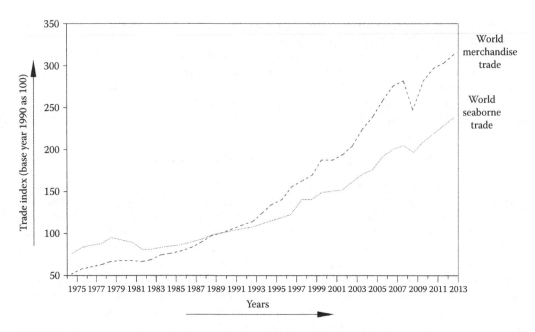

FIGURE 4.9
World trade development (including sea-borne trade) since 1975.

the requirement of ships at any time. If the availability of ships is more than the requirement, the price of a new ship is likely to decline. If the ship availability is not adequate, the demand for new shipbuilding is likely to be high, with ship pricing showing a higher trend. Thus, ship pricing follows the supply and demand relationship as per standard economic norms. The ship availability is not solely dependent on transport requirement but also on ship scrapping/demolition trend and technology demands for demolition (e.g. regulatory double-hull tankers eased out single-hull tankers). Since shipbuilding is a long process, the supply of new ships has a time lag with the demand. Figure 4.9 shows the sea-borne trade growth in the world in the last few decades (UNCTAD Report 2013). It can be observed that sea-borne trade had a dip in 2008–2009 due to the worldwide depression, but now it is showing and is likely to show growth in the near future.

The shipbuilding market has shown impressive growth following the demand for ships in the early part of this century, particularly for bulk carriers and tankers. Figure 4.10 developed from the data of the Shipbuilders' Association of Japan (2013) shows the jump in ship order book position from then on, peaking around 2009–2010. This peak was perhaps a bit unnatural and was not likely to last. Shipbuilding is known to be a cyclic industry moving between crests and troughs of ship demand and, as was expected, the demand for shipbuilding has reduced. Clarkson's market research group has collected and statistically analysed a large amount of shipping and shipbuilding data. Figure 4.11 (Stopford 2009) shows shipbuilding demand and deliveries. This figure also shows the excess delivery done in 2009–2010. Further, Clarkson's extrapolation of data indicates that shipbuilding is likely to grow again in the near future. Clarkson's data are collected for three main categories of commercial ships: tankers, bulk carriers and container ships. The prices for such ships have also been analysed and a ship price index has been established with 1988 price as 100. The variation of this index is shown in Figure 4.12 developed from the data published by Clarkson's research. It can be observed that the index has fluctuated more or

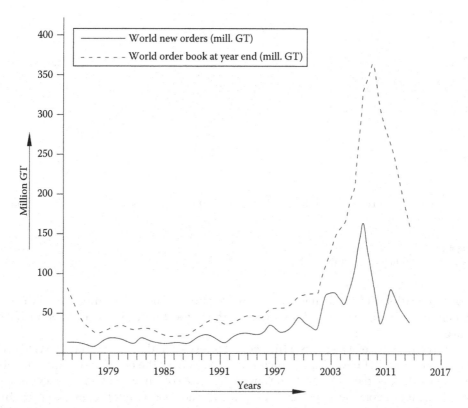

FIGURE 4.10
Shipbuilding order book position since 1975.

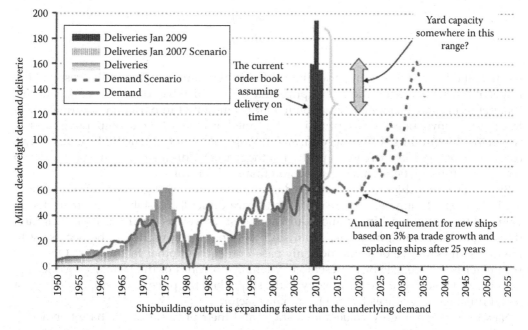

FIGURE 4.11
Demand of ships since 1950 projected up to 2035 and deliveries.

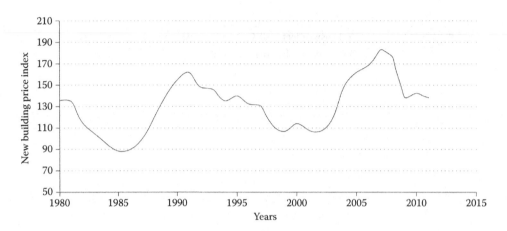

FIGURE 4.12
Clarkson's new building price index for ships.

less based on shipbuilding demand with a high value in 2007–2009 after which the index has fallen due to a fall in demand. In any case, the highest index has been around 170 and the lowest is 135 in 2012.

The contractual price of a ship is a function of international price standards based on the supply and demand of new ships, which must also satisfy the enterprise requirements of business growth based on shipbuilding cost. If the price of a new conventional commercial vessel (P_n) is under consideration having cgt_n, it should follow that the normal price trend dictated by the price index extrapolated from a similar known ship's (cgt_b) price (P_b) in recent past indicates

$$P_n = P_b * \frac{cgt_n}{cgt_b} * \frac{I_n}{I_b}$$

where I_n and I_b are shipbuilding price indices for the new ship and the similar vessel in the respective years of construction. This price may be modified based on financing patterns, contract terms and currency fluctuation if a large amount of imported items are used. In the case of conventional ships, price is determined at international competitive rates to which the shipbuilding cost must confirm. The shipbuilder may intentionally take such an order anticipating a loss to catch a portion of the market. This also compels the builder to reduce cost through better management and financial control.

But value-added ships and structures do not follow this pricing procedure since these are built in small numbers where price trends are difficult to establish. Further, due to value addition, the machinery and equipment cost as well as overhead cost, including design and integration, may vary from ship to ship. Therefore, it may be necessary to modify a known price of a ship not only based on cgt, but other factors also. One such parameter is the area factor represented by length * breadth of the factory or workshop. Another factor could be the technology of the ship. Then the price can be represented as a direct function of cgt * area factor * technology factor.

Naval vessels are highly value-added ships, and their price should be based on actual building cost estimation.

4.7 Design versus Tendering and Contract

Ships and marine structures are large products involving large financial investment; therefore, tendering and contract signing must be done with adequate care. Tender documents consist of a preliminary general arrangement drawing, brief specifications and the quoted price. The documents are compared against internationally competitive bids. Therefore, it is necessary that the feasibility study at the concept design stage must be done carefully even if extrapolated from a basic design. Due to large investment, the client is very often interested in confirming performance standards at this stage itself, which may require numerical analysis, including computational fluid dynamics and finite element analysis. The quoted price must be done with care so that it is not only competitive but also satisfies the needs of the enterprise.

At the contract-signing stage, no change in the main parameters or major equipment identified during the tender stage should be made. The technical documents of the contract should include the basic design drawings and detailed specification. Complete information on hull form, structural design, hydrodynamic characteristics and details of all systems and subsystems should be available at this stage, specifying all performance standards confirming to statutory rules and regulations. Changes in the contract or any additions/deletions from the contract should be discouraged after contract signing since the builder and the client have to agree on these through mutual discussion, resulting in delay and, in some cases, rework.

The financial part of the contract can be of two types: (1) fixed price contract subject to clauses of escalation due to undue and unforeseen rise in prices of major cost items and (2) 'cost plus' contract where the price is as per the actual building cost plus a predetermined profit. In this case the accounts of the builder are to be scrutinized by the client before any payment is made. This type of contract is negotiated only when a product is of a new type where estimating building cost could be difficult and erroneous. Since a marine product takes a long time to deliver, starting from contract signing, the payment of contractual price is made in stages so that the builder's expenditure is compensated at the right time. This means that building stages must be identified when a portion of the payment can be made. The identified stages may be some or all of these: contract signing, 50% steel erection, main engine installation, transfer to water (launching), dock trials, sea trials and delivery or other identified stages arrived at by mutual discussion.

Guarantee items in a marine product include all performance standards, including construction standards. During the sea trials, performance characteristics are recorded and compared with specified criteria. It must be ensured that the trial conditions are clearly specified since the sea is not stationary at any time. Construction standards are ensured by (normally) the classification society during design evaluation and construction. Regulatory performance is ensured by design checks and presence of flag state control authorities during the sea trials. The machinery and equipment, which are purchased items, are also guaranteed by the builder through an agreement with the supplier of such items. In any case, the builder provides a guarantee engineer on board for the specified guarantee period. Failing any of these technical criteria or default in construction time or default in stage payment, penalty is also specified in the contract. In the case of excessive deviation from the specified standards, the client has the right to reject the ship and claim refund of payment with penalty. Similarly, if the client is unable to take delivery of the product, he has to compensate the builder with adequate penalty.

The contract is also a legal document which specifies how the client and the builder can resolve disputes. If a building schedule is disrupted and delivery is delayed, the builder suffers heavily. The following are the reasons: (1) the real value of contractual money received by the builder at the delayed periods falls and there is a substantial loss, (2) loss due to payment of penalty due to delayed delivery, (3) disruption of build schedule of the shipbuilder due to delay in one ship and consequent expenditure under 'overheads' and (4) loss of goodwill of the shipbuilder in the market and loss of contracts thereof.

4.8 Engineering Economics Application to Ship Design

Any design of marine structure or vehicle must be evaluated with respect to its economic performance, and thus the best alternative should be selected from many engineering solutions based on different criteria (Buxton 1971). The operating environment for different marine structures and vehicles may vary widely. Any marine product can be economically evaluated once its operating conditions can be established. As an example, commercial ship is considered here for its economic performance.

4.8.1 Ship-Operating Economics

The ship owner's responsibility towards various items of ship-operating expenses can be under four major heads: (1) capital charges, (2) daily running costs, (3) voyage costs and (4) cargo expenses. Capital charges cover items such as capital repayment and interest as well as all expenses related to investment, profit and taxes. The full calculation of effective capital charges can be complex. Voyage costs cover fuel cost, canal dues and charges at ports of call. Voyage costs vary considerably from trade to trade and the ship's operating condition. Daily running costs are those incurred on a day-to-day basis whether the ship is at sea or at port, which includes crew wages and benefits, victualing, ship upkeep, stores, insurance, administration and management. Daily running costs are largely a function of type, size and flag of registration of the ship. Cargo is normally handled by stevedores at ports. Stevedore charges, cargo insurance and cargo claims due to pilferages and damage come under cargo expenses, which are truly related to voyage expenses.

In commercial shipping, the owner need not be the operator. In most cases, the ship is operated by the charterer, and depending on the type of charter and contractual terms, ship owner and ship operator share the expenses. The charter collects the freight charges and pays the owner the ship hire charges. Figure 4.13 shows the major expenses and how these are shared.

4.8.2 Application to Ship Design

Marine transport and service requirements must be developed into a series of feasible ship designs, which must then be evaluated for their technical and economic performance covering the following:

- Trade pattern and operating environment
- Range of feasible technical designs

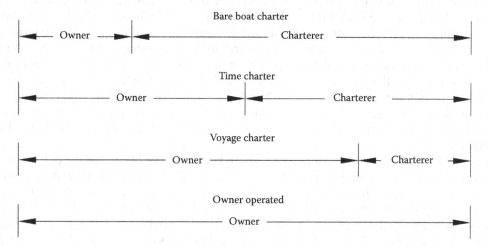

Capital charges	Daily running cost	Voyage cost	Cargo expenses
Loan repayment	Crew expenses	Fuel cost	Cargo handling
Loan interests	Maintenance and	Port charges	Cargo claims
Taxes	repair	Canal dues	
Return after tax	Stores		
(Depreciation)	Insurance		
(Profit)	Administration		

FIGURE 4.13
Components of ship-operating expenses and sharing between owner and operator.

- Estimation of building and operating costs and income
- Economic evaluation of alternatives

The ship designer's concern is the optimal design satisfying the operator/owner and the shipbuilder based on two fundamental principles:

1. *Specialized knowledge*: To design the optimal ship, extensive experience of the influence of different design features on the first cost is required. The designer should quantify accurately the cost of alternative hull proportions, materials, machinery, arrangements, etc.

2. *Commercial competition*: Through the ship designer, the shipbuilder must show that this design is not only the cheapest but also the most profitable. For industrial or naval vessels, 'most profitable' may be replaced with 'having minimum operating expenses'.

A modern approach aimed at improving the designs of ships requires good collaboration between the owner, the builder and the designer, who become partners is making an optimal ship. Based on market research and transport demand, concept evaluation should be done taking into account the owner's needs and experiences. From a number of feasible designs, an optimal design should be so chosen that it should be easy to build also. This selection should be done based on either mathematical optimization or judgement. Then the design should be firmed up after discussion with the builder and the owner. Contract and detailed design should follow from there.

4.8.3 Comparison of Alternative Designs

A typical situation faced by the designer is to compare alternative designs considering both technical and economic factors. The comparison of alternatives does not need to be based on the entire design. This may be done on individual features such as different cargo-handling systems, different propulsion systems or different materials for the piping systems. Such features are straightforward to analyse economically when they do not affect earning capacity. The alternative first costs and maintenance costs are evaluated in terms of annual cash flows and converted to present worth to find the system with the highest NPV.

In practice, most alternative designs differ not only in building and operating costs, but in performance, so that care must be taken to include second-order effects. For example, better cargo-handling gear may not only save on operating costs, but also reduce port time, thereby carrying more cargo per annum. The secret of success in comparing alternative designs is to obtain sufficiently realistic data and use an appropriate method of economic analysis.

One of the ways of doing economic evaluation is to calculate NPV over the entire life of the ship as a project. A simplified method of cash flow calculation would be to do a tabular calculation with columns indicating the various expenditure and income items, the last column being the present worth of the annual cash flow. The rows should represent the annual cash flow from 0th year, incremented by 1 until the Nth year, when the ship would be sold as scrap or as a second-hand vessel, this being an income. In a simplified manner, the variables can be calculated as follows:

For a typical commercial vessel, a round trip means that the ship is to start from a port, travel to another port (or more than one ports) and return to the same port.

Distance travelled in the round trip is the range in nautical miles. If the ship touches only one port in its voyage, range is twice the distance between the two ports, which is normally the case for bulk carriers and tankers.

Sea days per round trip (SD) = Range$/(24 * V)$ where V is the service speed in knots, which should be a realistic value taking into account weather, fouling (speed reduces gradually between two dry docking periods due to fouling), passage through narrow and shallow waterways, loading conditions resulting in variation in draught (particularly in ballast condition).

Port days per round trip (PD) = Number of ports of call after leaving origin port $*$ Average duration at each port. Port duration should allow for waiting time, berthing time, delays, etc.

$$\text{Number of round trips per annum}\,(\text{RTPA}) = \frac{(365 - \text{days ship out of operation})}{(SD + PD)}$$

Maximum payload (metric tonnes) = Deadweight − weight of consumables and non-cargo weights such as ballast. In volume-limited ships, maximum payload = maximum volumetric capacity/stowage factor.

Load factor (LF) is the ratio of the actual cargo carried per annum and the maximum cargo-carrying capacity, which is given by

$$\text{Load Factor} = \frac{\text{Actual tonne-miles per annum}}{\text{Maximum tonne-miles per annum}}$$

$$= \left(\frac{\text{Average cargo payload on loaded voyage}}{\text{maximum cargo payload}} \right)$$

$$* \left(\frac{\text{Average kn-miles steamed with cargo}}{\text{Total kn-miles steamed}} \right).$$

Fuel consumed at sea per day (t) = Service power * SFOC * ($24/10^6$) + diesel oil for electrical power at sea per day where SFOC is the specific fuel oil consumption in g/kW·h.

Fuel consumed at port per day (t) = Diesel oil for electrical power at port per day.

Total fuel consumed per round trip (FPRT) = Sea fuel * SD + port fuel * PD (Maximum fuel load carried will depend on the location of bunkering ports, prices, reserve fuel and bunker capacity of ship. Typical reserve capacity is 10%–20% of fuel carried or 4–6 days of steaming.)

- Cargo carried per annum (CCA) = Maximum payload * RTPA * LF
- Cargo charges (mainly cargo handling) = CCA * cost per unit
- Port charges per annum = Average charge per day at port * PD
- Canal dues per annum = Average canal dues for one passage * number of passages per annum (Normally, these two charges are based on GT and NT.)
- Daily running cost = Crew + maintenance + upkeep + miscellaneous (This component of cost is obtained from a shipping company's annual cost figures.)
- Fuel cost per annum = FPRT * RTPA * unit price of fuel (If the ship has a dual fuel system, separate fuel costs should be estimated.)
- Capital charges are to be calculated based on actual loan terms.
- Freight earning = CCA * freight rate per tonne or tonne mile

(If the distance between ports of call is same, the freight rate is per tonne. But if the distance between ports of call is not same, then the freight in one leg of journey will be different from another leg. It is possible to get average freight rate per tonne mile.)

- Annual cash flow must be discounted by the present worth factor and summed over N years to get the NPV. Figure 4.14 shows a typical cash flow calculation for a large ocean-going merchant vessel. It can be observed that ship operation is generally a long-term investment where, in initial years, there is a loss due to large repayment of capital and interest. But the situation changes to profit making once the capital is paid back.

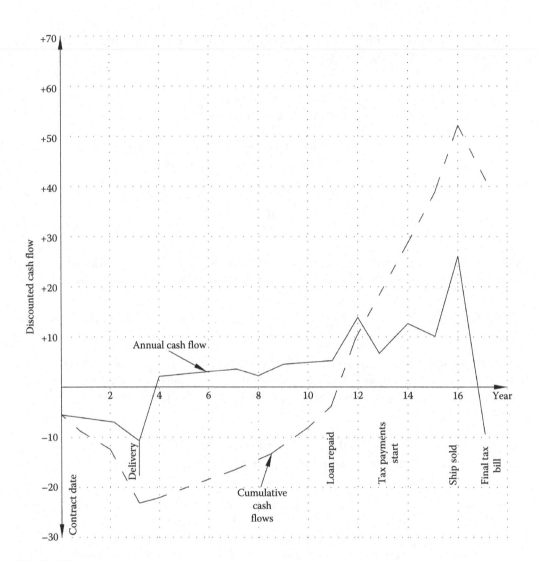

FIGURE 4.14
Annual and cumulative discounted cash flow.

4.8.4 Uncertainties in Ship Design

It can be observed that cash flow calculation is fairly complicated. It is difficult to express the economic criterion as a mathematical function of economic and technical variables. However, it is possible to express this in numerical form, as has been shown in Figure 4.14. Further, in such calculations, the economic variables may or may not be totally independent and their dependence may not be known exclusively. Some of these may be heuristic in nature. A cash flow calculation is a good way to use in such conditions. In any design activity, it is necessary to optimize the identified optimizing function expressed as a function of the design variables. Since the cash flow calculation is a highly nonlinear subject with rather difficult constraints, it may be difficult to go for a mathematical optimization process. In such a case, various calculation methods can be used to do sensitivity studies as well as to find one or a number of optimized solutions. A few methods are discussed as follows:

- The results of a techno-economic ship design process may be sensitive to changes in the data, because there may be uncertainty about many of the technical and economic parameters. For example, it is not possible to predict exactly the fuel price over the life of the ship, nor is it possible to predict other factors like port time, maintenance cost, load factor, etc. The simplest way of investigating such uncertainties is to repeat the calculation with different values of key parameters and assess how sensitive the results are to such changes. Figure 4.15 shows a typical presentation of such calculations with the economic measure of merit plotted against key parameters for different design solutions. Where the curves of alternate designs do not cross, the ranking is not changed, but where there is a crossover, the decision to be made is whether the operating situation is likely to be to the left or right of the crossover.

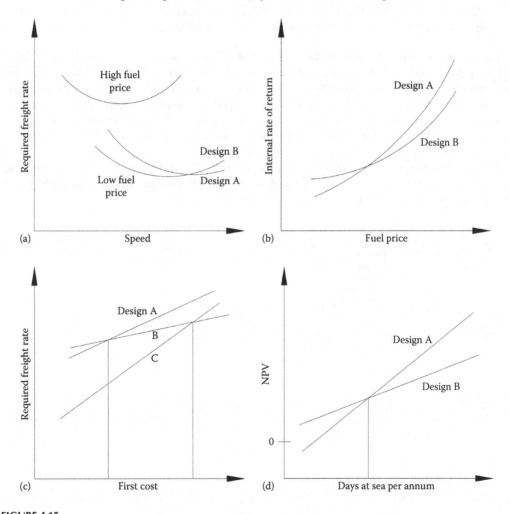

FIGURE 4.15
Decision-making in ship design based on economic criteria: (a) RFR vs. ship speed, (b) IRR vs. fuel price, (c) RFR vs. ship first cost, and (d) NPV vs. sea days per annum.

- It is also possible to make trade-offs on design decisions based on results of sensitivity calculations, e.g. how much extra first cost can one afford to pay to obtain a reduction in fuel consumption. The decrease in NPV from, say, a 10% increase in the first cost can be compared with the percentage decrease in fuel consumption needed to generate a corresponding increase in NPV.

- Another way to understand the effects of uncertainty of all the independent economic variables, such as fuel oil price, port days, maintenance cost or freight rate, is to model the probability density of each such variable by collecting large number of data from across the world. Then one can model the probability density of NPV by using the Monte Carlo simulation technique where one could generate a random number, get the value of the corresponding variable based on its probability density, get such values for all variables and calculate the NPV. The probability density of NPV can give the mean and the standard deviation of NPV, which can be used to take a decision on the selection of the design solution. If the data sample of each variable is very large and the distribution of each is well defined, one can use the central limit theorem to arrive at a normal distribution of NPV, its mean and variance.

For example, if crude oil price data are collected (U.S.$/barrel) and plotted to predict a future price based on extrapolation, it may be difficult, as seen in Figure 4.16. In this figure, 40 years' annual average crude oil price at four locations (until 2012) and international average price (until the beginning of 2015) are plotted, but as can be seen there is a spurt in fuel oil price after 2000 until about 2012 and then a sharp drop. It is expected that this price will stabilize in the short term. Therefore, these data can be used neither for extrapolation nor for predicting future spurts in fuel oil price.

An example of price of heavy fuel oil, types IFO 380 and IFO 180, collected for the 4-month period between January 2014 and April 2014 all over the world is modelled and the histogram plots are shown in Figures 4.16 and 4.17, respectively.

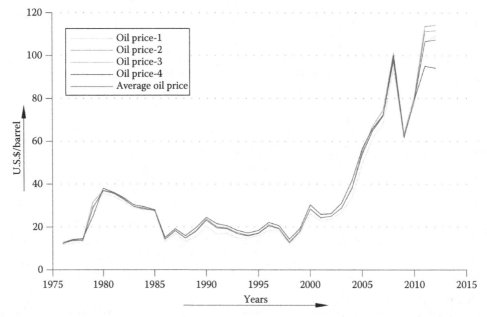

FIGURE 4.16
Annually averaged fuel oil price internationally and at four locations.

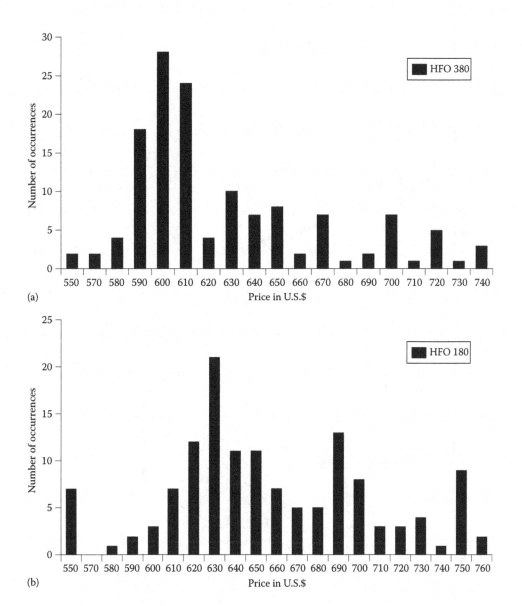

FIGURE 4.17
Histogram of international heavy fuel oil prices per ton for the period between January and April 2014: (a) IFO 380 and (b) IFO 180.

4.8.5 The Optimal Ship

Some examples of decisions on optimal ship design are as follows:

1. *Optimal ship size for a given speed*: For a bulk cargo or crude oil trade where there are no restrictions on ship size or cargo availability, the economies of scale in building and operating costs indicate that the optimal ship, in general, is the largest possible, offering the lowest transport cost. The situation is shown in Figure 4.18. The top half shows a curve of freighting cost (FC) per metric tonne against ship size. One particular freight rate (FD) is shown. The lower half shows the annual

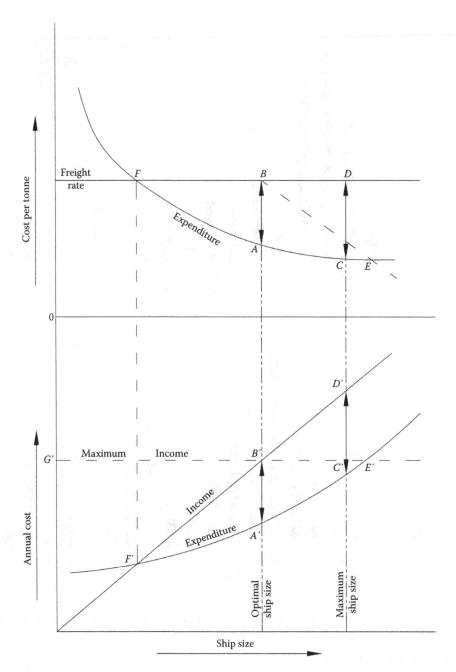

FIGURE 4.18
Optimal ship size for a constant speed.

expenses at present worth, which is obtained by multiplying unit cost FC with freight-carrying capacity. Maximum NPV is obtained at CD with the maximum permissible size of the ship for the trade. This size may be determined by a number of physical restrictions, particularly depth of water, canal restrictions, shallow water en route, etc. There may also be limitations on cargo availability. In this case, an upper bound is set on the freight income (shown as line G′E′). Here maximum return occurs at $A'B'$, and any increase above this optimal size merely increases expenditure, including capital charges, while income remains constant at $G'E'$. A similar effect is obtained if loading and discharging rate is slow compared with ship size. Port time normally increases with size, thus reducing the number of voyages, which may be compensated by providing faster cargo-handling facility.

2. *Optimal ship speed for a given size*: Figure 4.19 illustrates diagrammatically the effect of ship speed on the total cost and total income. Generally, increasing ship speed does not have a great effect on the hull and equipment part of the first cost apart from the secondary effects on dimensions due to reduced block coefficient to keep the payload constant. Likewise, crew cost and other components of daily running cost are not affected appreciably by change in speed. The propulsion power,

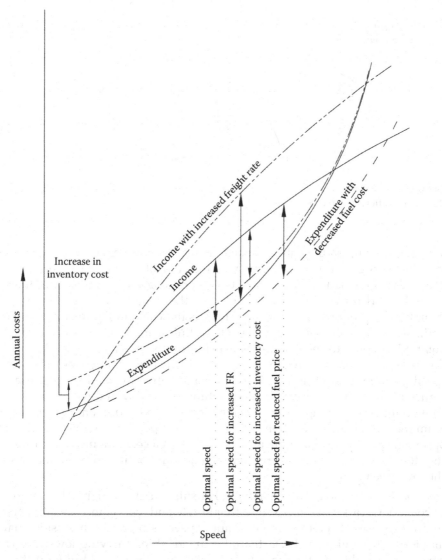

FIGURE 4.19
Optimal ship speed for a given payload.

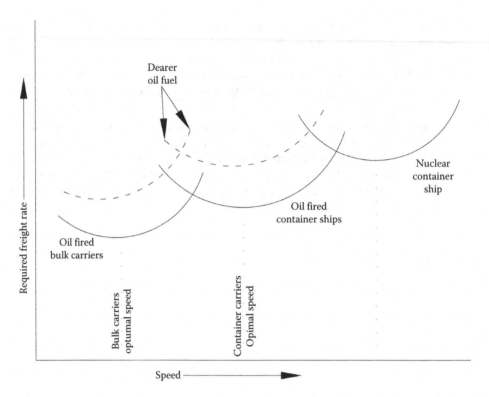

FIGURE 4.20
Optimal speed for different ship types.

however, goes up as the cube of speed, increasing fuel consumption and, therefore, fuel cost goes up approximately as the cube of speed. Machinery first cost
goes up roughly as the square of speed, increasing the capital cost of the ship.
Since the port turnaround time is not zero, the freight earned can be only less
than directly proportional to speed. Thus, as indicated in Figure 4.19, there is an
optimal speed which is a function of both technical and economic factors at which
point NPV (equal to income minus expenditure) is maximum. The NPV variation is lens shaped with regard to speed, which shows flatness at the optimum
speed position, indicating that it is not too sensitive to small speed variations.
Figure 4.19 also shows the variation of optimum speed with three other variables:
(1) with increase in freight rate, income increases indicating an increase in optimum speed; (2) with a reduction in fuel oil price, expenditure reduces pushing the
optimum speed point for an increase and (3) with a constant increased inventory
cost, the slope of the expenditure curve changes and optimum speed increases, as
shown in the figure.

3. *Optimal ship speed for different ship types*: For ships carrying high-value cargo such
   as containers, cars, passengers, etc., freight rate and, hence, income are high and
   so are expenses. Therefore, the optimum speed is high. So these ships have a
   fine form and high speed. On the other hand, ships carrying low-value cargo
   have a low freight rate; therefore, bulk carriers are slow-speed full form ships. In
   case the designer wishes to consider using an expensive propulsion plant such

as a nuclear-powered ship, the capital cost goes up and also running cost may increase. It is necessary to increase income by increasing the number of voyages. Therefore, optimum speed increases. Figure 4.20 indicates the optimum speed in all these cases. The curves in the figure are quite flat near the optimum point. The exact optimum point may move to the right or left during the ship's lifetime due to fluctuations in economic or operating conditions. This may also happen at the design stage due to small variations in engine power (diesel engines are normally available in steps of power) with the possibility of being slightly away from the optimum point. Due to the flatness of the curve, these small variations do not make appreciable variation in NPV.

Though only ship speed and size are discussed earlier, similar analysis can be done for many other variables and one can select them based on optimized vessel performance.

5

Vehicle Parameter Estimation

The first step in designing a marine vehicle is the selection of the main parameters which define the size of the vehicle. The size must ensure carriage of the requisite payload or deadweight during operation at sea according to the volume or capacity available in the ship. The size of the ship must also satisfy the technical performance requirements of the ship as defined in the mission requirements. It is difficult to do all scientific calculations at the beginning when the ship has not been fully defined, and one has to take recourse to statistical data and experience to ensure selection of parameters satisfying technical requirements. Thus, the parameter selection process has evolved over the last hundreds of years based on the data of ships built so far. This process can also be extended to determine the size of new types of vehicles and structures. This chapter discusses the parameter selection method as the first step of the vehicle design process.

Since ships and ship calculations have evolved over a large number of years, various nomenclatures of ship terms have also evolved, mostly on a scientific basis. Though the reader may be aware of many of these terms, it is necessary to state briefly the nomenclature used in ship design and calculations. It is possible that due to the changes in the conventional form and arrangements of ships, there could be some confusion in nomenclature, and in such cases, scientific description of various terms is necessary.

5.1 Ship Nomenclature

A ship's end which moves forward is known as the forward end or stem, and the rear of the ship where the propeller(s) and rudder(s) are housed is known as the aft(er) end or stern. Looking towards the forward end, the left side of the ship is the port side and the right side is the starboard side. Geometrically, ships are symmetrical about their longitudinal centre plane so that the shape of the vessel on the port side is the mirror image of the starboard side. The various terms for defining ship particulars are described below and are shown in Figure 5.1.

Generally, in ocean-going merchant vessels, the baseline coincides with the keel line (or the bottom line) of the ship. In smaller vessels such as tugs, trawlers or twin screw vessels, the keel line may be inclined with respect to the base line. The load water line (LWL) of the vessel, that is the line up to which the vessel sinks in normal fully loaded service condition, is parallel to the baseline. The height of LWL above the baseline is the draught (T) of the ship. The perpendicular to the baseline drawn through the point of intersection of the LWL and the stem contour is called the forward perpendicular (FP). Similarly, the perpendicular line passing through the centre of the rudder stock or the aft end of the rudder post, if there is a rudder post, or the centre of the rudder pintle, as in the case of a spade rudder, in the aft end of the ship is known as the aft(er) perpendicular (AP) of the

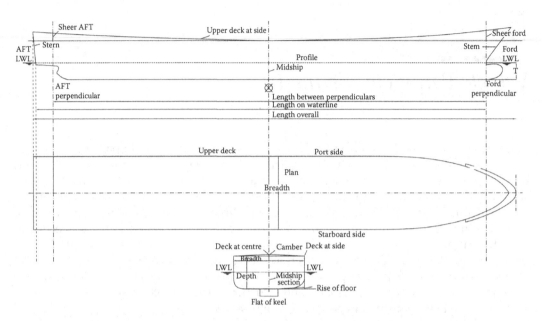

FIGURE 5.1
Ship particulars.

ship. In case the rudder stock position is not defined, as in the case of vessels having azimuthal or water jet propulsion system or in some cases of naval vessels, the AP is taken as the perpendicular through the point of intersection of the aft end of the ship and the LWL. The longitudinal distance between AP and FP is the length between the perpendiculars (L_{BP}) of the ship. The distance between FP and the intersection of the aft end of the ship with LWL is the length on water line (L_{WL}). The length between the extreme ends of the ship is the length overall (L_{OA}). It can be observed that $L_{OA} > L_{WL} \geq L_{BP}$. For normal single screw merchant ships, L_{BP} = 95%–96% of L_{WL} and L_{OA} depends on the forward and aft extent of the ship beyond AP and FP due to erections above water and projection of bulbous bow below water.

The cross section of the ship at the middle of the L_{BP} is the midship section, which is normally the maximum cross section of the ship. The maximum horizontal distance between the sides of the ship at LWL perpendicular to the ship centre line is the breadth (*B*) of the ship, which normally occurs at the midship section. The height of the ship's upper deck at side from the baseline (or keel line) along the length of the ship, normally at midship, is the depth (*D*) up to the upper deck. The upper deck of the ship may be horizontal with constant height from end to end or the deck may be raised towards the forward and aft ends from the midship. The raise at the FP above the midship deck is the forward sheer, and similarly the raise at the AP is known as the aft sheer. The sheer curve, both forward and aft, can be either straight or parabolic. Normally, if there is a sheer, sheer (ford) > sheer (aft) and often sheer (ford) is twice sheer (aft). The purpose of giving sheer is primarily to increase the internal capacity (volume) of the ship without altering the defined depth. If the upper deck is straight in the cross section, i.e. the deck at centre and the deck at side are at the same height, the ship is said to have no camber. If the deck at centre is raised above the deck at side, the rise is known as camber. The camber line, which is the cross-sectional line of the deck, could be straight or parabolic. Thus, in profile view the deck at centre could be higher than the deck at side in the event of a non-zero camber. The purpose of a camber is to give increased internal volume and also to provide a natural downward slope to the deck at side, providing facility for discharge of any liquid on deck. Thus, the under-deck volume of the ship is the total capacity of the ship, consisting of the under-deck cargo space; under-deck space for carriage of liquids such as heavy fuel oil, diesel oil, lubricating oil, fresh water and ballast water, any empty space, space for carrying solids other than cargo such as chain locker and stores, and engine room volume. The dimensions defined so far are the moulded dimensions, which are the dimensions based on internal measurements of the ship. The capacity or volume of the ship or any compartment enclosed by moulded dimensions is the moulded capacity. The volume of the structural members inside any compartment, such as transverse and longitudinal stiffeners, decks and girders inside the hull and similar other structures, is not utilized in carrying cargo. This volume, therefore, is deducted from the moulded capacity to get the volumetric carrying capacity of a particular compartment. The structural volume is normally taken between 3% and 5% of the moulded volume. The total enclosed volume of the ship, including that above the upper deck, multiplied by a coefficient, is the gross tonnage (GT), and the portion of this volume utilized for the carriage of cargo or passengers, multiplied by a coefficient, is the net tonnage (NT). GT and NT are the volumetric measures of the ship, which are calculated based on the Tonnage Measurement Rules as per the International Maritime Organization (IMO) resolution on the subject in 1969 and do not have any relationship with the weight measure of the ship.

If the ship floats at even keel, draught forward (T_F), draught aft (T_A) and draught at midship are same. If it is not so, then the vessel is said to have a trim by aft = $T_A - T_F$ or trim forward = $T_F - T_A$. The draught at which the ship is designed to operate is known as the design draught or T(design). The deck which is maintained water-tight and makes the whole ship a water-tight envelope is known as the freeboard deck. Freeboard is measured from this deck as the difference between the depth up to freeboard deck at midship and draught. The minimum freeboard required to be maintained is the statutory freeboard specified by the International Convention on Load Line (ICLL), 1966 under the aegis of IMO. The statutory maximum draught the ship can attain is the freeboard draught or T(freeboard), which is the difference between the moulded depth at freeboard deck and the statutory freeboard. In no case the ship is allowed to be loaded to a draught more than this. Structural scantlings, or geometrical particulars of structural items such as plates and stiffeners, are calculated based on scantling draught or T(scantling), which may be more than T(design), but need not be more than T(freeboard). Thus,

$$T(\text{design}) \leq T(\text{scantling}) \leq T(\text{freeboard})$$

The fullness of the ship is defined by the form coefficients. The midship area coefficient is the ratio of the midship area up to LWL and the rectangle enclosing this area. Thus,

$$C_X = \frac{A_X}{(B*T)}$$

The water plane area coefficient is the ratio of the area of the LWL and the enclosing rectangle.

$$C_{WP} = \frac{A_{WP}}{(L_{BP} * B)}$$

The block coefficient is the ratio of the moulded volume up to LWL, which is normally known as the volume of displacement, and the enclosing rectangular parallelepiped. The prismatic coefficient is the ratio between the moulded volume and the cylindrical prism of cross section as the midship area having a length L_{BP}. The vertical prismatic coefficient is the ratio between the moulded volume and the cylindrical prism of horizontal cross section at the LWL having a depth equal to the draught. Thus,

$$\text{Block coefficient}: C_B = \frac{\text{Moulded volume of displacement}}{(L_{BP} * B * T)}$$

$$\text{Prismatic coefficient}: C_P = \frac{\text{Moulded volume of displacement}}{(A_X * L_{BP})}$$

$$\text{Vertical prismatic coefficient}: C_{PV} = \frac{\text{Moulded volume of displacement}}{(A_{WP} * T)}$$

And then,

$$C_P = \frac{C_B}{C_X} \quad \text{and} \quad C_{PV} = \frac{C_B}{C_{WP}}$$

The moulded volume of displacement is the volume of water displaced by the ship considering the volume enclosed by the moulded hull lines, which is the hull form enclosed by the inner lines of the ship. Total buoyancy force is generated by the water displaced, which consists of the moulded volume of displacement and the displaced water due to the thickness of hull shell plates. This is known as the extreme volume of displacement, and the corresponding weight of water displaced is the buoyancy force known as displacement. However, it is difficult to estimate the volume of water displaced by the hull plates of the ship, and it is normally taken as 0.6%–0.8% of the moulded volume of displacement. The total buoyancy force is equal to the weight of the vessel consisting of the lightship weight and the deadweight. Lightship weight is the weight of the completed ship, not including any item not required for the completion of the ship components of which have been discussed in Section 4.5.1. Deadweight items include cargo weight, ballast water, weight of crew and passengers and their effect and consumables such as heavy fuel oil, diesel oil, lubricating oil, fresh water, boiler feed water, stores, linen and spares. Then

$$\text{Displacement} = \text{Lightship weight} + \text{Deadweight} = L_{BP} * B * T * C_B * \rho * (1 + s)$$

where
   ρ is the density of the liquid in which the vessel floats, being 1.025 in sea water and 1.00 in fresh water
   s is the coefficient for extreme displacement, being between 0.006 and 0.008

A higher value is taken for smaller ships.

5.2 Controlling Equations for Preliminary Estimation of Main Parameters

It is necessary to determine the size of the maritime product in terms of its main parameters as the first step in the design process. This has to be done at a stage when detailed information of the product is not available for scientific analysis of its performance, but at the same time, it is necessary to ensure that the technical and nontechnical performance will be achieved without much change in estimated parameters, as shown in Figure 3.7a and b. For a conventional ship, this may be a relatively easy process, but for a new ship type with different proportions or for a new platform, this stage is very important and has to be carried out intelligently and cautiously.

A ship carrying cargoes of high density where the freight earned is based on weight carried, such as bulk carriers or tankers, is known as a deadweight carrier where the governing equation is

$$\Delta = C_B \times L \times B \times T \times \rho(1+s) = \text{Deadweight} + \text{Lightweight}$$

Here T is the maximum draught permitted with minimum freeboard or the freeboard draught. Since such vessels are designed as double-hull vessels nowadays, the volume of the cargo space is not adequate to provide the maximum deadweight as per the freeboard draught. Therefore, the design draught may be less than the freeboard draught even though draught is one on the main parameters of selection in such vessels.

Ships required to carry cargo of light density, such as light grain in bulk, passengers or containers, must be designed to maximize the cargo volume or the under-deck volume of the ship. Such ships are known as capacity carriers and the governing equation is

$$V_h = C_{BD} \cdot L \cdot B \cdot D' = \frac{V_r - V_u}{1 - s_s} + V_m$$

where
D' is the capacity depth in $m = D + C_m + S_m$.
C_m = mean camber = $2/3 \cdot C$ for parabolic camber
or = $1/2 \cdot C$ for straight line camber where C: camber at $C \cdot L$.
S_m = mean shear = $1/6 * (S_f + S_a)$ for parabolic shear where S_f: forward shear and S_a: aft shear

C_{BD} is the block coefficient at moulded depth, and an empirical relationship with the block coefficient of the ship up to LWL can be given as

$$C_{BD} = C_B + (1 - C_B)\left[\frac{(0.8D - T)}{3T}\right]$$

V_h is the volume of ship in m³ below the upper deck and between perpendiculars.
V_r is the total cargo capacity required in m³.
V_u is the total cargo capacity in m³ available above the upper deck.
V_m is the volume required for machinery, tanks, etc. within V_h.
S_s is the fraction of moulded volume to be deducted as volume of structurals in the cargo space (say 0.05).

Here T is not the main factor though it is involved as a second-order term in C_{BD}, and the draught of such a ship is the design draught, which is less than the freeboard draught. The weight equation is to be satisfied for the design draught, and a slight alteration in draught does not affect the design since there is enough margin with respect to statutory draught.

In unitized cargo carriers, the designer is limited to changing the vessel dimensions by the dimensions of the cargo itself. A barge carrier may have its dimensions altered as per the dimensions of the barge it carries.

A container ship, similarly, can have ship length altered in the units of 20′, breadth in the units of 8′ and depth in the units of 8½′. Such vessels are known as linear dimension ships (Watson and Gilfillan 1977). Constraints on ship dimensions may be imposed by external factors limiting particular ship dimensions, which may be due to the following.

- Those imposed by canals and seaways in the designated route of the ship such as restrictions imposed by St. Lawrence Seaway, Panama Canal, Suez Canal, Dover and Malacca Straits, as discussed in Section 2.1.3
- Restrictions imposed by ports of call, particularly on draught and length
- Restrictions imposed by the shipbuilding facility

The first estimates of parameters and coefficients can be done in one or more of the following three methods: (1) collection of recent data and statistical analysis, (2) approximate statistical relationships or empirical formulae available in published literature and (3) extrapolating from a basic ship, which is a nearly similar ship. The three methods are discussed as follows.

5.3 Data Collection and Analysis for Parameter Estimation

The main dimensions of built ships can be collected and analysed, which may give the empirical relationships between various parameters, parametric ratios, deadweight, power, speed (Froude number), etc. These relationships may be utilized to obtain the first estimate of the main parameters. An example of such data collection and derived relationships between deadweight, length, breadth, depth and draught is as follows. This example is by no means exhaustive, and the designer has to generate his or her own statistical relationships based on his or her requirements.

$L_{BP} = 3.351 \cdot (\text{dwt})^{0.387}$ for capacity carriers such as grain, container, LNG and LPG carriers, etc. (Figure 5.2)

$L_{BP} = 6.667 \cdot (\text{dwt})^{0.308}$ for deadweight carriers such as bulk carriers, tankers, etc. (Figure 5.3)

Similarly, for capacity carriers

$$B = 0.148 \cdot L_{BP} + 2.587 \text{ (Figure 5.4)}$$

$$D = 0.095 \cdot L_{BP} - 1.199 \text{ (Figure 5.4)}$$

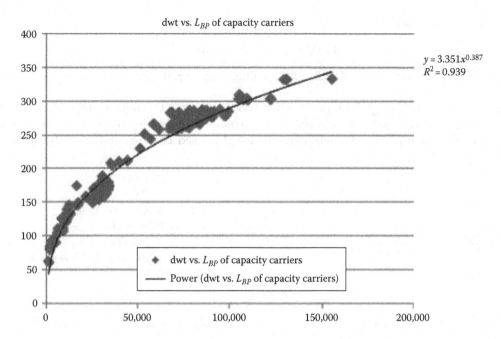

FIGURE 5.2
Deadweight–L_{BP} relationships of capacity carriers.

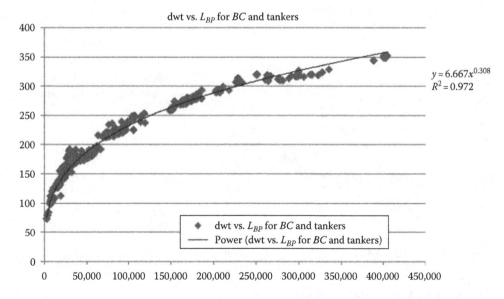

FIGURE 5.3
Deadweight–L_{BP} relationship of deadweight carriers – tankers and bulk carriers.

For deadweight carriers

$$B = 0.164 \cdot L_{BP} + 0.09 \text{ (Figure 5.5)}$$

$$D = 0.081 \cdot L_{BP} + 1.516 \text{ (Figure 5.5)}$$

From Figures 5.2 through 5.5 it can be observed that though the approximate relationship between the main parameters is obtained, variations could also be obtained from the data. In this example, deadweight carriers (weight-based designs) and capacity carriers

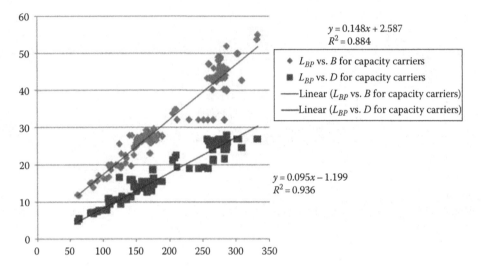

FIGURE 5.4
L_{BP}–breadth and depth relationship of capacity carriers.

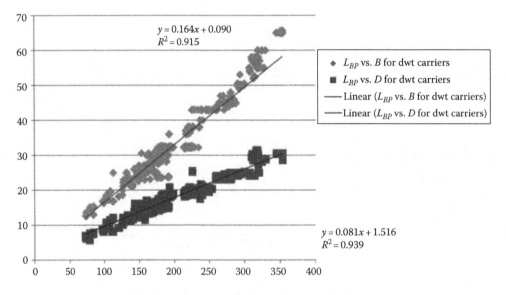

FIGURE 5.5
L_{BP}–breadth and depth relationship of deadweight carriers.

(volume-based designs) have different relationships. It is well known that weight-based designs follow maximum design draught principle. As per ICLL 1966, the statutory freeboard of tankers is the lowest and is termed Type A freeboard. All other ships require limiting draught confirming to a higher freeboard known as Type B freeboard. Certain bulk carriers can be assigned a somewhat lower freeboard designated as B-60 or B-100 freeboard. But due to recent regulations and trends for double-hull vessels and segregation of ballast spaces from cargo spaces, draught attainable is less than the statutory draught. All these give a new relationship between the depth and draught of deadweight carriers. Data collected indicate that

$$\frac{D}{T} = 1.385 \text{ for deadweight carriers with double hull.}$$

In the case of capacity carriers, this relationship is quite different since design draught is smaller than the statutory draught. Even within this dataset, liquefied gas carrying vessels carry much less cargo by weight than other capacity carriers. The data collected give D/T values as

$$\frac{D}{T} = 1.459 \text{ for capacity carriers such as container ships}$$

and

$$\frac{D}{T} = 2.137 \text{ for liquefied gas carrying vessels.}$$

If enough data are available for any different ship type such as vehicle carriers, offshore platform or any other marine product, an approximate size can be obtained with this kind of data analysis. One has to be careful about using the data collected by understanding the principles behind the relationships. Ship dimensions arrived at this way have to be modified to suit other requirements and constraints.

5.4 Approximate Semi-Empirical Relationships for Parameter Estimation

Various investigators have analysed a large amount of data available with them and have formulated semi-empirical relationships for parameter estimation and also estimation of performance as functions of parameters. The designer or reader can get this information from various well-established references such as Taggart (2003), Lamb (2003a,b), Schneekluth (1987, 1998), Watson (2009) and Munro-Smith (1975). However, these relationships are approximate and perhaps some of these are based on data which are old and have been superseded with new data. So these relationships can only be used with caution for initial estimations only and must be updated with scientific analysis. For this it is necessary to appreciate the interrelationship between various parameters and their contribution to vessel performance, some of which are mentioned in Table 5.1.

TABLE 5.1

Relationships between Parameters

Parameter	Depends On	Affected Vessel Characteristics
Displacement (Δ)	Deadweight or payload Volumetric capacity	Size of vessel
Length (L)	Speed Displacement	Resistance Manoeuvrability Longitudinal strength Hull volume Seakeeping behaviour Cost of vessel
Breadth (B)	Length ($L/B \leq 4$ for small vessel and up to 6.5 for merchant vessel)	Resistance Manoeuvrability Transverse stability Hull volume Cost of vessel
Depth (D)	Breadth ($B/D = 1.7$–2.5)	Longitudinal strength Transverse stability Freeboard Hull volume Cost of vessel
Draught (T)	Breadth and depth (Normally $2.5 \leq B/T \leq 3.75$, going up to 5.0 in heavily draught limited vessel)	Resistance Seakeeping behaviour Transverse stability Freeboard Displacement
Draught–depth ratio (D/T)	Type of freeboard Type of ship	Seakeeping behaviour
Length–depth ratio (L/D)	Type of ship (L/D between 10 and 14)	Longitudinal strength Hull deflection
Block coefficient (C_B)	Froude number (F_n)	Hydrodynamic behaviour Displacement
Midship area coefficient	C_B	Prismatic coefficient (C_P) Lengths of parallel middle body, entrance and run
Bilge radius	Breadth Midship area Flat of keel Rise of floor	Parallel middle body Quantum of production
Water plane area coefficient	C_B C_P Type of stern	Half angle entrance Wave resistance, stability

The form coefficients can be initially estimated by using the empirical formulae based on relationships mentioned earlier and given in references already cited, some of which are reproduced as follows.

Block coefficient:

$$C_B = 0.7 + \frac{1}{8}\tan^{-1}[25(0.23 - F_n)], \ F_n \text{ being the Froude number}$$

$$C_B = C - 1.68F_n \quad \text{where } C = 1.08 \text{ for single screw and } 1.09 \text{ for twin screw ships}$$

$$C_B = 1.18 - 0.69\frac{V}{\sqrt{L}} \quad \text{for } 0.5 \le \frac{V}{\sqrt{L}} \le 1.0$$

where V is in knots and L is in feet.

$$C_B = \frac{0.14}{F_n}\left[\frac{(L/B)+20}{26}\right] \text{ or } = \frac{0.14}{F_n^{2/3}}\left[\frac{(L/B)+20}{26}\right] \quad \text{where } 0.48 \le C_B \le 0.85, \text{ and } 0.14 \le F_n \le 0.32$$

$$C_B = -4.22 + 27.8\sqrt{F_n} - 39.1F_n + 46.6F_n^3 \quad \text{where } 0.15 \le F_n \le 0.32$$

5.4.1 Midship Area Coefficient

Midship area coefficient is given as a function of block coefficient as follows:

$$C_B = 0.55 \quad 0.60 \quad 0.65 \quad 0.70$$

$$C_M = 0.96 \quad 0.976 \quad 0.980 \quad 0.987$$

$$C_M = 0.977 + 0.085\,(C_B - 0.60), \text{ or}$$

$$= 1.006 - 0.0056\,C_B^{-3.56}, \text{ or}$$

$$= [1 + (1 - C_B)^{3.5}]^{-1}$$

5.4.2 Water Plane Area Coefficient

The water plane area has been expressed as a function of prismatic coefficient in the following manner by different designers.

Equation	Applicability/Source
$C_{WP} = 0.180 + 0.860C_P$	Series 60
$C_{WP} = 0.444 + 0.520C_P$	Small transom stern warships
$C_{WP} = C_B/(0.471 + 0.551C_B)$	Tankers and bulk carriers
$C_{WP} = 0.175 + 0.875C_P$	Single screw, cruiser stern
$C_{WP} = 0.262 + 0.760C_P$	Twin screw, cruiser stern
$C_{WP} = 0.262 + 0.810C_P$	Twin screw, transom stern
$C_{WP} = C_P^{2/3}$	Suggested by Schneekluth
$C_{WP} = (1 + 2C_B/C_m^{1/2})/3$	Suggested by Schneekluth
$C_{WP} = 0.95C_P + 0.17(1 - C_P)^{1/3}$	U-forms hulls
$C_{WP} = (1 + 2C_B)/3$	Applicable for average hulls
$C_{WP} = C_B^{1/2} - 0.025$	V-form hulls

5.5 Basic Ship Method of Parameter Estimation

The first step in this method is selecting a basic ship which is nearly similar to the ship required to be designed. Parameters are selected by choosing a basic ship such that $(V/\sqrt{L})$ and the ship type are same as the design ship, the dimensions are nearly same as the ship to be designed and detailed information about the basic ship is available. Then calculate

$$\text{Displacement} = L \cdot B \cdot T \cdot C_B \cdot \rho \cdot (1+s) \text{ for basic ship}$$

Choose the new ship's displacement from the relationship

$$(dwt/\Delta)_{basic} = \left(\frac{dwt}{\Delta}\right)_{new}$$

Select L, B, T and C_B to get the displacement of the new ship using difference equations. These equations are frequently used to alter the main dimensions for desired small changes in displacement or other characteristics. For example,

$$\Delta = L \cdot B \cdot T \cdot C_B \rho$$

$$\text{Or } \log \Delta = \log L + \log B + \log C_B + \log \rho$$

Assuming ρ to be constant and differentiating,

$$\frac{d\Delta}{\Delta} = \frac{dL}{L} + \frac{dB}{B} + \frac{dT}{T} + \frac{dC_B}{C_B}$$

So if a change of $d\Delta$ is required in displacement, one or some of the parameters L, B, T or C_B can be altered so that the aforementioned equation is satisfied.

5.6 Preliminary Performance Estimate

Once a set of parameters is selected, estimating its technical performance is necessary to ensure that the ship will perform within specified constraints. Since the design is at a very early stage and the details of shape, structural arrangements, general arrangement and machinery are not known, it is difficult to estimate performance on scientific basis and one has to take recourse to empirical formulae established based on the primary design variables based on scientific principles. Table 5.2 summarizes the performance characteristics and their dependence on various ship parameters, dependence being based on empirical formulae from references cited earlier and in the case of vibration, Todd (1961).

The most sensitive performance criteria given in Table 5.2 are perhaps the weight and the centre of gravity (CG) position (Watson and Gilfillan 1977, Schneekluth 1987, 1998) on which stability and power depend. It is necessary that the weight and CG as well as

TABLE 5.2

Ship Performance Characteristics and their Dependence on Ship Parameters

Performance Characteristics	Item	Depends on
Initial transverse stability	Vertical centre of buoyancy (KB)	Draught
		Vertical water plane area distribution
	Metacentric radius (BM_T)	Breath (B_2)
		Draught (T)
		Block coefficient (C_B)
	Metacentric height (GM_T)	Vertical centre of buoyancy (KB)
		Metacentric radius (BM_T)
		Vertical centre of gravity (KG)
Equilibrium and trim	Longitudinal metacentric radius (BM_L)	Breath (L^2)
		Draught (T)
		Block coefficient (C_B)
	Longitudinal metacentric height (GM_L)	$\simeq BM_L$
	Moment to change trim 1 cm (MCT 1 cm)	Displacement
		Length
		Longitudinal metacentric height
	Trim	Displacement
		Longitudinal centre of buoyancy
		Longitudinal centre of gravity
		Longitudinal centre of floatation
		MCT 1 cm
Light ship weight	Steel weight	
	Outfit weight	Length
		Breadth
		Draught
		Heavy outfit weight items
	Machinery weight	Break power
		Engine type and RPM
		Number of engines
		Auxiliary machinery items
Deadweight	Cargo weight	Volumetric capacity
		Cargo density
	Heavy fuel oil	Range of travel
		Service speed
		Break power
		Specific fuel oil consumption
		Density of oil
	Marine diesel oil	Total electrical power
		Diesel engine – number, power and RPM
		Power consumption at sea and point
		Specific fuel oil consumption
		Density of MDO
	Lubricating oil	Number and type of diesel engines
	Fresh water	Fresh water generator capacity
		Boiler feed water requirement
		Lives on board
	Crew and effects	Number of lives on board
	Provisions and stores	

(Continued)

TABLE 5.2 (*Continued*)

Ship Performance Characteristics and their Dependence on Ship Parameters

Performance Characteristics	Item	Depends on
Light ship centre of gravity	Longitudinal centre of gravity Vertical centre of gravity	Longitudinal lightweight distribution depth Length/breadth Length/depth Block coefficient
Deadweight centre of gravity	Longitudinal centre of gravity Vertical centre of gravity	Location of cargo holds and weight in each Location and weight of consumables Centroids of above items
Total weight and centre of gravity	Displacement Centre of gravity (LCG and VCG)	Lightship weight + deadweight Weight height, LCG and VCG Deadweight, LCG and VCG
Capacity	Moulded capacity	Length Breadth Depth Camber Sheer Cargo hold location
	Grain capacity	$\simeq 0.95 \times$ moulded capacity
	Ball capacity	$\simeq 0.90 \times$ grain capacity
Power	Effective power	Displacement Froude number Block coefficient Appendages Forebody shape Above water area
	Brake power	Number and type of propellers Shaft length and bearings Number and types of engines Gearing system
Seakeeping	Roll natural period	Transverse metacentric height Breadth
	Pitch natural period	Longitudinal metacentric radius Draught Length
	Heave natural period	Water plane area Length Vertical prismatic coefficient
Hull vibration	General	Types of ship Length Breadth Depth Draught Block coefficient Displacement
	Vertical vibration	Midship moment of inertia about transverse axis
	Horizontal vibration	Midship moment of inertia about longitudinal axis
	Torsional vibration	Midship polar moment of inertia
	Resonance with propeller excited vibration	Propeller RPM Propeller numbers of blades

buoyancy and centre of buoyancy (CB) are estimated correctly so that these do not change when a detailed design is carried out. Otherwise, if there is a negative change in stability performance or higher power requirement or cargo-carrying capacity, one has to start all over again from the beginning. Needless to mention that there could be a number of sets of design variables which satisfy design requirements at this stage and the designer may have to choose the best one based on some optimization criteria.

The empirical relationships for the estimation of form-based stability characteristics have been suggested by various authors cited before. A few of these are mentioned as follows.

5.6.1 Vertical Centre of Buoyancy, *KB*

$$\frac{KB}{T} = \frac{(2.5 - C_{VP})}{3} \quad \text{for hulls with } C_M \leq 0.9$$

$$\frac{KB}{T} = (1 + C_{VP})^{-1} \quad \text{for hulls with } 0.9 < C_M$$

$$\frac{KB}{T} = 0.90 - 0.36 C_M$$

$$\frac{KB}{T} = (0.90 - 0.30 C_M - 0.10 C_B)$$

$$\frac{KB}{T} = 0.78 - 0.285 C_{VP}$$

5.6.2 Moment of Inertia of Water Plane

Moment of inertia coefficient C_I and C_{IL} are defined as

$$C_I = \frac{I_T}{LB^3} \quad \text{and} \quad C_{IL} = \frac{I_L}{L^3B}$$

where I_T and I_L are transverse and longitudinal moments of inertia of the water plane, respectively. The formulae for the initial estimation of C_I and C_{IL} are as follows:

Equations
$C_I = 0.1216 C_{WP} - 0.0410$
$C_{IL} = 0.350 C_{WP}^2 - 0.405 C_{WP} + 0.146$
$C_I = 0.0727 C_{WP}^2 + 0.0106 C_{WP} - 0.003$
$C_I = 0.04(3 C_{WP} - 1)$
$C_I = (0.096 + 0.89 C_{WP}^2)/12$
$C_I = (0.0372 (2 C_{WP} + 1)^3)/12$
$C_I = 1.04 C_{WP}^2/12$
$C_I = (0.13 C_{WP} + 0.87 C_{WP}^2)/12$

Once I_T and I_L are estimated, metacentric radii can be calculated since,

$$B_{MT} = \frac{I_T}{\nabla} \quad \text{and} \quad B_{ML} = \frac{I_L}{\nabla}$$

Performance can be extrapolated from the available performance data of the basic ship. The difference equation mentioned in the earlier section can be comfortably used for this purpose. For example, one can assume the following functional relationships:

- KB is proportional to T.
- KG is proportional to D.
- BM_T is proportional to B^2/T.

Then using difference equation, it can be shown that

$$\frac{dBM}{BM} = 2 * \frac{dB}{B} - \frac{dT}{T}$$

Then new transverse metacentric height can be obtained.

Similarly, power requirement for the new ship with small variations of dimensions and having the same Froude number can be obtained by assuming that brake power is proportional to $\Delta^{2/3} \cdot V^3$.

Similar relationships for various weight items can also be used to determine new design ship values from the basic ship. The weight and CG of the new ship can then be established with some degree of certainty.

6

Stability of Floating Bodies

A floating body is said to be in equilibrium if resultant forces and moments acting on the body are zero. In case of a floating body in water, this is achieved because the downward force, weight of the body, is equal to the upward buoyancy force which is equal to the weight of the displaced water and the centre of gravity and the centre of buoyancy act in the same vertical or, for a ship, $LCB = LCG$. This equilibrium is stable if the body, displaced from its equilibrium position due to an external force or moment, comes back to its original position after the force or moment is removed. If the body moves away further from its original position, it is said to be in unstable equilibrium. If the body stays in the displaced position, it is said to be in neutral equilibrium. In case of a floating body like a ship or any floating platform, it has to be in stable equilibrium for survival. A submerged body could be in stable or neutral equilibrium for survival, whereas it must be in stable equilibrium for effective operation at sea. At design stage it is not enough to ensure stability in intact condition, but also in a probable damaged condition for its survival. It is necessary to understand and estimate the geometrical properties of the body so that its mechanics in water can be studied properly.

6.1 Bonjean Curves and Hydrostatics

In the standard Cartesian co-ordinate system, x axis is positive along the longitudinal direction along the centre plane of the body, y axis is positive towards port side and z is vertically upwards. The origin is the intersection of the base plane where $z = 0$, the centre plane with $y = 0$ and the midship plane $x = 0$. The body is defined by one or more three-dimensional continuous surfaces generally expressed as

$$y = f(x,z)$$

separated by boundary space curves between continuous surfaces which may be the deck line, the bow, stern and keel lines, knuckle lines etc. From this surface, the equations for the intersection planes can be obtained. The equation of a plane is

$$ax + by + cz = d$$

and if $d = 0$, the plane passes through the origin. The curves defining the body sections, the waterlines and the buttocks can then be defined as

$$\text{Body sections: } y = f(z) \text{ at } x = \text{constant}$$

$$\text{Water lines: } y = f(x) \text{ at } z = \text{constant and}$$

$$\text{Buttocks: } z = f(x) \text{ at } y = \text{constant.}$$

The section properties normally required for various calculations are the sectional areas and vertical moments of area at different draughts which can be obtained as

$$\text{Transverse sectional area upto draught } z \text{ at longitudinal position } x: A_{xz} = 2\int_{0\,to\,z} y * dz \text{ and}$$

$$\text{vertical moment of this area upto draught } z \text{ at longitudinal position } x: M_{xz} = 2\int_{0\,to\,z} y * z * dz$$

The midship area coefficient is then obtained as $C_M = A_0/(B \cdot T)$ where T is the design draught and A_0 is the midship area up to draught T. The graphical representation of sectional areas and vertical moments are commonly known as Bonjean and moment curves, respectively.

With a previously defined co-ordinate system, L_a is the aft end of waterline which is negative and L_f is the forward end of waterline. Taking all offsets in metre, area quantities in m² and density of water in tonnes per m³, the various waterline properties at constant-level trim draught z can be obtained as

$$\text{Water plane area: } A_{WP} = 2\int_{L_a\,to\,L_f} y * dx \quad \text{and}$$

$$\text{Centroid or longitudinal centre of floatation } LCF: LCF = 2\int_{L_a\,to\,L_f} \frac{x * y * dx}{A_{WP}}$$

$$\text{Tonnes per cm immersion: } TPcm = A_{WP} * \rho_{water}$$

$$\text{Water plane area coefficient: } C_{WP} = \frac{A_{WP}}{(L * B)}$$

$$\text{Transverse moment of inertia about body centre line: } I_T = \frac{2}{3} * \int_{L_a\,to\,L_f} y^3 * dx$$

If L_{BP} is considered for integration, then $L_a = -L_{BP}/2$ and $L_f = L_{BP}/2$. In a multi-body system, the transverse moment of inertia about the centre line of the multi-body system has to be obtained by shifting the own moment of inertia to the centre line. Typically, in a catamaran where the centre line is at a distance d from single-body centre line, transverse moment of inertia will be

$$I_T = 2 * (I_T \text{ about its own centre line + area of the single-body water plane } A_{WP} * d^2)$$

$$\text{Longitudinal moment of inertia about } LCF: I_L = \frac{2}{3} * \int_{L_a\,to\,L_f} x^2 * y * dx - A_{WP} * LCF^2$$

The volumetric properties of the underwater body can similarly be calculated up to any waterline z. The volume can be estimated as

$$\text{Volume of displacement upto draught } z: \nabla = 2 \int_{L_a \text{ to } L_f} \int_{0 \text{ to } z} y * dz * dx$$

$$= \int_{L_a \text{ to } L_f} A(x) * dx \quad \text{where } A(x) \text{ is the sectional area } A \text{ at location } x \text{ upto water line } z$$

or

$$= \int_{0 \text{ to } z} A_{WP}(z) * dz \quad \text{where } A_{WP}(z) \text{ is the water plane area upto water line } z.$$

$$\text{Block coefficient: } C_B = \frac{\nabla}{(L * B * T)}$$

$$\text{Prismatic coefficient: } C_P = \frac{C_B}{C_M}$$

$$\text{Vertical prismatic coefficient: } C_{VP} = \frac{C_B}{C_{WP}}$$

$$\text{Longitudinal centre of buoyancy upto draught } z: LCB = 2 \int_{L_a \text{ to } L_f} \int_{0 \text{ to } z} \frac{y * x * dz * dx}{\nabla}$$

$$= \int_{L_a \text{ to } L_f} \frac{A(x) * x * dx}{\nabla} \quad \text{or} \quad = \int_{0 \text{ to } z} \frac{A_{WP}(z) * LCF * dz}{\nabla}$$

$$\text{Vertical centre of buoyancy upto draught } z: VCB \text{ or } KB = 2 \int_{L_a \text{ to } L_f} \int_{0 \text{ to } z} \frac{y * z * dz * d}{\nabla}$$

$$= \int_{L_a \text{ to } L_f} \frac{M_{xz} * dx}{\nabla} \quad \text{or} \quad = \int_{0 \text{ to } z} \frac{A_{WP} * z * dz}{\nabla}$$

where M_{xz} is the vertical moment of section at x about base up to waterline z. The offset y is defined for the internal shape of the body and the shell plating is placed outside this hull form. Therefore, the dimensions defined so far are the moulded dimensions and the volume of displacement ∇ calculated previously is known as moulded volume. But in water, the dimensions should be measured from extremes to evaluate volume of displacement since shell plate volume also displaces water, thus making the volume of displacement greater than ∇. The shell plate volume is difficult to calculate and is normally taken as 0.6%–0.8% of moulded volume. Thus,

$$\text{Extreme volume of displacement: ext} = \nabla * (1 + s)$$

where $s = 0.006$–0.008, being on the higher side for smaller ships and vice versa.

The buoyancy force is equal to the weight of water displaced and thus the extreme displacement is given as

Extreme displacement: $\Delta = $ ext vol of disp $* \rho$

where $\rho = 1.0$ for *FW* and 1.025 for *SW* at 15°C.

Transverse metacentric radius for immersion upto water line z: $BM_T = \dfrac{I_T}{V}$.

Longitudinal metacentric radius for immersion upto water line z: $BM_L = \dfrac{I_L}{V}$.

Height of transverse metacentre above base: $KM_T = KB + BM_T$.

Height of longitudinal metacentre above base: $KM_L = KB + BM_L$.

All the previous particulars based on the geometry of the hull form can be calculated up to various waterlines z and plotted diagrammatically as a function of z. These properties are the hydrostatic particulars.

As can be observed, most of the calculations involve integration. If ship curves could be expressed mathematically as polynomials, one could attempt integration analytically. Unfortunately, ship curves are normally not analytically expressed. The curve to be integrated is generally defined numerically at pre-defined specific ordinates so that integration can be done using a straight line between two ordinates (trapezoidal rule), using a second-order parabolic curve between three equidistant ordinates (Simpson's first rule or 5-8-1 rule) or using a third-order parabolic curve between four equidistant ordinates (Simpson's second rule). These formulations are useful for manual calculations (Munro-Smith 1964, 1988, Comstock 1967, Tupper 2004). But with availability of high-speed computers, these calculations can be done very easily, quickly and very accurately. The three-dimensional ship body can be intersected by a plane generating two bodies on either side of the intersecting plane. The intersecting plane generates the intersecting line. It is easy to get the properties of intersecting line or the bodies under or above the intersecting plane by accurate numerical integration which is automatically processed in the available CAD software.

6.2 Stability at Small Angles

External forces and moments acting on a ship or floating offshore platform can be static in nature or dynamic, that is changing with time, and therefore, these could be steady or unsteady. Examples of such forces acting on a ship in a seaway trying to move it away from its equilibrium position could be wind forces, wave action on the body, turning of a vessel, collision, grounding or due to cargo movement such as grain shifting, etc. The forces could act in any direction, beam-wise, longitudinally or from quartering direction. Due to these actions, if the body inclines about x axis through an angle ϕ, it is known as heel and the corresponding stability is transverse stability. Similarly, longitudinal stability is considered when the vessel inclines through an angle θ about the y axis known as trim. Other disturbances, rotation about z axis or yaw and translational displacements due to surge, sway and yaw do not cause stability problems and are generally ignored while evaluating vessel stability. For a ship, the proportions of vessel dimensions are such that, for the same disturbance, heel is large compared to trim and so transverse stability must be estimated to ensure vessel safety in a seaway and longitudinal stability is not a critical item to be considered at design stage except in platforms and length and breadth are in equal proportions. For stability studies, generally dynamic effects are ignored and the vessel is considered to be poised in a position in static condition. Though the vessel heels and trims simultaneously due to external disturbance, it is customary to consider heel and trim separately and then combine the two to get the vessel's final equilibrium position.

If a vessel heels to a small angle $\delta\phi$, from Figure 6.1 it can be observed that the emerged volume having its centroid at g_1 is equal to the immersed volume having its centroid at g_2 causing the heel, and the righting moment is given as

$$\text{Righting moment} = \Delta * GZ$$

$$\text{and} \quad GZ = GM_T * \sin\delta\phi \approx GM_T * \delta\phi \quad \text{where } \delta\phi \text{ is in radians}$$

where M_T is the transverse metacentre which is the point of intersection of the verticals through the original centre of buoyancy B and the altered centre of buoyancy B_1 due to heel.

The distance GM_T is therefore important as an index of transverse stability at small angles of heel and it is called the transverse metacentric height. GZ is considered positive when the moments of weight and buoyancy tend to rotate the ship towards the upright position and GM_T is positive or M is above G. GM_T is negative when M is below G and the moment moves the ship away from its original position. Metacentric height GM_T is often used as an index of stability at concept design stage when preparation of stability curves

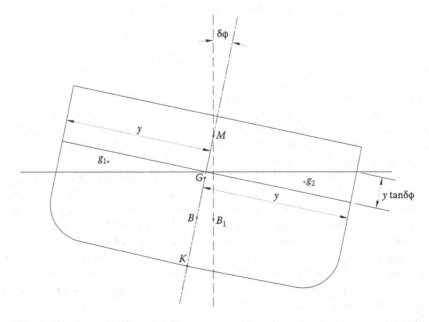

FIGURE 6.1
Transverse metacentric height.

for large angles has not been made. Its use is based on the assumption that adequate GM_T, in conjunction with adequate freeboard, will assure that sufficient righting moments will exist at both small and large angles of heel. GM_T can be calculated knowing the hydrostatic characteristics of the ship and the height of the ship's centre of gravity (CG) above base (K) as follows

$$GM_T = KB + BM_T - KG$$

where BM_T is the metacentric radius. For a floating body weight = buoyancy and thus, if the angle of heel is known and displacement or W is known, the transverse metacentric height can be determined as

$$GM_T = \frac{\text{Righting moment}}{\delta\phi}$$

where $\delta\phi$ is small (generally less than 7°) in radians. This principle is used during inclining experiments when the ship is heeled to an angle $\delta\phi$ by a known heeling moment (= righting moment), by moving a known weight w through a transverse distance d causing a heeling moment wd. Then,

$$GM_T = \frac{w*d}{\delta\phi}$$

Needless to say that GM_T should be positive or more than zero for the vessel to have positive initial stability. But having a very high GM_T is also not desirable since it would generate a small rolling period leading to faster rotational motion. The period of roll in still water, if not influenced by damping effects, is

$$\text{Period} = \frac{\text{const.} *k}{\sqrt{GM}} = \frac{c*B}{\sqrt{GM}}$$

where k is the radius of gyration of the ship about a fore-and-aft axis through its centre of gravity. The factor 'const. $* k$' is often replaced by $c * B$, where c is a constant obtained from observed data for different types of ships. c can be about 0.80 for surface types and 0.67 for submarines. In almost all cases, values of c for conventional, homogeneously loaded surface ships lie between 0.72 and 0.91.

The case of the ore carrier is an interesting illustration of the effect of weight distribution on the radius of gyration and, therefore, on the value of c. The weight of the ore, which is several times that of the lightship, is concentrated fairly close to the centre of gravity, both vertically and transversely. When the ship is in ballast, the ballast water is carried in wing tanks at a considerable distance outboard of the centre of gravity, and the radius of gyration is greater than that for the loaded condition. This can result in a variation in the value of c from 0.69 for a particular ship in the loaded condition to 0.94 when the ship is in ballast. For most ships, however, there is only a minor change in the radius of gyration with the usual changes in loading.

In the longitudinal plane, since ships are usually not symmetrical forward and aft, the centre of buoyancy at various even-keel waterlines does not always lie in a fixed transverse

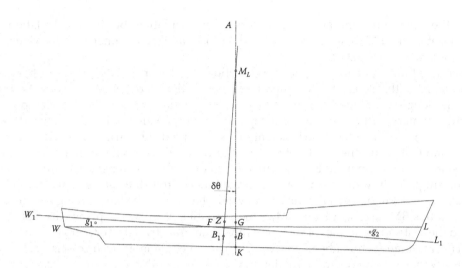

FIGURE 6.2
Longitudinal metacentric height.

plane, but may move forward and aft with changes in draft. For a given even-keel water-line, from Figure 6.2 it can be observed that if the vessel trims through a small angle $\delta\theta$ the immersed volume having its centroid at g_1 is equal to the emerged volume having its centroid at g_2 and this causes the shift of centre of buoyancy from B to B_1 and original and trimmed waterlines intersect at the centre of floatation F. M_L is the longitudinal metacentre which is the point of intersection of verticals through B and B_1. Then it can be shown that

$$GM_L = KB + BM_L - KG \approx BM_L$$

where
  BM_L is the longitudinal metacentric radius
  GM_L is the metacentric height

  and

$$\text{Trimming moment = righting moment} = \Delta * GM_L * \sin\delta\vartheta \approx \Delta * GM_L * \Delta\vartheta = \frac{\Delta * GM_L * Trim}{L_{BP}}$$

For a normal surface ship, the longitudinal metacentre is always far above the centre of gravity, and the longitudinal metacentric height is always positive. Moment to change trim by 1 cm is given as

$$MCT\,1\,\text{cm} = \frac{\Delta * GM_L}{(100 * L_{BP})} \approx \frac{\Delta * BM_L}{(100 * L_{BP})}$$

which is used to obtain the equilibrium position of the ship in the longitudinal plane due to a trimming moment.

    In cases, where substantial trim exists or when there is substantial change in water plane shape at normal trim, values for BM, KM and GM are likely to be substantially different from those calculated for the zero trim situations. It is important to calculate metacentric values for different trimmed conditions for the floating body. It may be difficult to do this in the manual or semi-automatic processes of calculation. But if the calculations are carried out using a high-speed computer, the exact stability parameters for various trim conditions can be estimated.

When a submarine with circular cross section is submerged, the centre of buoyancy is stationary with respect to the ship at any inclination and, therefore, both the transverse and longitudinal metacentre. To look at the situation from a different viewpoint, as the ship submerges, the water plane disappears, and the value of BM is reduced to zero. B and M coincide. The metacentric height of a submerged submarine is usually called GB rather than GM. Thus, the height of centre of gravity, KG, is of vital importance to transverse stability for submerged bodies where B has to be above G for stability under water.

If the single-body water plane can be located away from the centre of the multi-body system, the moment of inertia of the entire water plane, I_T can be large giving a large metacentric radius BM_T and so a large GM_T. This is possible if the water plane was distributed around the centroid. Examples of such cases are catamaran, trimaran or pentamaran vessels, SWATH vessels and semi-submersible vessels. The water planes in such vessels can be small giving advantages with regard to motion in waves along with high degree of stability. This facilitates large deck loading without compromising on metacentric height requirement.

6.3 Stability at Large Angles

At large angles of inclination, righting moment is still $\Delta * GZ$, but GZ is no more equal to $GM_T \cdot \sin \phi$ since M_T is defined only for small angles. To determine the moment of weight and buoyancy tending to restore the ship to the upright position at large angles of heel, it is necessary to know the perpendicular distance from the centre of gravity, through which the weight force W acts downwards, to the vertical line through the centre of buoyancy – shown as line AD in Figure 6.3 – through which equal upward force buoyancy acts. This distance, GZ, is referred as the statical stability lever, righting lever or righting arm.

6.3.1 Righting Lever of Floating Bodies

It is difficult to determine the GZ value for a vessel at any operating displacement at any angle of heel whenever required. Therefore, it would be convenient if GZ was available for various displacements and angles of heel for use whenever required. But G varies with loading condition and is not a pre-determined value. Therefore, this distance could be calculated from some standard reference point on ship centre line plotted as a set of curves, one for each angle of heel varying with displacement. Such a set of curves is generally referred as cross curves of stability. In Figure 6.3, weight W and buoyancy Δ act along the vertical through centre of gravity G in upright position and the weight acts along the vertical through G and buoyancy acts along the vertical through altered centre of buoyancy B_1 shown as line AD. Then, the righting lever is GZ and one could take as standard reference any point O on the ship centre plane or the point at intersection of keel and ship centre line, K. Then,

$$GZ = Q - OG * \sin \phi \quad \text{or} \quad GZ = KN - KG * \sin \phi$$

Commonly, KN curves are referred as cross curves of stability. GZ consists of two parts: KN or 'Q' as in Figure 6.3 and the height of G above base or KG. The first part can be calculated

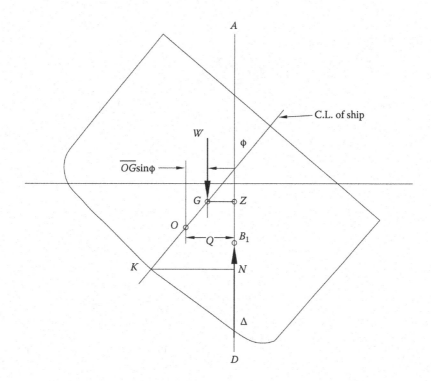

FIGURE 6.3
Righting arm.

without any reference to weight distribution if the geometry of the ship form is known. This part of stability is normally referred as form stability. The second part is known as weight stability. Figure 6.4 shows a typical diagram of cross curves of stability for a tug boat.

KN calculation essentially involves estimation of the volume of displacement and the three co-ordinates of centre of buoyancy – *LCB, VCB* and *TCB*. Ship form is a complex three-dimensional shape, not amenable to mathematical representation. This problem becomes more complex in this case since, with angle of inclination increasing, the deck may be immersed which has also to be modelled along with ship surface. Further, if some structures such as forecastle, hatch coamings, deck houses, etc. above the upper deck are also water-tight, they should be taken into account for calculation of *KN* since these contribute to buoyancy in case of immersion. Since ships are not symmetrical in fore and aft directions, with heeling the *LCB* is likely to shift from its original position. This imposes a trimming moment on the ship. In this altered position the *LCB* should be in the same vertical as the original *LCB* (or *LCG*). While doing numerical computation using a computer, it is convenient to take this into account and calculate the *KN* values at various angles of heel at the equilibrium trim condition known as 'free trim' condition. This can be done by a numerical trial and error method.

The *KN* values calculated at initial no-trim condition changes if there is an initial trim. The initial trim would depend on loading condition. It may be necessary to estimate cross curves of stability corresponding to various initial trimmed waterlines.

The statical stability curve is a plot of righting arm or righting moment against angle of heel for a given condition of loading. For any ship, the shape of this curve varies with the displacement, the vertical and transverse position of the centre of gravity, the trim and the

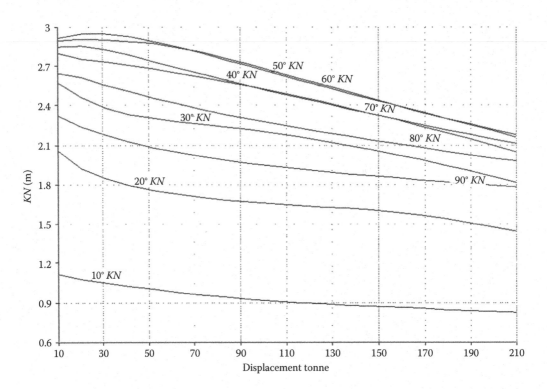

FIGURE 6.4
Cross curves of stability (*KN*) for a typical tug boat.

effect of free liquids. These values of the righting arm, plotted against angle of heel, form the statical stability curve. If the centre of gravity of the ship is not on the ship centre line due to uneven weight distribution on port and starboard sides, the ship would heel to one side initially and then come to a stable equilibrium position as shown in Figure 6.5a. If the *CG* of the ship was on the centre line, there would be no initial heel and *GZ* curve would be as shown in Figure 6.5b. If the vessel had no initial list, but had a slightly negative GM_T, the vessel would go away from the equilibrium position with slight disturbance and come to a stable equilibrium at an angle of heel known as angle of loll as shown in Figure 6.5c.

Static stability of any ship in any condition can be evaluated by superimposing various heeling arms resulting from specific upsetting forces (wind, turning etc.) on a curve of righting arms.

Although the statical stability curve, as the name implies, is the representation of the righting arm, *GZ*, or righting moment ($\Delta * GZ$) of a ship when in a fixed-heel attitude, the curve can be used to determine the work involved in causing the ship to heel from one angle to another against the righting moment. In the case of a ship, where the moment varies with the angle, if *M* is the moment at any angle of heel, ϕ, then the work required to rotate the ship against this moment through an angle $\delta\phi$ is *M*. $\delta\phi$ and the work required to rotate it from *A* to *B* is

$$\text{Work} = \int_A^B M d\phi$$

which is the area under the curve between *A* and *B* as shown in Figure 6.6.

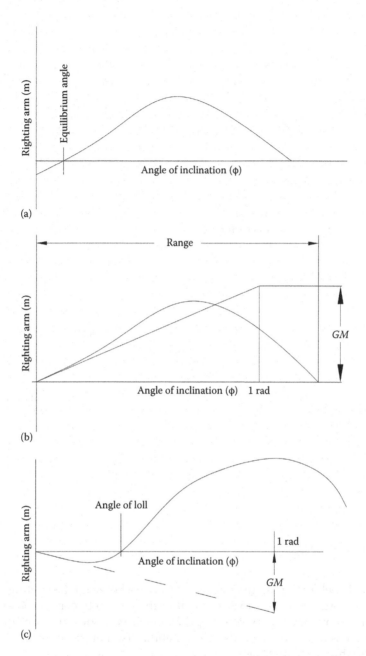

FIGURE 6.5
Statical stability curves in different conditions. (a) Static stability lever curve – *CG* off centre line. (b) Static stability lever curve – *CG* on centre line. (c) Statical stability lever curve – angle of loll.

If the heeling moments developed by the heeling forces are calculated for several angles of inclination, these moments may be plotted on the same coordinates as the statical stability curve. In Figure 6.7, curve *CAB* is the heeling-moment curve acting on the ship and curve *ADB* is the righting-moment curve. Note that both curves are extended to the left to show heel in the opposite direction. The dynamic stability can be estimated as area under *GZ* curve and it can be seen that a ship heeling from one side to the other from

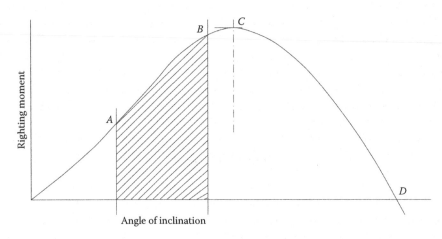

FIGURE 6.6
Work required to heel a ship.

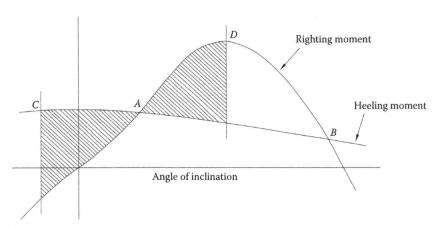

FIGURE 6.7
Heeling and righting moments.

point C to A will heel further up to D such that the area between the heeling-moment and righting-moment curves from C to A is equal to the area between the righting-moment and heeling-moment curves between A and D. The area under the righting lever curve between 0 and ϕ is referred as the dynamic stability (sometimes referred as dynamical stability) of the ship up to angle ϕ.

Heeling moment can be caused due to various external forces on the ship. Some of the common ones are given as follows:

1. If the curve labelled 'heeling moment' represents the moment of a beam wind, the moment will vary with the angle of inclination because of changes in the 'sail' area (area of the above-water portion) perpendicular to the wind direction, projected on a vertical plane, and in the vertical separation of the centroids of the wind pressure and the water pressure acting on the hull. This moment decreases as angle of heel increases as can be seen from Figure 6.8a.

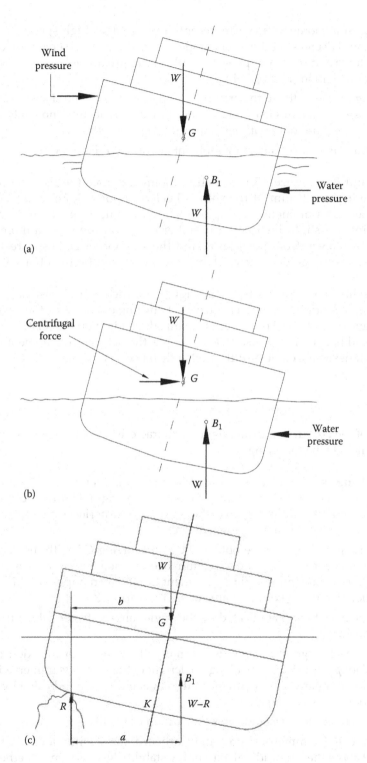

FIGURE 6.8
Effect of external heeling moment. (a) Effect of beam wind. (b) Effect of high-speed turn. (c) Effect of grounding.

2. If the heeling-moment curve represents the effect of high-speed turning, the moment will decrease at the larger angles, since the vertical separation of the centrifugal force and the water pressure will vary approximately as the cosine of the angle of inclination (Figure 6.8b).

3. A heeling moment due to the crowding of passengers to one side similarly varies as the cosine of the angle of inclination. In general, heeling moments vary with inclination because of variations in forces, levers or both.

4. Grounding also causes a heeling moment as shown in Figure 6.8c.

At points A and B in Figure 6.7 the heeling moment equals the righting moment and the forces are in equilibrium. If the ship is heeled to point A, an inclination in either direction generates a moment tending to restore the ship to position A. If the ship is heeled to point B, a slight inclination in either direction produces a moment tending to move the ship away from position B, and the ship either comes to rest in position A or capsizes. The range of positive stability is decreased by the effect of the heeling moment to point B.

When a heeling moment exists, as in Figure 6.7, the vertical distance between the heeling-moment and righting-moment curves at any angle represents the net moment acting at that angle either to heel or to right the ship, depending on the relative magnitude of the righting and heeling moments. As an example, the net righting moment (typically) in the case of a transverse weight shift through a distance d is

$$W * GZ - \omega * d * \cos \phi$$

The features of the GZ curve indicate stability characteristics of the vessel under consideration. These include the following:

1. At small angles, $GZ = GM_T * \sin \phi \approx GM_T * \tan \phi$, and therefore, initial metacentric height GM_T is equal to slope of the GZ curve at the origin (Figure 6.5b and c). If initial metacentric height is negative, the vessel may experience a sudden heel after which it may be stable at that angle of heel.

2. The righting moment can be obtained by multiplying GZ with displacement, or righting moment $= \Delta * GZ$. The maximum heeling moment the ship can withstand is given by the maximum righting moment the ship can impose and is, therefore, proportional to the maximum value of righting arm GZ.

3. The range of stability is indicated by the angle range between which the righting arm is positive.

4. Angle of deck edge immersion is normally at the point of inflexion of the GZ curve. Though, the deck edge along the length of the ship gets immersed at different angles; normally, it is within a small range and is well indicated by the point of inflexion of the GZ curve.

5. The area under the curve indicates the dynamic stability.

6. The ship being symmetrical port and starboard, the stability lever curve is symmetrical about the longitudinal axis, or the stability lever is same whether the ship heels to port or starboard.

6.3.2 Righting Lever of Submerged Bodies

In case of a submerged body like a submarine, since the body is completely submerged, the buoyancy force always acts through the centre of buoyancy B. If the body in inclined through an angle ϕ, the moment acting on the vessel is $\Delta * BG * \sin \phi$. It is clear that this will try to bring the body to upright position if B is above G. This is the condition of stable equilibrium. The righting arm follows a sine curve where

$$GZ = BG * \sin \phi$$

Thus, the stability lever curve is typical (Figure 6.9) and is dependent on the relative positions of B and G. To change the stability one has to change the CG only by flooding or emptying tanks.

6.3.3 Free-Surface Effect

The theoretical effect of free surface on metacentric height can be assessed by assuming that the weight of the liquid in each tank acts at the metacentre of the tank, because, at any small angle of heel, a vertical line through the actual centre of gravity of the liquid in the tank will pass through this point. This is equivalent to assuming that the weight of the liquid in each tank is raised from its centroid in the upright position to its metacentre, a distance of i_T/v. This increases the vertical moment of the mass of the ship by $(\omega/g)(i_T/v)$, where ω is the weight of the liquid. If the specific volume of the liquid, expressed as volume/mass, is designated as δ, then $\omega/g = v/\delta$ and the increase in vertical mass moment becomes

$$\frac{v}{\delta} * \frac{i_T}{v} = \frac{i_T}{\delta}$$

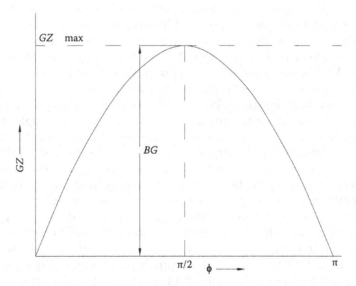

FIGURE 6.9
Righting arm of a submarine.

an expression which is independent of the quantity of liquid in the tank. Therefore, for any condition of loading, free surface may be evaluated for small angles of heel, by adding the values of i_T/δ for all tanks in which a free surface exists. If this summation, which is the increase in vertical moment due to the free surface, is divided by the ship's displacement, the result will be the rise in the ship's centre of gravity caused by the free-surface effect. This rise, called the free-surface correction, is added to KG, the height of the ship's centre of gravity above the keel, resulting in an equivalent reduction in the metacentric height. Hence, with displacement in mass units,

$$GM_{cor} = KB + BM - KG - \sum \left(\frac{i_T}{\delta * \Delta} \right)$$

Free liquid on a ship acts in the fore-and-aft direction in the same manner as in the transverse direction. For an intact ship with normal tankage, the effect of free liquid on trim is so small that it may be ignored. The trimming moment due to shift in CG of the tank liquid is small in comparison to the moment to change trim 1 cm and the centre of gravity is at the same height as the centre of buoyancy. On unusual craft, however, free liquids may have an important effect on trim.

When the tank is fully pressed or is empty, there is no free surface. The effect is felt due to free surface due to a partially filled tank. The subdivision of large tanks into two or more smaller tanks by dividing it breadth-wise may be an effective method of improving stability by suppressing the motion of free liquids.

6.3.4 Grain Shifting Moment due to Carriage of Bulk Dry Cargo

Bulk dry cargo, such as ore, coal or grain, may redistribute itself if the ship rolls or heels to an inclination greater than the angle of repose of the substance carried or earlier depending on the nature of cargo and accelerations involved due to rolling. Thus, a ship may start a voyage with the upper surface of such a cargo horizontal and with the cargo evenly distributed throughout the space. But if the ship rolls sufficiently to cause a cargo shift, a list results. A ship which has listed due to even a slight shift of cargo is open to the danger that it may later roll to increasing angles on the low side with further shifting of the cargo. Ships have been known to capsize from such progressive shifting of cargo.

Furthermore, all cargoes are directly influenced by the seaway-induced motions of the ship, which produce significant angular and lateral accelerations. In a rapidly rolling ship, such cargoes may shift even when the maximum angle of roll is less than the angle of repose of the cargo, because of the dynamic effects of rolling. Calculations using motion dynamics show that the accelerations involved in rolling produce a greater likelihood of cargo shifting when the cargo is located above the ship's CG (as in between deck spaces) rather than below (in the hold).

In bulk carriers, topside sloping tanks are fitted blocking the corner space to which grain could shift. Small hatches and sloping sides at the top of the compartment will reduce the danger of shifting cargo. For general cargo ships that may sometimes carry bulk cargo, it is essential to provide for fitting one or more longitudinal subdivisions in the holds and between decks to minimize the possibility of a shift of cargo in heavy seas. Such temporary subdivision bulkheads are called shifting *boards*. Usually they consist of wooden planks laid edge to edge in steel channels or equivalent fittings. In all cases it is essential to ascertain that adequate stability can be attained in operation to cope with any anticipated cargo, considering the restraints actually available. Of course, the ship operator is responsible for reviewing such factors prior to every voyage.

Grain has long been recognized as a dangerous cargo because of its tendency to flow or shift in the hold of a rolling ship. In the past, both national and international regulations relied heavily on the use of feeders from between decks to holds, which were intended to allow grain to flow downward to keep the hold full as the grain settled. Continued reports of grain cargo shifting, with some ship losses, led to a new investigation of the problem, which showed that even with feeders holds could not be assumed to be full and that shifting boards were still of great value in many cases. Therefore new grain regulations have been developed that change the emphasis from attempting to prevent grain shifting to making sure that the worst possible heeling moment does not exceed acceptable limits for each ship and loading condition.

6.4 Intact Stability Requirements

The International Maritime Organisation (IMO) has formulated rules under the Safety of Life at Sea (SOLAS) Convention for ensuring adequate intact stability for merchant ships during operation. The main requirements of properties of the GZ curve are as follows.

1. The initial metacentric height GM_0 should not be less than 0.15 m.
2. The righting lever GZ should be at least 0.20 m at an angle of heel equal to or greater than 30°.
3. The maximum righting lever should occur at an angle of heel preferably exceeding 30° but not less than 25°.
4. The area under the righting lever curve (GZ curve) should not be less than 0.055 m·rad up to $\phi = 30°$ angle of heel.
5. The area under the righting lever curve should not be less than 0.09 m·rad up to $\phi = 40°$ or the angle of flooding θ_i if this angle is less than 40°.
6. The area under the righting lever curve between the angles of heel of 30° and 40° or between 30° and ϕ_i, if this angle is less than 40°, should not be less than 0.03 m·rad.

Further, the regulations specify methods of estimation of loss of stability due to free surface in the tanks of the ship, amplitude and period of roll due to wind heeling moment and heel due to ship taking a turn. The regulations also specify satisfaction of stability criteria in all possible loading conditions – fully loaded, partially loaded, ballast and lightship conditions – both in departure and arrival states of the ship.

The intact stability requirements of passenger vessels specify that heel due to overcrowding of passengers to one side should not exceed 10°. Similarly, severe wind conditions and turning at speed should ensure roll or heel within specified limits. Grain-carrying ships should ensure limitation of heel due to grain shifting moment. IMO also specifies additional regulations for different ship types other than conventional merchant ships. Such ships include the following:

Cargo ships carrying large amount of timber deck cargo

Fishing vessels

Special purpose ships such as research vessels of $GT \geq 500$

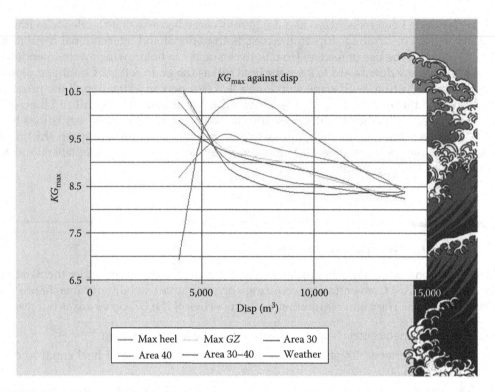

FIGURE 6.10
Limiting KG_{max} to satisfy SOLAS intact stability criteria.

Offshore supply vessels

Mobile offshore drilling units (MODU)

Pontoons

Dynamically supported vessels having a speed corresponding to Froude number ≥ 0.9; such vessels include hydrofoil craft, air cushion vehicles and surface effect ships.

Container ships

It can be observed that each of the SOLAS intact stability conditions can be satisfied only up to a certain value of KG (VCG) and if the KG value crosses this limit, the condition would be violated. The limiting KG_{max} value is likely to change based on the condition considered. Figure 6.10 shows the limiting KG_{max} curves satisfying various SOLAS conditions at various displacements for a feeder container vessel. The limiting envelop of the curves defines the maximum KG value the ship should not cross at any operating condition at that displacement.

6.5 Effect of Parametric Changes on Stability

Effect of minor changes of ship parameters on ship stability can be studied using difference equations as shown in Section 5.5.

6.5.1 Effect of Change of Breadth on Stability

It can be observed that if there is an increase in breadth without changes in other ship parameters including displacement, there will be a decrease in draught causing a slight reduction in *KB*, but substantial increase in *BM*. It can be seen in Figure 6.11 that, due to increase in breadth, there is an increase in the immersed volume which reduces the draught of the ship such that $W_2 + W_3 = W_1$. *KG* remaining constant, there is a net gain on *GM* or initial metacentric height. Reserve buoyancy of the ship increases due to increase in breadth and also freeboard. Thus, there will be an improvement in *GZ* or the righting arm curve. The most important parameter affecting stability is the breadth increase which improves stability to a large extent.

6.5.2 Effect of Change of Depth on Stability

If there is an increase in depth without changing the underwater body shape, *BM* and *KB* remain unchanged. However, the steel weight increases, and primarily, the steel weight of the portion at and above the original depth go up by an amount equal to the change in depth. This increases the *KG* of the lightship. The cargo *CG* also goes up due to increase in cargo volume in the upward direction. Thus, there is a net increase in the *KG* of the loaded ship which can be approximated to be proportional to depth change as a first approximation. Thus, there is a reduction of *GM* or initial metacentric height. This pulls down the *GZ* curve at small angles. But as the angle of inclination increases, the excess reserve buoyancy due to increase in depth comes into play and *GZ* increases. The deck edge immersion is delayed (see Figure 6.12).

6.5.3 Effect of Change of Form

Keeping the displacement same, if the bilge radius is increased, there will be an upward movement of underwater volume which results in an increased load water plane area (Figure 6.13). *KB* goes up and so does *BM*. Initial metacentric height improves. On the other hand, if the underwater volume is pushed down like a bulbous bow or reduction in

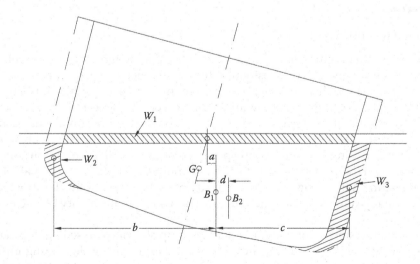

FIGURE 6.11
Effect of increase of beam.

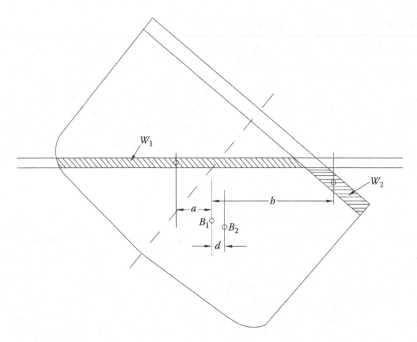

FIGURE 6.12
Effect of increase of depth.

bilge radius, *KB* reduces and so does *BM*, reducing the net initial stability. Providing more flare above water leads to only marginal increase in *KG* but increases reserve buoyancy (decreases reserve buoyancy in case of tumblehome) and, hence, an improvement in the *GZ* curve (Figure 6.13). Addition of water-tight erections above deck will have similar effects.

6.6 Discussion on Stability

It must be noted that accurate estimation of stability parameters depends largely on the estimation of the weight and *CG* position of the vessel. Inclining experiment at the time of delivery of the ship when it is nearly complete in all respects is the standard way to estimate the lightship weight and *CG* with minor corrections. But estimation of each of the deadweight items and its respective centroid has to be done with care whenever the ship is being loaded so that an accurate total weight and corresponding *CG* can be obtained. At a port during loading and unloading operations at intermediate conditions, there may be situations when the vessel has lost stability. Therefore, stability must be evaluated during partial loading/unloading and ballasting/de-ballasting operations. For this purpose, ships are normally provided with a loading instrument so that stability can be evaluated at any stage.

The stability assessment done so far is with the ship statically poised. This is never the case in a seaway. The ship rolls, pitches and heaves in a seaway depending on the sea condition. Also at any instantaneous position, weight, buoyancy and restoring moments are balanced with other external forces and moments. These factors affect stability.

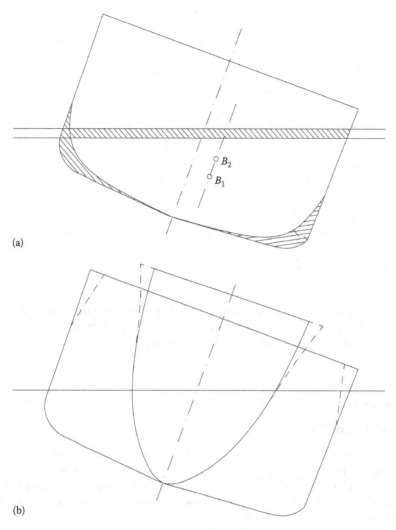

FIGURE 6.13
Effect of change of form. (a) Fining of the bilges. (b) Change of tumblehome and flare.

Waves may have a significant effect on static stability, particularly following or overtaking waves of approximately the ship's length. Righting lever curves can be drawn by superimposing offsets from the wave profile on the body plan used for the calculation of cross curves. In a computer calculation, the wave profile is used for input instead of a straight waterline where dynamic effects of rolling are generally excluded.

Figure 6.14 shows typical righting arm curves for a ship in a regular wave of the same length as the ship and height equal to $L_W/20$, with either wave crest or wave trough amidships.

In a dynamic condition of head or following sea, a ship could momentarily experience a heeling moment (could be due to loss of GM) and this could lead to parametric rolling which is one of the dynamic effects.

Beam seas, on the other hand, tend to impose a constant heel on the ship around which position the ship tends to roll as the wave passes the beam of the ship. Quartering seas impose an oscillating heel-and-trim (roll-and-pitch) combination on the ship.

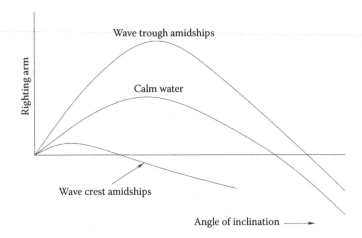

FIGURE 6.14
Righting arm of a ship poised on a wave.

In the absence of any method to evaluate stability instantaneously in a dynamic environment, the earlier procedure of evaluation associated with stability recommendations of IMO is in practice which has been accepted universally.

6.7 Damaged Stability

A ship may be damaged at sea due to various reasons such as grounding, collision, accidental opening of a sea water inlet valve or an underwater explosion. This causes water entry into compartments which have been damaged occupying spaces which do not hold cargo and replaces the liquids that may be there in such compartments. The unwanted entry of water has the following consequences:

- Sinkage and trim in the final equilibrium condition with loss of freeboard and reserve buoyancy. Could be severe enough causing capsize due to loss of complete reserve buoyancy or progressive flooding through openings on upper deck.
- Loss of transverse stability causing loss of static and dynamic stability leading to capsize.
- Loss of longitudinal stability and excessive trim leading to plunge.
- Damage to cargo.

For this reason, it is necessary to contain the extent of water entry due to damage to outer hull. Therefore, a ship is divided into a number of compartments by providing transverse water-tight bulkheads so that the water entry is only in the volume enclosed within the fore and aft bulkheads of the damaged area. If the complete damaged compartment is at a level below the intact waterline, the entrained water would occupy the entire space of the compartment and this may be treated as an added weight to the ship. There is no free-surface effect of the stability of the ship since the compartment

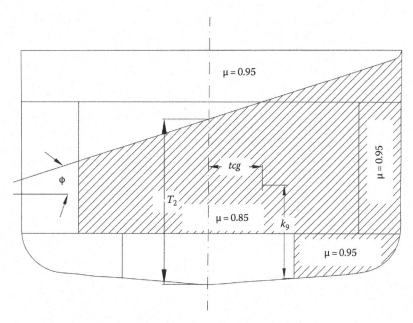

FIGURE 6.15
Section through flooded compartments.

would be full of water. However, in a partially flooded compartment, free-surface effect of a flooded compartment needs to be taken into account only if one uses the added weight method of calculation as shown later in this section. For this reason, ships are often provided with water-tight longitudinal bulkhead limiting water entry and free-surface effects. In modern ships there are vertical side tanks as in container ships, tankers and bulk carriers or sloping side tanks as in bulk carriers and double bottom tanks. In loaded condition, the ballast tanks are empty and are the first casualty of damage. On the other hand, in ballast voyage the hold space may be flooded. Further, the position and longitudinal extent of damage determines whether longitudinally one or more compartments have been damaged and also how much unsymmetrical flooding takes place. Thus, the extent and location of damage plays a major role in determining equilibrium and stability in damaged condition. Figure 6.15 shows a damaged ship section indicating unsymmetrical flooding with assumed permeability of various compartments which are empty. There is a shift of the transverse centre of gravity (TCG) due to unsymmetrical flooding.

For doing stability calculations in damaged condition, the loading (Δ, LCB, LCF etc.) in the original intact condition and the corresponding equilibrium draught and trim must be known. Then a particular compartment or a group of adjacent compartments may be damaged based on the extent and location of damage. The final equilibrium waterline inside the damaged compartment and outside is the same. There are two alternative methods of calculation of final damaged equilibrium position and stability (Comstock 1967). In lost buoyancy method, damaged compartment(s) is considered open to the sea and, therefore, does not contribute to the total buoyancy of the ship and ship's weight (or displacement) and CG remains constant. In added weight method, water entering into the damaged compartment is considered as a weight added to the ship, the body remaining intact. Therefore, weight and CG position change and so do buoyancy and CB. For stability

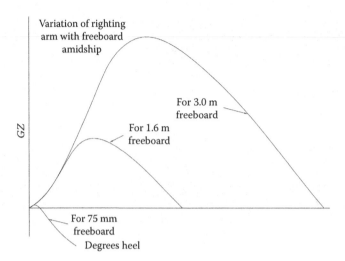

FIGURE 6.16
Variation of righting arm with freeboard amidships.

calculation, in this procedure, free-surface effects of the damaged compartment have to be considered since this compartment remains a part of the ship.

When a compartment is damaged, the rate of water entry into the damaged compartment depends on the amount of tear of the outer hull. So the vessel may take some time to come to final equilibrium position when the water levels outside and inside the compartment will be the same. During this period, there is only partial water entry into the compartment. In this condition, water level inside the damaged compartment is parallel to but below the outside water level. The stability in this condition may be even worse than in the final equilibrium position. It is necessary to investigate damage stability in partially flooded condition also.

There is a loss of stability due to damage and Figure 6.16 shows a typical intact GZ curve and damaged GZ curve due to flooding and loss of freeboard. It can be seen easily that dynamic stability is greatly impaired and the vessel's capability to withstand external heeling moment reduces. In some cases, the initial metacentric height may be negative giving an angle of loll (Figure 6.5c) when the vessel heels to an angle for stable equilibrium. Even then both static and dynamic stability are greatly reduced. In extreme cases of excessive flooding, such as car deck in a RORO ship, the loss of stability may be total leading to capsize (Figure 6.16 – case of 75 mm freeboard).

If there is unsymmetrical flooding due to damage of side compartments on one side, the vessel attains an initial heel along with loss of stability (Figures 6.5c and 6.15). If the compartment is far away from midship, there is also large trim. Thus, in trimmed and heeled condition, deck (particularly forward deck) may come under water when openings on deck may take in water causing progressive flooding and ultimately the vessel may capsize. To avoid such a situation, side tanks are very often cross-connected so that damage of one side may flood both the port and starboard compartments leading to symmetric flooding leading to higher sinkage and trim, but less heel. Of course, the intermediate stages of flooding may be critical in such cases since flooding of the cross-connected compartment will have a time lag depending on the cross section of the connecting pipe.

6.8 Safety and Subdivision

Damage of a compartment or a group of compartments is dangerous and unsafe for survival of life and property. Therefore, IMO (SOLAS) has been involved in formulations of rules to be followed during the ship design stage itself. Also other organisations such as the U.S. Coast Guard and various navies (defence) have formulated different rules regarding damage stability. Under IMO, the International Load Line Convention (ILLC), international regulations for carriage of dangerous (noxious) chemicals in bulk (IBC Code) and the Pollution Prevention Rules (MARPOL) have stipulated damage stability requirements of ships including damage extent and stability in damaged condition. Some rules also specify bulkhead positions for safety. The most comprehensive rules have been framed under SOLAS convention under IMO.

The prescriptive rules for safety rarely work in practice. Occurrence of damage, its extent and location are probabilistic in nature. IMO has realised this and the previously formulated subdivision rules (not discussed here) have now been superseded with probabilistic damage stability rules for all ocean-going ships as per the reference Resolution MSC 194 (80), Annex 2, Chapter II-1, Part B-1 of IMO.

The probabilistic criterion is based on the concept that a ship damaged in an accident will have a probability that the damage will have a certain extent, a probability that the damage will occur at a certain position along the length of the ship and a probability that the ship will survive such damage. These three probabilities have been determined by an analysis of accidents involving nearly 300 ships.

The rules specify that the deepest subdivision load line is the waterline corresponding to the summer load line of the ship. The partial load line is the waterline corresponding to a draught equal to the light draught plus 60% of the difference between the summer draught and the light draught. Based on these definitions, the required subdivision index R is calculated based on subdivision length and number of passengers, which indicates that the vessel will remain afloat if attained subdivision index A based on actual subdivision of the ship is greater than R or $A \geq R$.

The attained subdivision index A is calculated as follows:

$$A = 0.4A_s + 0.4A_p + 0.2A_l$$

where each of the partial indices A_s, A_p and A_l is calculated as follows:

$$A = \sum p_i s_i$$

where

p_i accounts for the probability that ith compartment or group of compartments may be flooded, disregarding any horizontal subdivision

s_i accounts for the probability of surviving the flooding of the ith compartment or group of compartments

A_s, A_p and A_l correspond to indices obtained for damage cases corresponding to three loading conditions corresponding to mean draughts d_s, d_p and d_l, where d_s is the deepest subdivision draught, d_l is the light service draught corresponding to the lightest possible loading and d_p is the partial subdivision draught which is

$$d_p = d_l + 0.6(d_s - d_l)$$

The rules specify the method of calculation of p for each damage condition. It is desirable to calculate p not only for single compartments but also for adjacent compartments taken two, three or more at a time, so that the highest value of the attained subdivision index A is obtained.

A compartment or a group of compartments is assumed to be flooded from the initial conditions. The GZ curve after flooding is determined for all cases. The value of s is determined from the GZ curve and heeling moments due to passengers, wind or launching of survival craft as

$$s = \min\{s_{intermediate}, s_{final}, s_{mom}\}$$

where
> $s_{intermediate}$ is the probability of survival of intermediate flooding stages
> s_{final} is the probability of survival in final equilibrium condition
> s_{mom} is the probability of survival against external heeling moments

If there is a horizontal subdivision in the compartment, the probability that the upper space will be flooded is taken into account by multiplying the value of s by a factor v calculated by a formula depending upon the height of the subdivision. In the calculation of GZ after flooding, values of permeability are to be taken as given earlier.

The attained subdivision index is obtained by adding up the values of $p_i s_i$ for each individual compartment and group of compartments. It is advantageous to consider as many compartments and groups of compartments as possible to obtain the highest value of A, but those groups of compartments that will definitely give $s = 0$ need not be considered.

7

Hydrodynamic Design

A marine structure or vehicle interacts with the marine environment in which it works and operates such that the mission requirement is adequately satisfied. The technical aspects of design must include considerations which are primarily hydrodynamic in nature. A vehicle design also includes the requirement of mobility of the vehicle in water. Marine vehicle design has evolved over hundreds of years, and ship designs have refined adequately to cope with changes in technology. This technical design process has served as a guide to design fixed and floating structures such as oil rigs and other platforms at sea, suitably altering technical requirements based on operational issues. This chapter discusses various issues related to hydrodynamics affecting ship design, which may be extrapolated to designing other sea structures.

7.1 Resistance

The resistance R of a ship is the force which opposes the forward motion of the ship at a constant speed V in a straight line in still water. The power required to overcome this resistance is called the effective power P_E, and $P_E = R * V$, resistance R being in kN, speed V in m/s giving P_E in kW. The speed of the ship is often given in knots, which is 1 nautical miles/h or 0.5144 m/s. Understanding ship resistance is important because (1) it is necessary to determine the effective power at the design speed so that an appropriate propulsion system may be selected for the ship, (2) it is desirable to design the hull form so as to minimize its resistance subject to the various design constraints and (3) to understand the flow so that appendages could be designed properly.

7.1.1 Components of Total Resistance

The total resistance of a ship is due to several causes, and the phenomena involved are extremely complicated. This has been discussed by many investigators as stated in Saunders (1957), Comstock (1967), Lewis (1989b), Schneekluth and Bertram (1998) and others. It is, therefore, usual to simplify the problem by regarding the total resistance to be composed of several components independent of each other and to disregard the possible interaction between the different components. For a ship moving at the surface of water, the total resistance is composed of the resistance of the above-water part of the ship (air and wind resistance or aerodynamic resistance) and the resistance of the underwater part of the ship (hydrodynamic resistance). The hydrodynamic resistance, which constitutes the largest component of total drag, is mainly the bare hull resistance which increases due to various appendages such as rudders, bilge keels, stabilizer fins, sonar domes, etc.

The total resistance (hydrodynamic, ignoring aerodynamic drag) of the bare hull moving at a constant speed in calm water can be divided into two main components. When the hull moves in water, the motion is resisted by the viscosity of water and thus experiences viscous resistance R_V. The motion of the hull at or near the surface also generates waves at the surface, and this gives rise to wave (or wave-making) resistance R_W. Thus, the total resistance R_T is given by the sum of these two components as

$$R_T = R_V + R_W.$$

Looking at the drag in a different way, the total resistance to forward motion is the sum of the longitudinal components of stresses tangential to the surface (friction), R_F, and normal to the surface (pressure), R_P:

$$R_T = R_F + R_P.$$

The viscosity of water also alters the pressure distribution around the hull and thereby causes an increase in the pressure resistance. The part of the pressure resistance due to the viscosity around a 3D form is called the viscous pressure resistance R_{VP} given by

$$R_{VP} = R_V - R_F = R_P - R_W$$

Then,

$$R_T = R_V + R_W = (R_F + R_{VP}) + R_W = R_F + (R_{VP} + R_W) = R_F + R_P$$

The viscous pressure resistance is usually a small component of the total resistance. However, if the hull is excessively curved at the stern and there are large waterline or buttock line slopes or discontinues, the flow separates from the hull surface and gives rise to eddies or vortices. This results in a significant increase in the viscous pressure resistance. The additional resistance due to separation of flow and the generation of eddies is called separation drag or eddy resistance.

The frictional resistance R_F is the sum total of the *frictional* resistance of a 2D surface of infinite aspect ratio (surface of zero pressure gradient), R_{F0}, and an additional component due to the 3D effect on friction, commonly known as friction form effect, and taking k as the form factor, $R_F = (1 + k)R_{F0}$. When the waves generated by the ship have high wave slope, like in full form ships moving even at slow speed, waves break and the drag is manifested as viscous resistance. Thus, if one measured the energy content of the wake behind the ship experimentally, the drag would consist of the normal viscous resistance and the drag due to wave breaking, and the remaining part of the total drag can be measured by estimating the energy content in the waves, which is known as wave pattern resistance. Figure 7.1 shows the various components of hydrodynamic resistance to the forward motion of a body in water, which are discussed subsequently.

A detailed study of the different components of ship resistance is necessary to understand the complex phenomena involved and to design the hull form of a ship to minimize the resistance of the ship. However, for many practical purposes, it is sufficient to divide the total bare hull resistance into two components: (1) frictional resistance and (2) remaining components lumped together as residuary resistance, which is mainly wave resistance.

The total-resistance coefficient is a function of Reynolds number and Froude number such that viscous resistance or, mainly, frictional resistance is a function of Reynolds number and the residuary resistance, mainly the wave-making resistance, is a function of Froude number. Thus,

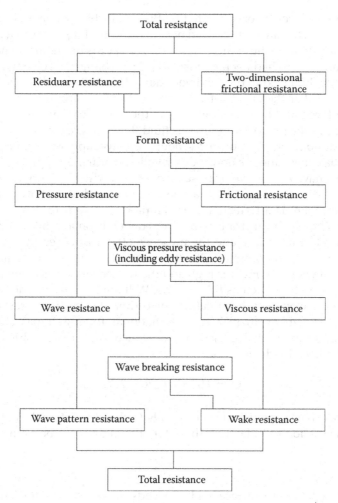

FIGURE 7.1
Resistance components of bare hull in calm water.

$$C_T = f(R_n, F_n) = C_F + C_R$$

$$\text{where } C_F = f_1(R_n) \quad \text{and} \quad C_R = f_2(F_n)$$

where

$C_T = R_T/(1/2 * \rho * S * V^2)$ is the total resistance coefficient

$C_F = R_F/(1/2 * \rho * S * V^2)$ is the frictional resistance coefficient and

$C_R = R_R/(1/2 * \rho * S * V^2)$ is the residuary resistance coefficient;

S is the wetted surface, proportional to L^2

$R_n = VL/\nu$ is the Reynolds number, named after Osborne Reynolds known for his experiments on viscous fluids among other things

$\nu = \mu/\rho$ is the kinematic coefficient of viscosity

$F_n = V/\sqrt{gL}$ is the Froude number

As per the boundary layer theory initiated by Prandtl in 1904 when a viscous fluid flows past a solid boundary, the layer of the fluid next to the boundary sticks to it ('no slip' condition), and the velocity of the fluid increases from zero at the boundary to nearly the value it would have had if there had been no viscosity. This change in velocity takes place in a narrow layer of the fluid next to the solid boundary. This layer is called the thin boundary layer. It is assumed that the effects of viscosity on the flow around a body are confined to the boundary layer, and that the flow outside the boundary layer is that of an inviscid fluid. This simplifies the problems of viscous fluid flow to a great extent.

At low Reynolds numbers, the flow in the boundary layer appears to take place in a series of thin layers or 'laminas', and the flow is described as laminar. At high Reynolds numbers, the fluid particles have a mean velocity superposed on which are small random velocity fluctuations in all directions and such a flow is called turbulent flow. As the Reynolds number increases, there is a transition from laminar flow to turbulent flow. The critical Reynolds number at which this transition occurs depends on a number of factors, including the roughness of the surface and the presence of disturbances such as eddies in the flow approaching the solid boundary. The flow around a ship is almost always turbulent because the ship Reynolds number is high and the wetted surface is comparatively rough.

A number of researchers such as R.E. Froude, William Froude, Hughes and Schoenner have worked on developing a formulation for estimating frictional resistance of ships based on Reynolds number. In 1957, the International Towing Tank Conference (ITTC) decided that in all future work, the frictional resistance coefficient for ships and ship models would be calculated by the formula

$$C_F = 0.075(\log_{10} R_n - 2)^{-2}.$$

This frictional resistance coefficient corresponds to the 2D frictional drag coefficient C_{F0} and applying the friction form factor k, the 3D frictional drag can be estimated as

$$C_F = (1 + k)C_{F0}.$$

Ship surface is not a smooth surface, and this leads to an increase in frictional drag. The effect of roughness is usually taken into account by adding a roughness allowance ΔC_F to the frictional resistance coefficient. A commonly used value is $\Delta C_F = 0.4 \times 10^{-3}$. However, one may also use the following formula:

$$10^3 \Delta C_F = 105\left(\frac{k_s}{L}\right)^{1/3} - 0.64$$

A standard value of the equivalent sand roughness of a newly painted steel hull is $k_s = 150 \times 10^{-6}$ m (150 μm), but lower values are now routinely obtained by modern shipbuilding techniques and paint technology. During the service of the ship, the hull surface becomes progressively rough due to damage to the paint coating, corrosion and erosion of the surface and 'fouling' by marine organisms that attach themselves to the hull, resulting in increased resistance. This makes it necessary to dry-dock the ship at intervals to clean and repaint the hull. The rate of fouling depends on a number of factors such as the time spent in port and at sea, and the time spent in temperate waters and in tropical waters. Empirical allowances are sometimes used to allow for the increased resistance due to fouling, e.g. a drop in speed of ¼%/day in temperate waters and ½%/day in tropical waters at constant power.

In an inviscid fluid, the pressure distribution around a curved body is such that there is no resistance. The effect of viscosity in the fluid causes a gradual decrease in the pressure around the body in the direction of flow compared to the pressure distribution in the inviscid flow, and this results in the component of resistance called the viscous pressure resistance. If the body is streamlined, the viscous pressure resistance is small and need not be considered separately but included in form resistance. With a body having a large curvature in the afterbody, i.e. a body which is 'bluff' and not streamlined, the particle may come to rest and its flow is reversed by the adverse pressure gradient. Fluid particles moving in the reverse direction meet the particles moving from forward to aft, pushing them away from the surface of the body, and an eddy is created between the surface of the body and the flow moving from forward to aft, and a vortex is shed. The extent of boundary layer separation and the magnitude of eddy resistance depend on a number of factors apart from the shape of the curved surface. Separation is more likely to occur in laminar flow and low Reynolds numbers than in turbulent flow and high Reynolds numbers. A high hydrostatic pressure reduces separation.

In the case of bluff hull forms, the free surface flow ahead of the bow becomes irregular and complex even at very low Froude number, usually leading to breaking waves at the bow. The resistance associated with wave breaking has been the subject of extensive investigation. Bow wave breaking is considered to be due to flow separation at the free surface, and it can generally be avoided by requiring that the tangent to the curve of sectional areas at the forward perpendicular is not too steep.

The movement of the hull through water creates a pressure distribution, i.e. areas of increased pressure at bow and stern and of decreased pressure over the middle part of the length. Since the free surface is a surface of constant pressure, this pressure difference generates waves. There is greater pressure acting over the bow, as indicated by the usually prominent bow wave build-up, and the pressure increase at the stern, in and just below the free surface, is always less due to the build-up of the boundary layer at the stern. The resulting added resistance corresponds to the drain of energy into the wave system, which spreads out at the stern of the ship and is continuously recreated. This is the so-called wave-making resistance. The result of the interference of the wave systems originating at bow, shoulders (if any) and stern is to produce a series of divergent waves spreading outwards from the ship at a relatively sharp angle to the centre line and a series of transverse waves along the hull on each side and behind in the wake.

The presence of the wave systems modifies the skin friction and other resistances, and there is a very complicated interaction among all the different components. Submerged bodies just below the surface of water also create wave systems and, therefore, experience wave-making resistance. However, as the depth of submergence increases, wave making reduces.

An early idea of the ship wave pattern was given by Lord Kelvin (1887–1904) by considering a pressure point travelling over the water surface. The Kelvin wave pattern consists of

1. A transverse wave system and
2. A divergent wave system

The meeting point of the transverse and divergent waves is a high point. The transverse waves move in the same direction of the ship and with the same speed. If the wave length is λ, then

$$\lambda = \frac{2\pi V^2}{g}$$

or

$$\frac{\lambda}{L} = 2\pi \left[\frac{V}{\sqrt{gL}} \right]^2 = 2\pi F_n^2$$

where L is the length of the ship. Since the divergent waves move at an angle θ to the direction of the ship, the speed of these waves is ($V \cos \theta$) and hence the wave length λ' is

$$\lambda' = \frac{2\pi (V \cos \theta)^2}{g} = \lambda \cos^2 \theta$$

The wave-making resistance increases with ship speed. But since this is the integration of the longitudinal pressure components developed by the wave system, the increase in wave resistance is undulatory in nature. When there is a crest in the wave profile in the forepart and a trough in the aft part, the wave-making resistance is high. But when there are crests near both fore and aft ends, the longitudinal pressure components in the fore and the aft tend to cancel and this resistance increase is reduced. Therefore, based on ship length and Froude number, there are humps and hollows in the wave resistance curve. If n is the number of wave crests in the ship length L, the hollow and hump speeds can be shown to occur at F_n given in Table 7.1.

Normally, the first bow wave crest occurs around the quarter of a wavelength aft of the bow. For high-speed vessels (say, planing craft), e.g. $F_n = 1.5$, the wavelength is more than 14 times the ship length and the first wave crest occurs at about 3.5 times the ship length behind the bow. Therefore, at high speeds, the water surface along the ship length is almost horizontal.

Up to speeds corresponding to $F_n \leq 0.27$, the length of the marine craft spans two or more waves, changes in draught and trim are small and the drag is predominantly frictional. As the speed increases, wave-making resistance increases and above $F_n = 0.36$, it increases at a very fast rate. At $F_n = 0.40$, when the ship length equals the wavelength, the wave resistance is maximum and virtually forms a barrier to the speed of displacement vessels. This is primarily because the increased velocities around the hull form result in negative pressure causing the stern to settle deeply in water and trim by stern. If the boat is to be driven in the high-speed displacement mode, i.e. $0.40 \leq F_n \leq 0.95$, it is necessary to change the stern shape to reduce separation drag and also reduce the build-up of negative pressure. This is achieved by designing a wider, flatter and broader stern than before. Then the wave or residuary resistance barrier is crossed and wave resistance is no longer an important factor. The frictional resistance, however, remains a dominant factor. At these speeds, the flat bottom of the aft body may generate some lift force, which may support some weight. Around this speed, some

TABLE 7.1

Humps and Hollows in Wave Resistance

n	Hollow Speed		Hump Speed	
	λ/L	F_n	λ/L	F_n
1	4/1	0.798	4/3	0.461
2	4/5	0.357	4/7	0.362
3	4/9	0.266	4/11	0.235
4	4/13	0.221	4/15	0.206

lift is generated and this range is also known as 'semi-planing' region. At high speeds, length loses its importance as a principal parameter for resistance, and weight, which requires to be supported by buoyancy, becomes important. A volume Froude number $F_{n\nabla}$ is defined as

$$F_{n\nabla} = \frac{V}{\sqrt{(g\nabla^{1/3})}}$$

At speeds higher than those corresponding to $F_n = 0.95$, the bottom and aft should be designed for planing, i.e. the lift generated at the boat bottom should support the weight and the boat's centre of gravity must rise up so that there is an effective reduction in wetted surface and, hence, frictional resistance. Flow is made to separate at the side as well as at the stern. This is the fully planing region when the residuary resistance increases very slowly with speed. Wave resistance is almost negligible with some drag due to spray. The development of flow from displacement mode to fully planing mode is discussed in detail by Savitsky (1964, 1985). If the vessel is supported by hydrofoils, at high speeds, the lift generated by submerged or surface-piercing hydrofoils can support the entire weight of the ship and the vessel comes out of water, thus reducing viscous drag and limiting it to drag of foils and struts connecting the vessel and foils. Figure 7.2 shows the changes in resistance pattern with increasing speed in displacement mode, planing mode and hydrofoil-borne mode. Larsson (2010) has shown diagrammatically the percentages of components of resistance of surface ships, which is reproduced in Figure 7.3.

When two wave systems meet together, a resultant wave system is created. Mathematically, the resultant wave height can be obtained by linear superposition. Simply stated, if two wave crests meet, a higher wave is generated, and if a wave crest meets the trough of another

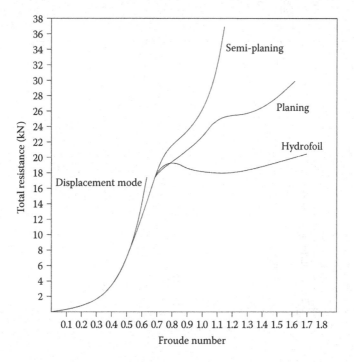

FIGURE 7.2
Resistance trends for different ship types.

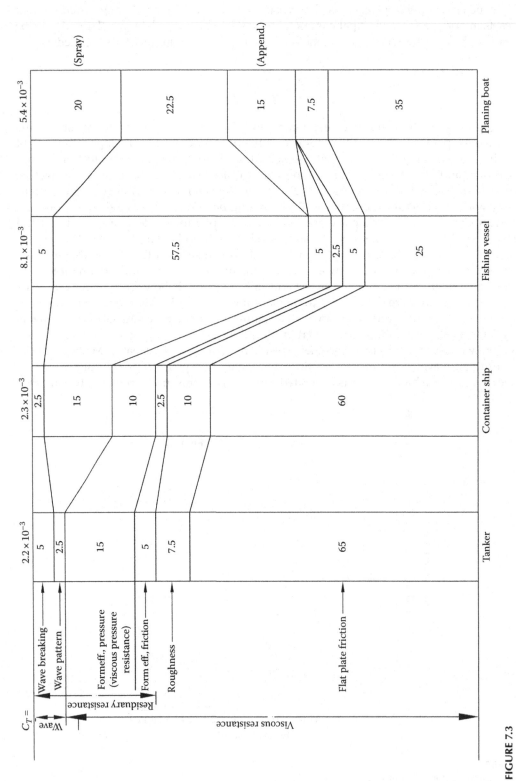

FIGURE 7.3
Percentages of resistance components of surface ships.

wave, a relatively low-height wave system is generated. An example of application of this principle is the bulbous bow of a ship where the forward wave crest due to bow and the wave trough due to the immersed bulb interact to reduce wave-making resistance. In a multihull ship such as a catamaran, trimaran or pentamaran, the wave system in between the hulls is superposed. If this superposition is such that the internal wave system is reduced in height, the total wave resistance becomes less than the sum of the wave resistance of individual monohulls. This nature of superposition is dependent on the hull separation. Therefore, in a catamaran vessel, hull separation is very important. A properly chosen hull separation can reduce the total resistance considerably. If the displacement volume of each hull of a twin-hull vessel is pushed below the waterline such that the waterplane becomes thin, one obtains a small waterplane area twin-hull (SWATH) vessel. Because of the thin waterplane, the waves generated are small, and judicious hull separation distance can reduce waves further. Hence, resistance becomes predominantly viscous. Care must, however, be taken to reduce flow separation. But due to an increase in wetted surface, the frictional resistance is high, and so these vessels are generally not driven at high speed.

Surface ships are normally designed for moving at a level trim condition. Even if there is a resultant vertical component of pressure distribution around the hull form, the resulting trim in running condition is negligible and its effect on resistance can be ignored. However, in ballast and light-load conditions, there may be considerable aft trim, which may cause flow separation and increased drag. So if a surface ship has to spend a large portion of its life in partly loaded or ballast condition, the hull form design should be done such that there is no increased drag due to trim. On the other hand, slight trim of a planing vessel is advantageous for generating lift, which must be provided at the design stage itself.

A ship sailing on a smooth sea and in still air experiences air resistance, but this is usually negligible, and it may become appreciable only in high wind. Although the wind speed and direction are never constant and considerable fluctuations can be expected in a storm, constant speed and direction are usually assumed. Even in a steady wind, the speed of the wind varies with height above the sea. For consistency, therefore, the speed V_0 is quoted at a datum height h_0 of 10 m. Near the sea surface, the wind is considerably slower than at and above the datum height. Variation of speed V with height h was given in Section 2.3 as

$$\frac{V}{V_0} = \left(\frac{h}{h_0}\right)^{0.2}$$

where
h_0 is the datum height
V_0 is the mean wind speed at the datum height

The axial wind force (wind resistance) is given in terms of a coefficient C_{XA}, which is expressed as

$$C_{XA(\Psi_A)} = \frac{\text{Axial force at relative wind angle } \Psi_A}{0.5\rho_A A_{TA} W}$$

where
A_{TA} is the transverse projected area of the ship
W is the relative wind speed experienced by the ship

The axial wind force coefficient C_{XA} is a function of the relative wind angle ψ_A, and typically it varies between +0.8 and −0.8 as ψ_A varies from 0 at head wind condition to 180° at following wind condition. This force is generally insignificant except when the ship is 'stopped' in a wind or during low-speed manoeuvring. The wind side force is computed on the basis of the lateral (side) projected area A_{LA} and is given by the expression

$$C_{YA(\Psi_A)} = \frac{\text{Side force at relative wind angle } \Psi_A}{0.5\rho_A A_{LA} W}$$

The variation of C_{YA} with the relative wind angle is generally more or less sinusoidal, and the maximum value of about 0.8 occurs near 90° (beam wind). The yaw moment generated by the wind is given by the expression

$$C_{NA(\Psi_A)} = \frac{\text{Moment at relative wind angle } \Psi_A}{0.5\rho_A A_{LA} L_{OA} W}$$

where L_{OA} is the length overall and the moment coefficient is

$$C_{NA(\psi_A)} = x_{A(\psi_A)} C_{YA(\psi_A)},$$

where x_A is the centre of pressure, which typically varies between +0.3 and −0.3 of the ship's length.

The principal appendages in ships are the bilge keels, rudders, bossings or open shafts and struts. All these items give rise to additional resistance. The drag due to appendages is very complex and depends on the appendage location (extent of turbulence or Reynolds number effect), shape (stream line shape), appendage orientation (separation of flow if not oriented along the streamlines of bare hull) and its wetted surface.

In a ship with a transom stern, a part of the wetted surface is perpendicular to the direction of motion or nearly so. This gives rise to a resistance component contributed by the transom stern and is called transom resistance. This component is primarily due to the separation of flow at the stern.

In certain circumstances, the motion of a body in a fluid produces a force normal to the direction of motion. This is called *lift*. When lift is generated, there is an associated resistance or drag known as *induced drag*. Some types of high-speed marine crafts depend on the generation of lift for supporting their weight in motion, and in these crafts, induced drag is a component of resistance. In some marine crafts, the motion of the craft generates spray and this may give rise to *spray resistance*, particularly if the spray strikes the hull. A ship may continuously take in large quantities of air or water from outside for some internal purpose. This air or water, assumed to be at zero velocity outside the ship, is forced to acquire the velocity of the ship when taken into the ship. The rate of change of momentum of this fluid gives rise to *momentum drag*.

7.1.2 Shallow-Water Effects

The resistance of a ship is quite sensitive to the effects of shallow water. In the first place, there is an appreciable change in the potential flow around the hull. If the ship is considered at rest in a flowing stream of restricted depth, but unrestricted width, the water passing below it must speed up more than in deep water, with a consequent greater reduction in pressure and increased sinkage, trim usually by the stern, and increased resistance.

If in addition the water is restricted laterally, as in a river or canal, these effects are further exaggerated. The sinkage and trim in very shallow water may set an upper limit to the speed at which ships can operate without touching the bottom. The second effect is the changes in the wave pattern which occur in passing from deep to shallow water. When the water is very deep, the wave pattern consists of the Kelvin transverse and diverging waves, the pattern for a pressure point being contained between the straight lines making an angle α of 19° 28 min on each side of the line of motion.

The velocity of surface waves in water depth h is given by the expression

$$V_C^2 = \left(\frac{g\lambda}{2\pi}\right)\tan h \frac{2\pi h}{g\lambda}$$

As the depth h decreases, and the ratio h/λ becomes small, $\tan h\,(2\pi h/\lambda)$ approaches the value $2\pi h/\lambda$, and for shallow water, the wave velocity is approximately given by the expression

$$V_C^2 = gh$$

For values of V less than about $V = 0.4\sqrt{gh}$, the wave pattern is enclosed between the straight lines having an angle α = 19° 28 min to the centre line as for deep water. As V increases above this value, the angle α increases and approaches 90° as V approaches $\sqrt{gh}$. The pressure point is now generating a disturbance, which is travelling at the same speed as itself, and all the wave-making effect is now concentrated in a single crest through the pressure point and at right angles to its direction of motion. The pattern agrees with the observation on models and ships when running at the critical velocity in shallow water. The whole of energy is transmitted with wave, and the wave being called a wave of translation. When V exceeds $\sqrt{gh}$, the angle α begins to decrease again, the wave system being contained between the lines given by $\sin\alpha = \sqrt{gh}/V$. It now consists only of diverging waves, there being no transverse waves or cusps. The two straight lines themselves are the front crests of the diverging system, and the inner crests are concave to the line of advance instead of convex as in deep water (Figure 7.4).

Speeds below and above $V = \sqrt{gh}$ are referred to as subcritical and supercritical, respectively. Nearly all displacement ships operate in the subcritical zone, the exceptions being fast naval ships. As depth of water decreases, the speed of the wave of given length also decreases. Thus, to maintain the same wave pattern, a ship moving in shallow water travels at a lower speed than in deep water, and humps and hollows in the resistance curve occur at lower speeds than in shallow water. The ship speed loss is

$$\delta V = V_\infty - V_h,$$

where
V_∞ is the speed in deep water
V_h is the speed at a depth h

When the ship is operating in shallow water and in restricted channels, the corresponding speed loss depends on the hydraulic radius R_H of the channel defined as

$$R_H = \frac{\text{Area of cross section of channel}}{\text{Wetted perimeter}}$$

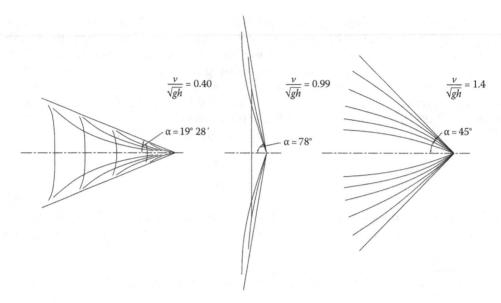

FIGURE 7.4
Effect of shallow water on wave pattern.

For a rectangular channel of width b and depth h

$$R_H = \frac{bh}{b+2h}$$

When b becomes very large, $R_H = h$, and this corresponds to the previous case of shallow water of unrestricted width. An analysis of shallow-water effects and a procedure for estimation of speed loss in shallow water have been suggested by Schlichting (1934) and discussed by Lewis (1989b). Shallow-water effects on resistance and the corresponding effective power estimation are required for the design of vessels moving in rivers, channels or waters of restricted cross section.

7.1.3 Methodical Series

The resistance characteristics of a hull form depend on a set of main hull form parameters as well as on local form characteristics. Ignoring the effects of local shape changes on resistance, if one could express resistance, particularly residuary resistance or wave resistance, as a function of a set of main parameters, it would be possible to estimate effective power at early stages of design. A methodical series of ships or ship models is a series whose resistance data are available through model experiments by systematic variation of certain hull parameters. A number of such methodical series are available in literature. In each series, one or more parent hull forms are chosen and their main features are identified. Then, by a process of systematic variation of these parameters, new hull forms are generated. Scale models of all these forms are then tested in a towing tank and the residuary resistance component (= total resistance – frictional resistance) is then presented as a function of these parameters. The following are a few of the standard series data available in literature:

- *Taylor's series*: Such a well-known standard series is the Taylor series developed by Admiral Taylor in the 1930s in David Taylor Research Centre (Experimental Model Basin at the time). The original parent hull was patterned after a British cruiser *Leviathan*. This series was later examined further by Gertler (Taylor 1943, Gertler 1954) and extended by Graff et al. (1964).

- *Series 60*: This is a very popular series developed by the Society of Naval Architects and Marine Engineers in cooperation with the ATTC (Todd 1963). It is based on the parents of the single-screw merchant vessels with variations of LCB, and apart from resistance, it also records trim effects and some propulsive coefficients. It has a narrower range than Taylor series. Sabit (1972a) has done a regression analysis of the resistance data and given the regression coefficients which can be readily used for effective power prediction of a vessel with different hull parameters.

- *BSRA series*: This has resulted from a long series of tests with single-screw merchant ship hulls. It was developed by the British Ship Research Association in the 1960s (Lackenby and Parker 1966). A comparison between parameters of Taylor series, Series 60, and BSRA series for applicability for resistance calculation is shown in Table 7.2. Here, $C_\Delta = W/(0.01L)^3$ and L_p is the length of the parallel middle body as the percent of the length between perpendiculars and LCF is the distance of longitudinal centre of floatation of load water line from midship as a percentage of length. The terms in the last row are propulsion-related factors. Sabit (1971) has given the regression coefficients for resistance estimation of BSRA series based hull forms.

- *SSPA series*: This series was developed by the Swedish State Shipbuilding Experimental Tank in the 1950s and includes data for high-speed twin-screw cargo liners, fast single-screw cargo ships, tankers and single-screw cargo ships (Edstrand and Lindgren 1955–1956, Edstrand et al. 1956, Freemanis and Lindgren 1957–1959). Regression coefficients for cargo ships of the SSPA series have been suggested by Sabit (1972b).

- *MARAD series*: This is a full form cargo ship series published by Roseman (1987).

TABLE 7.2

Range of Applicability of Standard Series for Resistance

Variables	Taylor C_P, L/B, B/T, C_Δ	Series 60 C_B, L/B, B/T, C_Δ, LCB	BSRA C_B, B/T, C_Δ, LCB, L_P
$V/\sqrt{L}$	0.5–2.0	0.4–1.0	0.4–0.85
C_B	0.4396–0.8018	0.60–0.80	0.65–0.85
L/B	3.92–17.25	5.5–8.5	6.89–7.27
B/T	2.25–3.75	2.5–3.5	2.12–3.95
C_Δ	26.5–221.6	68–302	114–385
LCF	0	−2.5 to 3.5	−0.5 to 4.05
L_P	0	0	0–50
Propulsion	None	$\eta_D, \omega, t, \eta_R$	ω, t, η_R

- *NPL series*: These are round-bottom fine form hull forms developed by the National Physical Laboratory, England, containing data for coastal vessels and high-speed displacement and semi-planing forms with Froude number going up to 1.0 (Bailey 1976).

Apart from these, there are a number of systematic series resistance tests for different types of crafts, which are available in literature. These include coaster vessels (Todd 1931–1942), trawlers (Ridgely-Nevitt 1963), semi-planing hull forms (Compton 1986) and planing hull form series 62 and 64 (Clement and Blount 1963, Yeh 1965). Almeter (1993) lists a number of other planing craft series such as the Norwegian series (Werenskiold 1970), the Dutch series (Keuning and Gerritsma 1982) and others.

In designing a new ship, it is necessary to estimate the effective power for engine selection. Resistance data of methodical series provide a method for extrapolating the resistance based on the design parameters from the series data. However, it is implicit in this method that the accuracy of the prediction depends on the hull form confirming to the hull shape of the series form. If the hull form is quite different, the prediction is likely to be inaccurate. Another use of series hull form prediction method is to compare the effective power prediction for the design ship from different series data and select the optimal series from which the lines plan could be generated. But with the advent of better design techniques discussed later, series prediction methods have become redundant since better hull forms are being developed giving a much better performance than that obtained from a series form.

7.1.4 Resistance Estimation by Statistical Method

If the resistance data of a large number of vessel models are available (say in a towing tank), it is possible to analyse that data numerically and establish a regression model of bare hull resistance as a function of a number of variables defining the hull form. Similar exercise can be carried out for appended hull or bare hull with appendages. If speed trial data are available, it is also possible to model the excess trial drag over and above the extrapolated appended hull resistance. Such an attempt has been made, and a convenient method of evaluating resistance using statistical models has been presented by Holtrop and Mennen (1978, 1982) and Holtrop (1984). The proposed method has been used successfully to estimate the required power at the early stages of design. Guldhammer and Harvald (1974) have given a set of curves based on statistical data to estimate the residuary resistance and, hence, effective power. Similar statistical prediction methods have been suggested by Doust and O'Brien (1958–1959) and Doust (1962–1963) for trawler forms, by van Oortmerssen (1971, 1973) for small ships, Savitsky (1985) and Blount (1976, 1994) and Hadler (1966) for planing craft. Moor and others have given effective power prediction methods for single-screw ships (Moor and Small 1960) and twin-screw ships (Moor and Patullo 1968). Moor (1971) have also suggested optimum standards of ships with regard to resistance until 1971, which have been superseded with better forms subsequently.

7.1.5 Resistance Estimation of Submersibles

A submarine operating on the surface behaves as does any surface craft. It is subjected to the same force phenomena when moving through the water. Prior to World War II, the configuration of submarines was the result of a compromise between surface and subsurface operation. Their hull form was a concession to the existence of wave-making resistance.

They were long and narrow, having an *L/B* ratio of about 11.5, with a centre of volume approximately amidships. The bow configuration was a modified surface ship's bow, and there was considerable flat-deck surface with many unstreamlined appendages. The eddy-making resistance submerged was considerable, because the craft was essentially a modified surface ship capable of submerging for short periods. The use of nuclear power has permitted the modern submarine to become a true subsurface ship. It is no longer dependent on the surface for oxygen to supply the engines. The present hull shapes are completely clear of appendages, except for the necessary control surfaces at the stern, the propeller and the streamlined sail enclosure. The basic configuration of the bare hull is that of a body of revolution whose *L/B* ratios range from 8 to 11 for attack submarines and from 11 to 13 for ballistic-missile submarines. The modern submarine experiences no wave-making resistance whatsoever when submerged more than three diameters from the free surface.

In order to make powering estimates for submarines, the resistance components are usually divided into those for the bare hull (subscript *BH*) and those of the appendages (subscript *AP*). The following symbols are useful in calculating the total resistance of a full-scale submarine by several different methods:

If $C_{F_{BH}}$ is the frictional resistance coefficient of the bare hull; $C_{R_{BH}}$ is the residual (eddy) resistance coefficient of bare hull, generally assumed independent of Reynolds number and $C_{V_{BH}}$ is the viscous-resistance coefficient of bare hull, which is equal to the total-resistance coefficient for a deeply submerged bare hull, then

$$C_{V_{BH}} = C_{R_{BH}} + C_{F_{BH}}$$

or alternatively,

$$\frac{C_{V_{BH}}}{C_{F_{BH}}} = 1 + 0.5\frac{B}{L} + 3\left(\frac{B}{L}\right)^3$$

where
B is the maximum beam or diameter of the submarine and *L* is the length.
The effective power of a fully submerged submarine is given as

$$\text{Effective power} = \frac{1}{2}\rho V^3 [(C_{V_{BH}} + C_A)S_{BH} + \sum C_{V_{AP}}S_{AP}]$$

Where C_A is the model-ship correlation allowance or roughness allowance for full-scale resistance estimates made without model tests
S_{BH} is the wetted surface of bare hull
$C_S = S_{BH}/\pi BL$ is the ratio of bare hull wetted surface to that of the outside of a cylinder with the same length and beam
$C_{V_{AP}}$ is the viscous-resistance coefficient of the various appendages, generally of the order of 0.5 times $C_{V_{BH}}$, depending on the length, shape and Reynolds number of the appendage
S_{AP} is the wetted surface of the various appendages, such as the fairwater (sail), rudder and planes, etc. generally totalling 0.1–0.2S_{BH}.

These definitions and symbols can be combined in the equation for the effective horse-power of a fully submerged submarine.

Generally, hull frictional resistance accounts for about 55%–60% of total resistance, viscous form effect is about 7%–8%, appendage drag is about 25%–30% and ship model correlation allowance accounts for 9% of the total resistance of a submerged submarine.

It is possible to do a paint flow test to show the direction of flow around the underwater ship body, which is useful for designing appendages.

7.1.6 Experimental Fluid Dynamics

In carrying out experiments to study ship resistance, it is usual to measure only the total resistance of a model geometrically similar to the actual ship by towing it in a towing tank. However, techniques have been devised to determine experimentally some individual resistance components for ship models. Frictional resistance can be determined by measuring the tangential stress at several points on the surface of the ship model and integrating the resulting stress distribution. The pressure resistance can be similarly determined by measuring the pressures on the hull surface. The wave resistance can be determined by calculating the rate at which the energy of the wave system generated by the ship model is increasing, since the work done by the wave resistance is theoretically equal to the energy of the waves generated by the ship. The energy of the waves is determined by the measurement of wave heights in the wave pattern behind the model. The resistance determined from the wave pattern in called *wave pattern resistance*, and this is slightly different from the wave resistance because of the effect of viscosity on the waves, wave breaking and other causes. The effect of viscosity is to cause the body moving in a viscous fluid to impart a momentum to the fluid in the direction of motion. The rate of change of this momentum is theoretically equal to the viscous resistance. This change of momentum is determined by measuring the velocities at several points in the *wake* (the disturbed fluid behind the ship model); the resistance calculated in this way is called *wake resistance*. However, for design purposes, generally, measuring total resistance of a ship model is enough to determine the effective power of a geometrically similar ship moving with a forward speed in calm water

Three conditions of similarity must be satisfied in carrying out a resistance experiment with a ship model:

1. *Geometrical similarity*, which requires that the ratio of any two dimensions in the model must be equal to the ratio of the corresponding dimensions in the ship.
2. *Kinematic similarity*, which requires that the ratio of any two velocity components in the flow around the model and the corresponding velocity components in the flow around the ship must be equal, i.e. the flow patterns around the model and the ship must be geometrically similar.
3. *Kinetic similarity*, which requires that the ratio of any two forces acting on the model must be equal to the ratio of the corresponding forces on the ship.

For a geometrically similar model, kinematic and kinetic similarities can be maintained if Reynolds number and Froude number are similar for the model and the ship. Then

$$C_{TM} = C_{TS} \quad \text{if } R_{nM} = R_{nS} \text{ and } F_{nM} = F_{nS}.$$

where the subscripts M and S refer to the model and the full size ship, respectively. However, this is not possible to achieve. Therefore, if geometrically similar ships ('geosims') move at speeds such that their Froude numbers are equal, their wave (residuary) resistance coefficients are also equal. This division of the total resistance R_T into frictional resistance R_F and residuary resistance R_R was first proposed by W. Froude, who also stated what he called the *law of comparison*:

The residuary resistances of geometrically similar ships are proportional to their displacements if their speeds are proportional to the square roots of their lengths,

$$\text{i.e.} \ \frac{R_R}{\Delta} = \text{constant} \quad \text{if } \frac{V}{\sqrt{L}} = \text{constant for geometrically similar ships}$$

where Δ and L are the displacement and length of the ship, respectively. This is now called the Froude law. Speeds of geometrically similar ships proportional to the square roots of their lengths are called *corresponding speeds*. A more modern approach is to call the two components viscous resistance and wave resistance. Noting that in geometrically similar ships, the wetted surface S is proportional to the square of the length L and the displacement volume ∇ is proportional to the cube of the length,

$$F_n = \text{constant implies } V_m = \frac{V_s}{\sqrt{\lambda}} \quad \text{where } \lambda = \frac{L_s}{L_m}$$

And

$$C_W = \text{constant implies } R_W = C_W \cdot \lambda^3$$

The Froude law may be used for the determination of the resistance of a ship from the measured resistance of its geometrically similar model provided that a method can be found to determine the viscous resistances of the model and the ship:

The model total resistance R_{TM} is measured at a speed V_M.

The model viscous resistance R_{VM} at the speed V_M is calculated by some independent means (ITTC friction line and form factor).

The model wave resistance at the speed V_M is obtained as $R_{WM} = R_{TM} - R_{VM}$.

The ship wave resistance at the corresponding speed is obtained using the Froude law:

$$V_S = V_M * \sqrt{\frac{L_S}{L_M}} \quad R_{WS} = R_{WM} * \frac{\Delta_S}{\Delta_M}$$

The ship viscous resistance R_{VS} at the speed V_S is calculated. Normally, this is taken as

$$R_{VS} = 1/2 * \rho * S_S * V_{S^2} * C_F,$$

$$C_F = (1+k) * C_{F0} \text{ and}$$

$$S_S = S_M * \lambda^2$$

where C_F, C_{F0} and k denote the frictional resistance coefficient of the ship, the 2D frictional resistance coefficient of the ship and the form factor, respectively. Here, the total frictional resistance is taken as equal to the total viscous resistance.

The total resistance of the ship at the speed V_S is obtained as $R_{TS} = R_{VS} + R_{WS}$.

If the vessel has a large number of appendages, the model should be tested for resistance with and without appendages. Then,

Appendages drag = appendaged hull drag − bare hull drag.

Extrapolation of appendages from the model to the full scale is complex and must be done with care. Normal merchant ships have a few appendages and extrapolation error can be ignored. However, naval vessels and submersibles have large appendages and drag due to appendages is high. The total-resistance coefficient for the ship is then increased to take into account trial conditions.

$$C_{Ts} = C_{Ts} + C_A + C_{AA}$$

where
C_{AA} is the coefficient due to wind resistance
C_A is the so-called correlation allowance taking into account ship surface roughness

A standard value of this allowance is 0.0004. This effective power can be estimated as follows:

$$\text{Effective power} = R_{Ts} * V_s = 1/2\rho * S_S * V_s^2 * C_{Ts} * V_s = 1/2\rho * S_S * V_s^3 * C_{Ts}$$

The stream flow pattern around the bare hull is often required to be known for designing appendages, locating them and orienting the same in the direction of flow. If appendages are not aligned properly, the drag increases substantially and due to large flow separation, there could be vibration of the appendage and subsequent joint failure. This vibration could also extend to the ship structure leading to discomfort and equipment failure. Normally, paint flow test or tests with wool tufts are carried out and the flow pattern is recorded.

Experimental fluid dynamics or hydrodynamic tests are the best way to predict the effective power and decide on the brake power of the propulsion plant. However, this is an expensive and time-consuming process. Experimentally developing an optimized hull form from the resistance point of view would mean testing different hull forms for comparison and then selection. This is a very expensive exercise. Normally, a reputed tank with a large experimental database can advise a client on hull form for better resistance characteristics. But a single hull form testing can always leave a doubt in the mind of the designer if he could have done better.

7.1.7 Computational Fluid Dynamics

The flow around a ship hull form is a mathematically complex problem involving solution of nonlinear equations of motion subject to kinematic and kinetic boundary conditions and the radiation condition. The nonlinear nature arises due to the complex 3D ship body geometry and the 3D free surface, which is unknown before a solution is arrived at. Assuming the flow to be incompressible, irrotational and potential in nature, a flow solution can be arrived at by discretizing the entire fluid space and solving the equation of motion. The advent of modern high-speed computers has given rise to many computational fluid dynamics (CFD) techniques to solve the irrotational potential flow problem. The resultant flow gives pressure distribution around the body from which the wave resistance can be computed and stream line flow characteristics can be studied. Viscous flow calculation is more involved. One way to do this calculation is to compute the potential flow and velocities which can form the starting point of viscous flow calculation. Numerous research workers have attempted a complete viscous flow solution using CFD techniques generally known as Reynolds averaged numerical solution (RANS). Today, advanced commercial software are available, which are good flow solvers. Using suitable CFD packages, it is now possible to predict wave resistance using potential flow theory and total resistance using a numerical RANS solver. It is then possible to estimate the form factor as discussed before. It is also possible to get the flow pattern around the ship. Barring a few software bugs, the CFD techniques can now predict resistance of displacement crafts over the normal operating range. Even round-bottom vessels at high speed corresponding to a Froude number up to 1.0 can be modelled and studied using CFD. Studies conducted at the Indian Institute of Technology (IIT), Kharagpur, have indicated that resistance predictions can be done at model scale, which compare favourably with experimental results giving a difference of maximum 6%–7%. Figure 7.5 shows the comparison of CFD estimate of resistance with experimental data. Figure 7.5a compares the data of CFD computation with experimental data from the experimental tank of Insean, Italy, for the standard vessel, which is a model of KRISO very Large Crude Carrier 2 or KVLCC2. Figure 7.5b compares the data for a round-bottom boat up to a Froude number of 0.5 and Figure 7.5c compares the data of an NPL round-bottom high-speed form running up to a speed corresponding to a Froude number of 1.0, experimental data for both these models being obtained from conducting experiments at the hydrodynamics tank at IIT Kharagpur. It can be observed that the percentage error is nearly the same throughout the test range for various ship hull forms. Though more accurate predictions are necessary for numerical tanks to replace model test tanks, these CFD tools can be successfully used to design hull forms and appendages even at the concept design stage. Sha (2014) has demonstrated that appended hull resistance can be predicted using CFD to the same accuracy level as the bare hull prediction (Figure 7.6), the comparison being made with two different methods of CFD analysis, volume of fluid (VOF) method and RANS solver. Sha (2014) has also shown the use of CFD tools for designing a gooseneck bulb

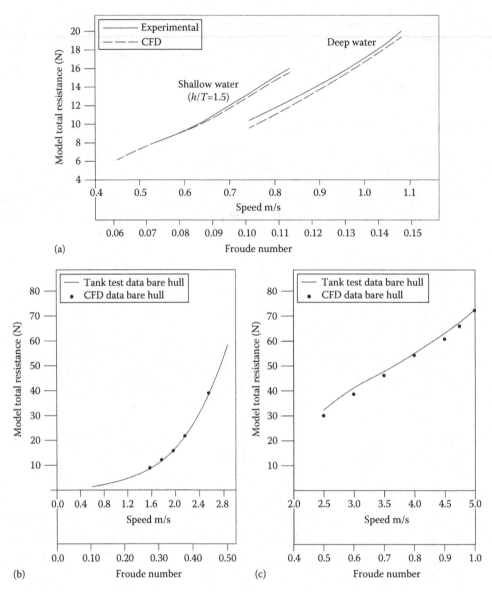

FIGURE 7.5
Comparison of total resistance obtained from CFD calculations and model tests for different vessels: (a) KVLCC, (b) round-bottom high-speed boat and (c) NPL round-bottom boat.

for a slender hull of about 90 m length with a block coefficient of about 0.5. Figure 7.7a shows the original and bulbous hull models, and Figure 7.7b shows the improvement in resistance due to bulb as proven in towing tank experiments. Since the experiments were conducted for both the models for the same draught condition, the bulbous form had about 2% higher displacement and, to make the comparison independent of displacement, Telfer coefficient $C_{TL} = R_T * L/(\Delta * V^2)$ has been used in the presentation. The design methodology of a hull form can use both CFD and experimental fluid dynamics tools as shown in Figure 7.8.

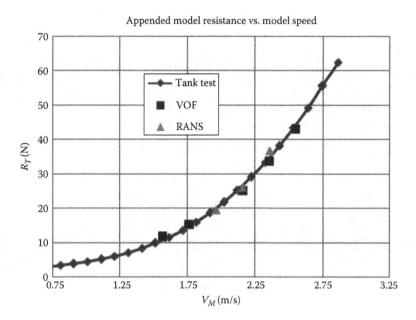

FIGURE 7.6
Improvements in bare hull resistance.

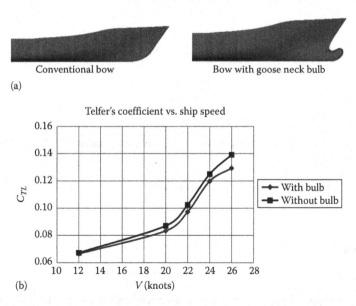

FIGURE 7.7
Comparison of appended hull resistance obtained from CFD calculations and model tests. (a) Forward end of normal hull and modified hull with gooseneck bulb. (b) Resistance improvement with bulbous bow from model experiments.

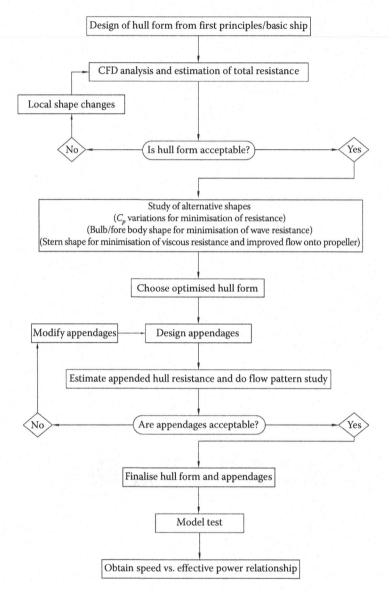

FIGURE 7.8
Flow chart for design of hull form using CFD and experiments.

7.2 Propulsion

Conventionally, a ship is propelled by a screw propeller, which is a hydrodynamic device converting the torque delivered to it from the propulsion engine to a forward thrust moving the ship forward. A screw propeller consists of a number of blades attached to a hub or boss, as shown in Figure 7.9 (Ghose and Gokarn 2004). The boss is fitted to the propeller shaft through which the power of the propulsion machinery of the ship is transmitted to

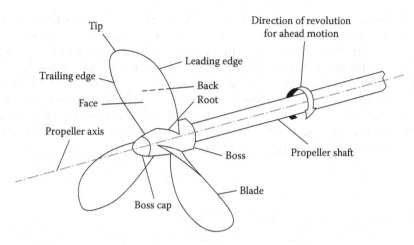

FIGURE 7.9
Three-bladed right-handed propeller.

the propeller. When this power is delivered to the propeller, a turning moment or torque Q is applied, making the propeller revolve about its axis with a speed ('revolution rate') n, thereby producing an axial force or thrust T causing the propeller to move forward with respect to the surrounding medium (water) at a speed of advance V_A.

The main parameters of a propeller are as follows:

N: Revolutions per minute or n: revolutions per second

D: Propeller diameter

P/D: Pitch ratio

A_P, A_E, A_D: Projected, expanded and developed blade areas, respectively

A_0: Propeller disc area = $\pi \cdot D^2 / 4$

A_E/A_0: Expanded area ratio

The performance of a propeller advancing in undisturbed fluid (in our case, water) is known as 'open-water' performance. But a propeller is normally fitted at the stern of a ship where it operates in water disturbed by the ship as it moves ahead. The performance of the propeller is thus affected by the ship to which it is fitted, so that the same propeller performs slightly differently 'behind' different ships.

The axial velocity of water onto the propeller is known as speed of advance, V_A, which is the speed with which propeller moves forward in undisturbed fluid. Non-dimensionalizing propeller thrust, torque and speed of advance, one can write the thrust and torque coefficients and advance coefficient of a propeller as

$$K_T = \frac{T}{\rho n^2 D^4} \quad K_Q = \frac{Q}{\rho n^2 D^5} \quad J = \frac{V_A}{n \cdot D}$$

The thrust and torque coefficients along with the advance coefficient J give the open-water efficiency:

$$\eta_O = \frac{TV_A}{2\pi n Q} = \frac{K_T}{K_Q} \frac{J}{2\pi}$$

These quantities are generally used to describe the open-water characteristics of a propeller. When a propeller is fitted at the stern of a ship, it operates in water disturbed by the ship during its forward motion. There is a mutual interaction between the hull and the propeller, each affecting the other. This 'hull–propeller interaction' has three major effects on the performance of the propeller and the ship taken together as compared to their behaviour considered individually: wake, trust deduction fraction and relative rotative efficiency.

The water behind the ship is already disturbed due to flow around the ship and thus the water at the propeller disc has a forward speed v, which is known as the wake of the ship. The speed of advance of the propeller is then given by

$$V_A = V - v = (1-w)V$$

The wake fraction w can, therefore, be written as

$$w = \frac{(V - V_A)}{V}$$

When a propeller produces thrust, it accelerates the water flowing through the propeller disc and reduces the pressure in the flow field ahead of it. The increased velocity of water at the stern of the ship and the reduced pressure cause an increase in the resistance of the ship. Then the thrust required to propel the ship forward at a constant speed must be more than the calm water resistance R_T. Since this increase in resistance is an effect due to the propeller, it is convenient to define thrust deduction in terms of thrust deduction fraction as

$$t = \frac{(T - R)}{T}$$

The thrust deduction fraction is related to the wake fraction, a high wake fraction usually being associated with a high thrust deduction fraction. The thrust deduction fraction also depends to some extent on the rudder if placed in the slipstream of the propeller. The hull–propeller interaction factors w and t give a component of propulsive efficiency known as the hull efficiency, which is the ratio of effective power to thrust power and is given by

$$\eta_H = \frac{(1-t)}{(1-w)}$$

The hull efficiency is usually slightly more than 1 for single-screw ships and slightly less than 1 for twin-screw ships. One may note that for a high hull efficiency, the wake fraction w should be high and the thrust deduction fraction t low. If the propeller were placed at the bow of the ship instead of the stern, it would be advancing into almost undisturbed water so that the wake fraction would be much lower than if the propeller were at the stern. On the other hand, the ship would be advancing into water disturbed by the propeller, and since the propeller accelerates the water flowing through it and increases its pressure, there would be a greater increase in the resistance of the ship, and hence a higher thrust deduction fraction, due to propeller action when the propeller is fitted at the bow than when it is fitted at the stern. Therefore, a propeller at the stern results in higher hull efficiency than a propeller at the bow.

Since the propeller in the open-water condition works in undisturbed water whereas the propeller behind the ship works in water disturbed by the ship, the efficiencies of the propeller in the two conditions are not equal. If T_o and Q_o are the thrust and torque of a propeller at a speed of advance V_A and revolution rate n in open water, and T and Q are the thrust and torque at the same values of V_A and n with the propeller operating behind the ship, the relative rotative efficiency η_R, which is the ratio of propeller efficiency in behind condition to open-water condition, is then given by

$$\eta_R = \frac{T}{T_o}\frac{Q_o}{Q}$$

$$= T/T_o \text{ with torque identity or } Q_o/Q \text{ with thrust identity.}$$

The relative rotative efficiency is usually quite close to 1, lying between 1.00 and 1.10 for most single-screw ships and between 0.95 and 1.00 for twin-screw ships.

7.2.1 Power Transmission

The power delivered to the propeller is produced by the propulsion plant of the ship and is transmitted to the propeller usually by a mechanical system, or sometimes by an electrical system. The propulsion plant or main engine may be a steam turbine, a gas turbine or a reciprocating internal combustion (diesel) engine. A turbine runs at a very high speed, and it is necessary to reduce the speed by using a speed-reducing device such as mechanical gearing. High-speed and medium-speed diesel engines also require speed-reducing devices. On the other hand, a low-speed diesel engine is generally directly coupled to the transmission shaft. The power produced by the main engine is transmitted to the propeller, after speed reduction if necessary, through a shafting system consisting of one or more shafts supported on bearings. The shaft on which the propeller is mounted is called the tail shaft, propeller shaft or screw shaft and is supported by bearings in a stern tube. There is also a thrust bearing to transmit the propeller thrust to the ship hull. In an electrical propulsion drive, the main engine drives an electric generator and the electrical power is transmitted by cables to an electric motor which drives the propeller through the propeller shaft. The brake power is the power output of the engine and is carefully measured at the engine manufacturer's works along with the other operating parameters of the engine and is slightly less than the indicated power measured at the cylinder. The power produced by a steam turbine or a gas turbine is usually determined by a torsion meter fitted to the shafting connecting the turbine to the propeller through the gearbox. The power determined by measuring the torsion of the shaft is called the shaft power P_S. A torsion meter may also be used to determine the shaft power when the ship has a diesel engine. The shaft power varies with the location of the torsion meter on the propeller shafting, being slightly higher when the torsion meter is fitted close to the engine than when it is fitted close to the propeller.

The power that finally reaches the propeller is the delivered power P_D, and this is related to the propeller torque Q:

$$P_D = 2\pi n Q$$

n being the propeller revolution rate per second. The delivered power is somewhat less than the brake power or the shaft power because of the transmission losses taking place between the engine and the propeller, i.e. in the gearing and the bearings in mechanical transmission, and in the generator, cables, motor and bearings in an electrical propulsion drive. The propeller thrust is transmitted to the ship hull through the propeller shaft to the ship structure, particularly the strong seatings of the thrust block. This causes the ship to move at a speed V overcoming its resistance R_T with effective power P_E. Figure 7.10 shows a schematic arrangement of the propulsion system of a ship, and the power available at different points of the system. The brake power P_B or the shaft power P_S, depending on whether the main engine is a diesel engine or a steam or gas turbine, may be regarded as the input to the propulsion system and the effective power P_E as its output. If one wishes to focus only on the hydrodynamics of the system, then the delivered power P_D is taken as the input.

The efficiency of the system as a whole or the overall propulsive efficiency or overall propulsive coefficient is then

$$\eta_{overall} = \frac{P_E}{P_B}\left(\text{for diesel engine}\right) = \frac{P_E}{P_S}\left(\text{for turbines}\right)$$

If one considers only the hydrodynamic phenomena occurring outside the hull, the input to the propulsion system is the delivered power P_D, and the propulsive efficiency, known as the quasi-propulsive coefficient (QPC) is given by $\eta_D = P_E/P_D$. The efficiency of power transmission from the engine to the propeller is called shafting efficiency

$$\eta_{overall} = \eta_D \cdot \eta_S \quad \text{where } \eta_S = \frac{P_D}{P_B}\text{ or }\frac{P_D}{P_S}$$

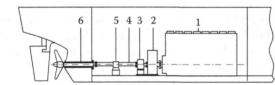

1. Engine
2. Reduction gearing (optional)
3. Thrust bearing
4. Shaft
5. Bearing
6. Stern tube

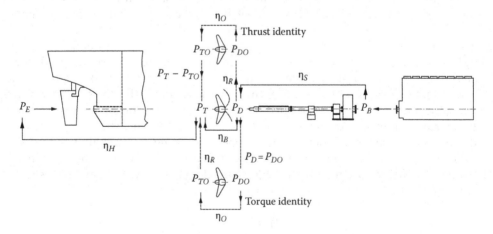

FIGURE 7.10
Schematic arrangement of the propulsion system of a ship. (From Ghose, J.P. and Gokarn, R.P., *Basic Ship Propulsion*, Allied Publishers Pvt. Ltd., India, New Delhi, 2004.)

The losses taking place in the transmission of power are usually expressed as a percentage of the brake power or shaft power. In installations in which the engine is directly connected to the propeller by shafting, the transmission losses are usually taken as 3% when the engine is amidships and 2% when the engine is aft. With mechanical reduction gearing or electric propulsion drives, the transmission losses are higher, 4%–8%. The QPC can be expressed as

$$\eta_D = \eta_0 \cdot \eta_H \cdot \eta_R$$

Once the speed required is known, the effective power and, hence, the thrust required from the propeller can be estimated. Knowing the QPC, the torque required by the propeller can be estimated and this must be provided at the particular propeller rotation rate by the engine. The engine, the propeller and the hull must match exactly to work effectively. The engine cannot be overloaded by propeller demand of higher power at a given RPM nor by the propeller demand of power at RPM higher than the engine-rated RPM. While designing a propulsion system for a tug or trawler or a supply vessel where additional thrust may be required at a low speed, the propeller may be required to provide a thrust higher than the thrust for providing a forward speed only. Since, in such condition, the forward speed is low, the propeller demand of power for the required thrust also changes and this may be possible at a reduced RPM. The engine must be able to provide this without overloading. The propeller design conditions change depending on the operating condition. Care must be taken to ensure that there is no propeller engine mismatch in other operating conditions.

7.2.2 Cavitation

Cavitation can occur in the flow of any liquid. But it is necessary in the context of marine vehicles to consider cavitation only in water – both fresh water and sea water. Some properties of water are of special significance when considering propeller performance: density, viscosity, vapour pressure and surface tension. Cavitation is particularly affected by vapour pressure because water is often assumed to cavitate when the pressure at a point becomes equal to the vapour pressure. Surface tension is important because the formation of a cavity in water results in the generation of a surface. Another important parameter influencing cavitation is the presence of dissolved gases in water.

Vapour pressure depends on temperature. When temperature rises sufficiently so that the vapour pressure becomes equal to the atmospheric pressure, the molecules escape from the water very rapidly, bubbles (cavities) are formed inside the water and these then rise to the surface. This is the process of boiling. A similar situation occurs if instead of increasing the temperature, the pressure of the air above the surface of water is decreased. When the pressure of the air falls below the vapour pressure, bubbles are formed and the water 'boils' without being heated. This is the phenomenon when the water is not flowing.

When water flows, there is a relation between the velocity and pressure, an increase in velocity being accompanied by a decrease in pressure. If the velocity increases sufficiently so that the pressure at a point becomes equal to the vapour pressure, the water will vaporize at that point and form a cavity. In theory, the pressure in water cannot fall below the vapour pressure because the water will then boil or cavitate and bubbles or cavities of water vapour will be formed. The vapour pressure of water depends on its temperature, and the values for fresh water are given in Table 7.3. Vapour pressure of sea water is some 3.3% less than the vapour pressure of fresh water at the same temperature.

TABLE 7.3

Vapour Pressure of Fresh Water

Temperature (°C)	Vapour Pressure (kN/m²)
0	0.6108
10	1.2271
20	2.3369
30	4.2414
40	7.3746
50	12.3348
60	19.9173
70	31.1557
80	47.3563
90	70.1077
100	101.3253

Formation of cavities also depends on surface tension since the vapour cover is the surface separating two media: liquid and gas. Surface tension of fresh water at 15°C is 0.07348 N/m and that of sea water is 0.07425 N/m. It must be noted that these values can change considerably due to the presence of contaminants such as oil in the water.

Cavitation can start at pressures significantly higher than the vapour pressure if the water contains dissolved gases. Sea water contains mainly nitrogen, oxygen and argon as well as traces of many other gases. The amount of dissolved gases depends on the temperature and the salinity. More gases are in solution at lower temperatures and at lower salinities. At large depths, the water is saturated with most gases (except oxygen). Near the surface, the oxygen content is normally between 0.1% and 0.6% but may increase due to photosynthesis by marine plants. At lower levels, the oxygen content is usually less because of consumption by living organisms and in oxidation processes. It is generally understood that boiling and cavitation occur when the ambient pressure at a point in a liquid becomes equal to the vapour pressure. If the ambient pressure is lower, the boiling point is reduced, while if the pressure is higher, the boiling point is increased.

For the purpose of cavitation, the Euler number or cavitation number is usually expressed in the following form:

$$\sigma = \frac{p - p_v}{(1/2)\rho V^2}$$

where

p and V are a suitably defined reference pressure and reference velocity, respectively

p_v is the vapour pressure of water.

Burrill (1943) introduced a thrust loading coefficient τ_c and a cavitation number $\sigma_{0.7R}$, where

$$\tau_c = \frac{T/A_P}{(1/2)\rho V_{0.7R}^2} \qquad \sigma_{0.7R} = \frac{p_0 - p_v}{(1/2)\rho V_{0.7R}^2}$$

Here

    T is the propeller thrust

    A_P is the projected blade area

    ρ is the density of water

    $V_{0.7R}^2 = V_A^2 + (0.7\,\pi n D)^2$, where V_A is the speed of advance, n is the propeller revolution rate and D is the propeller diameter

    p_v is the vapour pressure

    p_0 is the total static pressure at the shaft axis (= hydrostatic pressure + atmospheric pressure).

The Burill cavitation chart defines three limiting lines: one for warship propellers with blade sections of a special type, one for normal merchant ship propellers and one for tug and trawler propellers, giving the upper limit of τ_c for avoiding harmful cavitation for different values of $\sigma_{0.7R}$. These charts are used for propeller design till today.

Another criterion which is popular with propeller designers is one due to Keller (Ghose and Gokarn 2004).

$$\frac{A_E}{A_O} = \frac{(1.3 + 0.3Z)}{p_0 - p_v} \frac{T}{D^2} + k$$

where

    A_E is the expanded blade area

    $A_O = \pi D^2 / 4 = $ Disc area

    Z is the number of blades

    k is the constant

According to Keller, $k = 0$ for high-speed twin-screw ships with transom sterns such as naval vessels, $k = 0.1$ for twin-screw ships of moderate speeds with cruiser sterns and $k = 0.2$ for single-screw ships.

In the detailed design of propellers, i.e. when deciding details such as blade section shape and pitch distribution, cavitation considerations play a very important part. The pressure distribution along the chord at each radius, the effect of the angle of attack and its variation in a nonuniform wake, the camber and thickness of the blade sections are some of the factors that must be considered with respect to cavitation when designing propellers using detailed design methods.

Since an excessive lift coefficient in a blade section results in cavitation and lift is proportional to the product of lift coefficient and blade section chord, an increase in the chord reduces the lift coefficient for a given lift and hence reduces cavitation. This is confirmed by Burill charts as well as Keller formulae. High suction peaks near the leading edge give rise to sheet cavitation. High blade section camber may result in bubble cavitation. Unduly low pitch ratios may lead to face cavitation. An increase in thickness chord ratio may allow an increase in the range of angles of attack within which there is no cavitation. A reduction in blade loading at the tip may reduce or eliminate tip vortex cavitation. Decreased loading near the root may be necessary to avoid hub vortex cavitation. If the wake field in which the propeller works has excessive circumferential variations of axial and tangential velocity components, it is difficult to design a propeller which will be completely free of cavitation. Either the hull form must be suitably modified and propeller blade clearances from the hull increased, or some cavitation accepted in parts of the blade revolution.

In single-screw ships particularly, some cavitation is inevitable and the design effort is directed towards minimizing the harmful effects of cavitation.

Cavitation in a propeller generally has adverse effects: thrust breakdown, reduced efficiency and damage to the propeller. Since the propeller blade moves across a variable pressure zone experiencing low pressure while on top and high pressure when at the bottom, there could be a possibility of cavitation on top with the cavitation bubbles bursting as the blade moves down. This gives rise to vibration and noise at a frequency = rotation rate * number of blades. Therefore, in propeller design an attempt is made to eliminate the possibility of cavitation. The measures taken to eliminate cavitation usually tend to decrease the efficiency of the propeller, and it is necessary to obtain the best compromise between eliminating cavitation and maintaining high efficiency. This may involve designing the propeller to operate in its design condition with a small amount of cavitation (e.g. 5% back cavitation).

Although minimizing cavitation is generally one of the goals of propeller design, the design operating conditions of the propeller (high revolution rate and speed of advance, high delivered power and high thrust with restricted propeller diameter, limited depth of immersion) may make it impossible to keep cavitation within acceptable limits. In such a situation, alternative design solutions such as supercavitating propellers, surface-piercing propellers (SPPs) or waterjet propulsion may have to be explored.

At an overall ('macro') level, cavitation in a propeller occurs basically because of excessive thrust loading (thrust divided by blade area) at too high speeds and insufficient depths of immersion. Therefore, possible design measures to reduce the possibility of cavitation may include:

- Decreasing the thrust
- Distributing the thrust among a larger number of propellers
- Increasing the blade area by increasing the diameter or the blade area ratio or both
- Reducing the ship speed or the propeller revolution rate or both
- Increasing the depth of immersion of the propeller

Some of these measures may not be practicable in many design situations. The most practical solution to the problem of minimizing propeller cavitation which a propeller designer can adopt is usually to select an appropriate blade area ratio.

A more modern approach (Eppler and Shen 1979) is propeller blade section design based on nonlinear theories to distribute pressure on blade surface reducing cavitation possibility. Compared to the NACA sections traditionally used in propeller design, the Eppler–Shen sections have a non-cavitating range of angles of attack, which is several degrees greater, with the position of maximum camber being closer to the trailing edge.

7.2.3 Selection of Screw Propeller Parameters

The design of screw propellers has been dealt with extensively by a number of designers, manufacturers and researchers. Comprehensive methods of propeller design have been dealt with by a number of authors, some of whom are O'Brien (1962), Ghose and Gokarn (2004), Carlton (2007) and Molland et al. (2011). The propeller diameter D is restricted by the geometry of the propeller aperture at the stern and position of the rudder. Generally, the velocity field around a propeller is highly nonuniform both in axial and circumferential directions. Each propeller blade section moves in this nonuniform wake field generating variable thrust at the angle of orientation from 0° to 360° as it rotates a full circle. Thus, variable impulse pressures are transmitted to water in a cyclic manner by each blade at a frequency equal to the rate of rotation. If a propeller has z number of blades, the propeller excitation frequency is $z*n$. If this impulse impacts with the stern structure of the vessel, there is a possibility of vibration at the same frequency, which may affect ship operation and crew comfort. This is aggravated if there is a possibility of cavitation. If this propeller-excited vibration frequency resonates with the hull vibration frequency of any mode, there could be serious vibration problems. This is generally avoided by drawing the hull resonance diagram at the concept design stage itself by estimating the vibration frequencies and selection of propeller rotation rate and number of blades. Figure 7.11 shows a typical hull resonance diagram.

To reduce the vibration amplitude due to propeller excitation, the clearance between hull, propeller and rudder must be adequate such that hydrodynamic impact forces are minimized. Classification societies recommend minimum geometric clearance as shown in Figure 7.12a. However, hydrodynamic clearance must also be adequate as shown in Figure 7.12b. Frequently, propeller blades may be given a rake, i.e. tilting to the aft, so that the hull propeller clearance can be increased slightly. Often, to reduce vibration, propeller blades may be skewed, sometimes highly skewed, so that the amplitude of vibratory

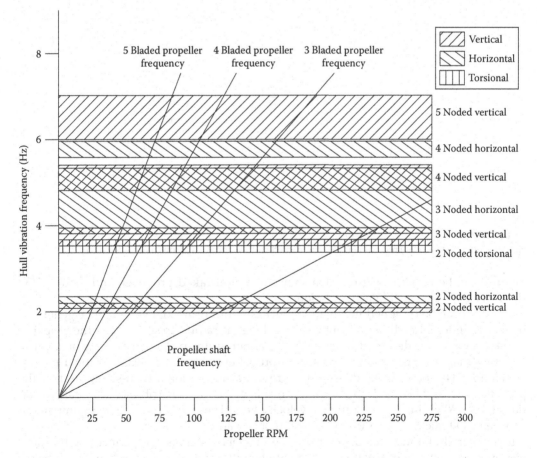

FIGURE 7.11
Hull resonance diagram.

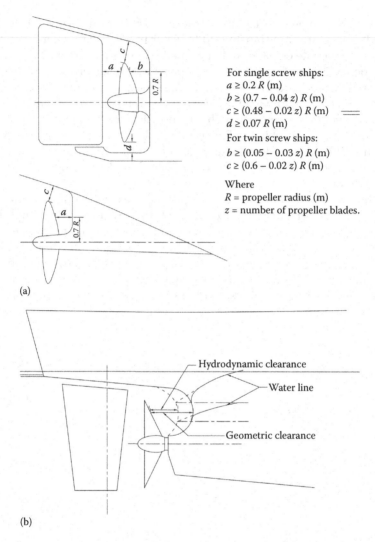

For single screw ships:
$a \geq 0.2\ R$ (m)
$b \geq (0.7 - 0.04\ z)\ R$ (m)
$c \geq (0.48 - 0.02\ z)\ R$ (m)
$d \geq 0.07\ R$ (m)

For twin screw ships:
$b \geq (0.05 - 0.03\ z)\ R$ (m)
$c \geq (0.6 - 0.02\ z)\ R$ (m)

Where
R = propeller radius (m)
z = number of propeller blades.

FIGURE 7.12
Hull propeller clearance: (a) recommended geometric clearance and (b) hydrodynamic clearance.

impulse can be reduced. Figure 7.13a shows a typical raked propeller, and Figure 7.13b shows a highly skewed blade orientation.

The stern can be designed in a manner such that uniform axial wake can be generated in the circumferential direction, but varying along the radius. In such a case, the propeller blade section can be designed in an optimal manner for the particular velocity of advance V_A rather than average velocity and such a propeller is known as wake-adapted propeller. Figure 7.14a shows wake distribution around a conventional twin-screw stern at the propeller disc, and Figure 7.14b shows improvements of wake distribution (see wake variation at $0.64r/R$) giving a more uniform velocity field. These figures also show comparison between CFD calculations and experimental observations.

It is generally known that screw propeller efficiency increases with increase in diameter and reduced rate of rotation, n, compatible with the engine. Therefore, the designer is always tempted to provide an optimized propeller with the largest possible diameter.

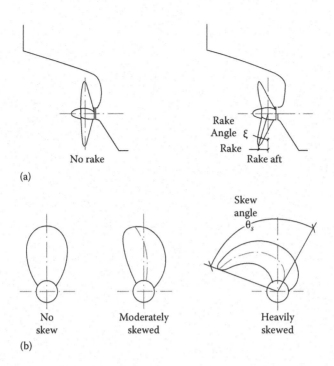

FIGURE 7.13
Propeller rake and skew: (a) rake and (b) skew. (From Ghose, J.P. and Gokarn, R.P., *Basic Ship Propulsion*, Allied
Publishers Pvt. Ltd., India, New Delhi, 2004.)

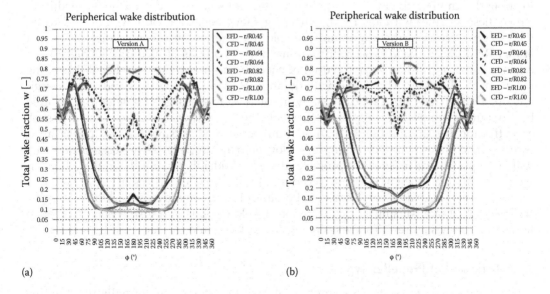

FIGURE 7.14
Wake field in the propeller disc: (a) wake field – original stern, version A and (b) wake field – modified stern,
version B.

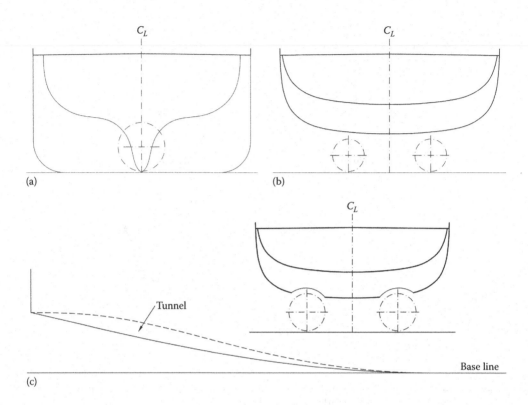

FIGURE 7.15
Screw arrangements at the stern: (a) single screw, (b) twin screw and (c) screw in tunnel.

Merchant ships are generally propelled by slow-speed diesel engines with direct drive. Keeping this in view, the marine engine manufacturers have developed slow-speed long-stroke diesel engines with low RPM, delivering a few megawatts of power. In such cases, the designer provides the maximum possible diameter to absorb the delivered power at that RPM. But if the vessel propulsion machinery consists of medium- or high-speed engines, the designer has some freedom of choosing an optimum propeller RPM suiting the available diameter. This is particularly applicable to small vessels and large vessels with medium-speed engines. Occasionally, in small vessels, a larger diameter may be fitted by providing a tunnel in the way of the flow onto the propeller. Care has to be taken in such cases that the flow around the stern is not unduly disturbed causing increase in resistance, excessive and undesirable trim or improper flow onto the propeller. If the vessel is provided with multiple screws, wake in the propeller disc is considerably reduced and the propeller-induced vibration problem is also reduced. But in this case, it is necessary to ensure that the appendages such as *A*- or *P*-brackets, bossing, etc. are properly aligned to flow so that these do not vibrate, causing crack generation at welded joints or transmission of vibration to ship's aft body. Figure 7.15 shows general single-, twin- and tunnel-screw arrangements.

7.2.4 Selection of Propeller Type

When vessels operate mostly in a single-load condition, these are normally fitted with a fixed pitch propeller (FPP) where the propeller consisting of the propeller boss and blades is manufactured in one piece and the pitch of the blades cannot be changed during its life.

Such a propeller works efficiently at the design load condition and less efficiently at other load conditions. Though propeller open-water test and self-propulsion test techniques are well established, such tests are not a standard technique for conventional ship propeller design work. There have been a number of methodical series propeller test data available, which have been successfully used for FPP design.

The Gawn or the AEW (Admiralty Experimental Works) 20 in. methodical series (Gawn 1953) consists of propellers which have elliptical developed blade outlines and segmental blade sections. The parameters which are systematically varied in the Gawn series are the pitch ratio and the developed blade area ratio. The major particulars of the Gawn series propellers are given in Table 7.4 and the performance characteristics are available in literature.

Another noteworthy methodical series of propellers is the B-series of MARIN, also known as the Troost, Wageningen or NSMB B-series (Oosterveld and Oossanen 1975). The B-series has been developed over several years as an asymmetric wide-tipped blade outline, and aerofoil sections at the inner radii changing gradually to segmental sections at the blade tip. The range of the variation of the blade area ratio depends on the number of blades and is given in Table 7.5 along with the other main particulars. All the propellers have a constant pitch, except for the four-bladed propellers, which have the pitch reduced by 20% at the blade root. The performance characteristics are available in literature.

Such a large amount of data for both Gawn series and B-series has been fitted to regression models, and the values of K_T and K_Q have also been put into the form of polynomials:

$$K_T = \sum_{i,j,k,l} C_T(i,j,k,l) J^i \left(\frac{P}{D}\right)^j \left(\frac{A_E}{A_O}\right)^k Z^l$$

$$K_Q = \sum_{i,j,k,l} C_Q(i,j,k,l) J^i \left(\frac{P}{D}\right)^j \left(\frac{A_E}{A_O}\right)^k Z^l$$

TABLE 7.4

Particulars of Gawn Series Propellers

Number of blades	$Z = 3$
Pitch ratio	$P/D = 0.60 - 2.00$
Blade area ratio (developed)	$A_D/A_O = 0.20 - 1.10$
Blade thickness fraction	$t_0/D = 0.06$
Boss diameter ratio	$d/D = 0.20$

TABLE 7.5

Particulars of B-Series Propellers

Z	P/D	A_E/A_O	t_0/D	d/D
2	0.5–1.4	0.30	0.055	0.180
3	0.5–1.4	0.35–0.80	0.050	0.180
4	0.5–1.4	0.40–1.00	0.045	0.167
5	0.5–1.4	0.45–1.05	0.040	0.167
6	0.5–1.4	0.45–1.05	0.035	0.167
7	0.5–1.4	0.55–0.85	0.030	0.167

The regression coefficients C_T and C_Q for various values of i, j, k and l are given in Ghose and Gokarn (2004), which can be conveniently used not only for performance prediction but also for propeller design as per the particular series.

If the ship stern has been designed for uniform circumferential wake, one could design a wake-adapted propeller using circulation theory. Methods of designing wake-adapted propellers have been given initially by Lerbs (1952) and van Manen (1957). With CAD and CFD techniques, it has been possible to model propeller blades and get its open-water performance characteristics. Figure 7.16 shows the comparison of calculated and predicted (B-series) propeller performance characteristics, which are excellent. The modern trend is to use CFD analysis for stern design and use CFD techniques for optimum propeller design.

In the case of vessels which are required to provide extra thrust over and above the thrust required for a forward speed, such as tugs and trawlers providing pull at a low speed, the propeller may be housed in a nozzle designed to provide higher thrust at low J values. The Kaplan series propellers are nozzle propellers, or propeller housed inside a fixed nozzle, where the blade shapes are suitably altered so as to be efficient inside a nozzle. The data are a function of J inside a nozzle having an aerofoil section of type 19A. Data are available in Ghose and Gokarn (2004) for three-bladed propellers with BAR 0.65, four-bladed propellers with BAR 0.55 and 0.70 and five-bladed propellers with BAR 0.75.

Cavitation is a major problem in propeller design as has been discussed in the previous section. Cavitation tunnel is an experimental facility for cavitation studies. In this facility, the total pressure (=atmospheric pressure + hydrostatic pressure) is required to be scaled down or for generating the required cavitation number. Thus, the facility is sophisticated and doing such experiments is an expensive affair. Whereas conventional vessel design is done based on the data from standard literature, for specific vessels such as naval vessels or specialized vessels, cavitation tests are resorted to for better propeller design.

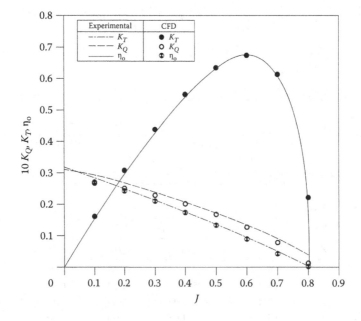

FIGURE 7.16

Comparison of CFD-based propeller open-water performance characteristics with series data for a B-series propeller.

Supercavitating propellers are a type of propellers in which the entire blade is covered with a cavity formed due to cavitating conditions and the effective blade outline changes to that with the cavity. This is particularly useful where the propellers are heavily loaded, and there is a possibility of cavitation, particularly in high performance vehicles. Newton and Radar (1961) have given a methodical series propeller performance data known as the Newton Radar Series. Table 7.6 gives the parameters of the Newton Radar Series.

Surface piercing propellers (SPP) are propellers fitted to low-draught vessels where the propeller projects out of water partially, and while the blade rotates, it takes an air cavity along with it while going into water and while coming out of water, the entrained water is dropped. Misra et al. (2012) have developed a methodical SPP series and conducted model experiments. The test results and a design procedure based on these results have also been presented. The parameter variation of this series is shown in Table 7.7, where h is the portion of the blade diameter out of water.

Figure 7.17 shows typical blade sections of different propeller types, Figure 7.18 shows different blade outlines and Figure 7.19 shows typical performance characteristics of different propeller types discussed earlier.

A problem with FPPs is that these propellers are inefficient at different load conditions experienced by ships running in full load and ballast conditions for sufficiently long periods, tugs and trawlers working in free running and towing conditions and naval vessels operating in different speed conditions. In such cases, a controllable pitch propeller (CPP) may be used where the pitch of each propeller can be altered from within the ship simultaneously to adjust to the load. This also allows the thrust direction to be reversed without altering the engine rotation. In such a case, pitch ratio selection is not that critical.

There are many alternative propulsion devices as reported in Ghose and Gokarn (2004). Paddle wheel is one of the oldest propulsion devices where a wheel is fitted with radial blades at regular angular intervals and it is rotated in water thereby generating reactive thrust on the boat that houses the wheel. Normally, river vessels used paddle wheels, two of these fixed on the port and starboard sides of the vessel near the midship or somewhat aft locations. Use of these wheels is limited now.

TABLE 7.6

Particulars of Newton Radar Series propellers

Number of Blades, Z		3	
Blade area ratio, A_D/A_O	0.48	0.71	0.95
Pitch ratio, P/D	1.05	1.05	1.04
For the different values of A_D/A_O	1.26	1.25	1.24
	1.67	1.66	1.65
	2.08	2.06	2.04
Cavitation number, σ	0.25 to a value corresponding to atmospheric pressure		

TABLE 7.7

Particulars of the Surface-Piercing Propeller Series

Number of Blades		4		
Blade area ratio	0.45	0.60	0.75	
Pitch ratio	0.80	1.00	1.20	1.40
h/D (%)	30.00	40.00	50.00	70.00

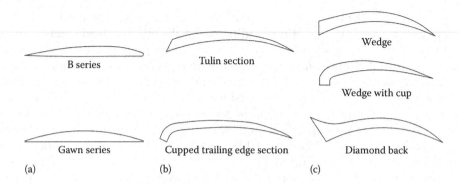

FIGURE 7.17
Typical blade section shapes of different propeller types: (a) fixed pitch non-cavitating, (b) supercavitating and (c) surface piercing.

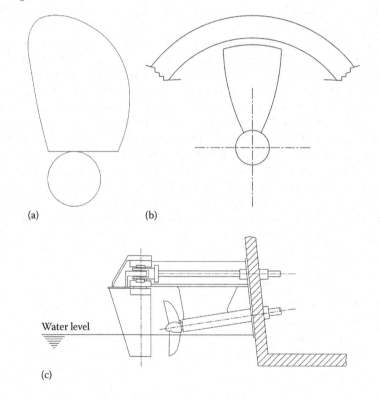

FIGURE 7.18
Typical blade outlines of different propeller types. (a) B-series propeller, (b) Kaplan series propeller, and (c) Surface-piercing propeller.

Sometimes two propellers are mounted on the same shaft and used for generating more thrust than what one screw would have generated. Such propellers are known as tandem propellers. If the two screws are mounted on two concentric shafts and are made to rotate in opposite directions, the one behind the other uses the energy dissipated to water by the forward screw and thus increases the efficiency of the combined system. Such a system is called contra-rotating propellers. Sometimes, to increase the efficiency of the screw propeller, minor devices are used around the propeller. These include vane wheels behind

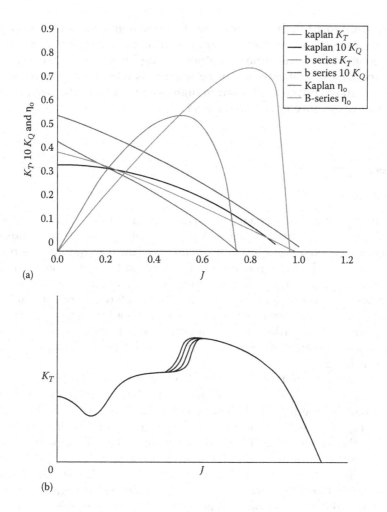

FIGURE 7.19
Typical open-water performance characteristics of different propeller types. (a) B-series and Kaplan series propellers and (b) Surface-piercing propeller.

the propeller, using propeller blades with end plates at the blade tips, a ring fixed to and aft of the boss, etc. Sometimes it may be convenient to have the propeller driving engine (electrical) mounted inside a 'pod' at the aft of which the propeller is attached. Then the propeller can be hung from the aft deck unlike being driven by a shaft through the stern. In this case, the podded propeller is on a 'Z' support system, which can be rotated to any angle, directing the generated thrust to any direction. In a small vessel, a similar arrangement is made with outboard motors. In tugs and supply vessels, there is a similar drive system called 'azimuth drive'. In such systems, the propellers combine the action of propulsion and steering, thus eliminating the need of a separate steering system. Cycloidal propulsion is a system where the drive shaft is vertical to which vertical blades are attached in a circumferential manner at a radius. The pitch of all blades can be controlled individually. By altering the pitch of the blades, it is possible to generate thrust in any direction, thus providing thrust in any required direction. This is known as the Voith Schneider propeller, which combines propulsion and steering in one unit. A waterjet or pump-jet propulsor consists of a tube housed at the stern of the vessel housing an impeller which throws water from forward to aft of the vessel at the aft end and generates a forward thrust. The direction of the waterjet can be moved to any direction from 0° to 180°, making it possible to move the vessel aft with least time lag without changing engine shaft rotation direction. This device works most efficiently when the waterjet is released at the free surface level. Therefore, this is most conveniently used in high-speed crafts, particularly planing boats where the vessel rises above its still water level.

7.3 Seakeeping

A seaway consists of a unidirectional and occasionally multidirectional random sea coupled with mild, moderate or heavy wind. A moving ship, as a system, is thus excited by this input, and its response is highly complex and nonlinear in nature. The response can be observed physically in the following manners: Rigid body motion with six degrees of freedom – three linear: surge in longitudinal (or x) direction, sway in transverse (or y) direction and heave in vertical (or z) direction and three rotational: roll (about longitudinal or x-axis), pitch (about transverse or y-axis) and yaw (about vertical or z-axis), large and varying wave load on hull girder, impact load on the fore body causing slamming, whipping and vibration, shipping of spray and green seas on board, loss of speed, effect on stability and capsizing.

Roll motions combined with lateral wind loads can cause dangerous heeling angles or even capsizing. Following waves can cause a considerable reduction in the transverse stability and unacceptable large roll angles can be the result. This can happen with fast ships in following waves with wavelengths equal to about the ship length. This is due to the reduction in the frequency of the encounter of the wave and ship. This phenomenon is known as parametric rolling. Excessive motion can cause discomfort to crew due to large motion amplitude and large vertical acceleration. Ships with large initial metacentric height, leading to small rolling periods, can be very inconvenient with high vertical acceleration if the point under consideration is away from the roll axis, which is, in most cases, the centre line of the ship passing through the centre of gravity or near that. With increase in motion, it becomes difficult to control and steer the ship properly, leading to loss of life, cargo and ultimately loss of ship due to capsizing. Large and continuously varying wave load subjects the ship girder to varying bending, shear and torsional stresses causing fatigue leading to structural damage. Slamming, or impact load on the forefoot of the ship, can cause substantial forebody damage and vibration. In a seaway, speed loss is either involuntary due to added resistance in waves or voluntary, i.e. to reduce the harmful effects of motion and wave loads or to reduce the probability of shipping of green seas and spray or, lastly, to increase the probability of survival at sea. Sometimes, the master of the ship may divert the ship from the desired route to avoid a predicted harmful seaway, thus increasing the voyage time and expense.

Till recently the response of a ship in a seaway did not affect the ship design practice very much. But with increase in complexity of equipment and machinery, competition in ship operation and demand of crew for better working environment and comfort, a lot of attention is being paid for good seakeeping behaviour of a ship in a seaway at the design stage. Briefly, a designer tries to design a ship with good seakeeping qualities such that the vehicle can perform the intended task safely and economically (with minimum delay) in the roughest of seas it is likely to encounter.

The designer's tools in his effort to design a sea-kindly ship are the various theoretical, experimental and statistical prediction methods available at his or her disposal. It is not intended here to do an exhaustive survey of all the theoretical prediction methods, experimental techniques or statistical data available in this subject, which have been done very thoroughly by the various Seakeeping Committees of ITTC (1978–2011) and others (Comstock 1967, Bhattacharya 1978, Lewis 1989c). The various prediction methods are touched subsequently in so far as these only aid the practical ship designer.

7.3.1 Ocean Waves and Ship Motions

Ocean waves constitute a major part of the environment of sea-going ships at sea and are also the cause of ship's oscillating motion at seas. Ocean waves are characterized by their irregularity, both in time and in space. The ocean waves and their statistical representation based on sinusoidal waves have been described in Section 2.5. Ship motions induced by ocean waves are also irregular. However, according to the principle of superposition, the irregular motions of a ship in ocean waves can be described as the linear superposition of the responses of the ship to all the wave components, which are regular and have different lengths, amplitudes and propagating directions, where the amplitudes are assumed small.

The motion of a ship in a unidirectional irregular seaway can be obtained quite simply by assuming that the ship's response to a regular wave is proportional to the wave amplitude (linearity assumption where both the seaway and the response are moderate) and there exists response amplitude operator (RAO), which is the ratio of the response amplitude to regular wave amplitude for a particular encounter frequency ω_e. So first a suitable wave energy spectrum representing the seaway is chosen. In the absence of any preferred spectrum, ITTC spectrum could be used. Next, the energy spectrum is modified to a base of frequency of encounter ω_e, which is based on wave frequency ω, ship speed V and the ship heading angle ($\mu = 0$ for following sea and 180° for head seas) and then,

$$\omega_e = \omega - \frac{\omega^2}{g} * V \cos \mu$$

$$S_\zeta(\omega_e) = \frac{S_\zeta(\omega)}{[1-(4\omega_e/g)*V\cos\mu]^{1/2}}$$

$$= \frac{S_\zeta(\omega)}{1-(2\omega/g)*V\cos\mu}$$

such that the area under the new energy spectrum curve does not change from that under the original energy spectrum, i.e. the total energy content of the sea remains same irrespective of the ship's heading. It is possible that $S_\zeta(\omega_e)$ may have a singularity at certain ω_e (say ω_{emax}) when the spectrum can be handled in two ranges, i.e. from 0 to ω_{emax} and ω_{emax} to ∞. The deterministic ship response amplitude ξ_i should yield the RAO against ω_e for the ship:

$$\xi_i = RAO * \zeta_i$$

Assuming that the response of a vessel in a seaway is the sum of its responses to the individual wave components comprising the seaway, response spectrum can be obtained as

$$S_\xi(\omega_e) = S_\zeta(\omega_e) * RAO^2$$

The information that can be gained about waves from a wave spectrum can similarly be gained for response from a response spectrum, i.e. the mean and significant motion amplitudes, the most probable maximum value, bandwidth, etc.

The motions which may be required to be obtained this way are heave, pitch and roll motions, which are the direct responses. Surge, sway and yaw are motions without any restoring force or moment and, therefore, are not oscillatory in nature. These are normally not considered in ship motion studies. One can observe that heave and pitch are longitudinal motions, whereas roll is transverse motion. It is customary to consider that longitudinal and transverse motions are independent of each other and can be studied separately.

The derived responses include vertical acceleration due to coupled pitch and heave or roll motion, shipping of water on deck and slamming based on relative motion of the bow with respect to the water surface, added power requirement in waves or speed loss, etc. Wave-induced loads generate large stresses on ship structure, but since this is not a motion problem, it is not discussed in the present section.

7.3.2 Prediction of Seakeeping Behaviour

In ship design development, the designer has to ensure that motions experienced by the ship while in service are within acceptable limit without causing risk to life and property and with minimum loss of speed in rough seas. Though the designer might have used broad design guidelines in parameter selection at the concept design stage to ensure good behaviour in a seaway, it may become necessary to predict behaviour for the developed form at the basic design stage and take necessary steps, such as local/regional/global parameter or shape change to ensure good behaviour in rough seas. Seakeeping behaviour can be estimated either numerically, experimentally or statistically from a large amount of experimental data.

7.3.2.1 Numerical Estimation

It is difficult to solve the equations of motion of the vessel with some external exciting force and predict the ship behaviour as a function of time or do a time domain computation. So, normally, frequency domain analysis is carried out to find out the RAO for a particular frequency, and thus the RAO as a function of encounter frequency is established from which the mean and significant motion amplitudes can be estimated. Even the frequency domain analysis for the 3D ship surface with a constantly undulating free surface is difficult to perform. The 'strip theory' method is a 2D method, extended to three dimensions, which has yielded good results matching with model tests and full-scale trials (Salvessen et al. 1969–1970, Bhattacharya 1978, Lewis 1989c). In this theory, the following assumptions are made:

- The length of the ship is much greater than the beam or draft and the beam is much less than the wavelength.
- The ship hull is considered rigid.
- The speed of the vessel is moderate and there is no lift due to forward speed.
- Ship motions are small and linear with respect to wave amplitude.
- Water depth is much greater than wavelength so that the deep-water wave approximations may be applied.
- Froude–Krylov hypothesis is applicable, which states that the hull has no effect on the incident waves.

CFD techniques are now being developed for full 3D solutions of motion equations for estimating motion characteristics (ITTC 2008, 2011). Commercial packages for such calculations are available in the market.

Roll motion affects ship performance in the following ways: Transverse accelerations due to roll may cause interruptions in the tasks performed by the crew. Under extreme conditions, it may even prevent the crew from performing tasks at all. Vertical accelerations induced by roll at locations away from the ship's centre line can contribute to the development of seasickness in the crew, affecting their performance by reducing comfort. A combination of the acceleration level and its average frequency of oscillation dictates the amount of comfort aboard. The human acceptance of this ship motion–related comfort depends on acceleration level, frequency, time of exposure to such acceleration and acclimatization of the crew and passengers to such motion. Figure 7.20 shows the human acceptance limit, MSI, for different amplitudes and frequencies of acceleration where MSI is 'motion sickness incidence' defined as the percent of people who would become physically ill if exposed to motion of certain characteristics in a given time interval. In this diagram, two ranges are shown, one for

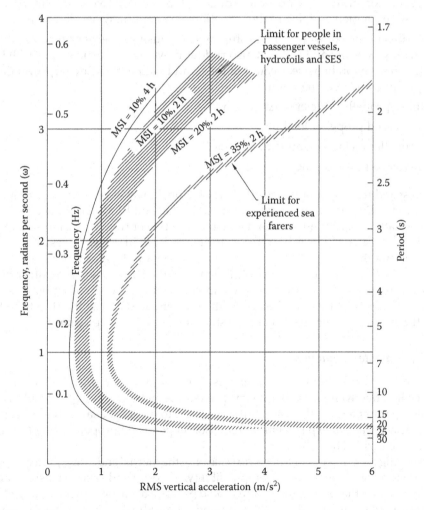

FIGURE 7.20
Tolerance levels based on MSI as a function of vertical acceleration and frequency.

vessels which carry unacclimatized passengers such as passenger vessels or research vessels and one for other vessels such as naval vessels and other merchant ships where people are experienced seafarers (Lewis 1989c). Large roll angles limit the capability to handle equipment on board. This is important for vessels performing launching or recovering operations.

Roll motion prediction requires special attention. Rolling can be high, and linearization assumptions may not hold. So nonlinear rolling motion analysis could be carried out instead. This is made more complicated by roll damping due to waves as well as due to viscosity of water.

Speed loss in an irregular seaway can be involuntary or voluntary. Involuntary speed loss can be due to the following:

1. Added resistance due to wind on hull and superstructure in headwind condition already discussed in Section 7.1.1.
2. Added resistance due to waves can be due to the following reasons:
    a. Direct wave action, which can be visualized as energy dissipation due to the oscillation of ship in calm water, due to phase shift between ship motion and wave excitation and due to the reflection of oncoming waves from ship body.
    b. Indirect effect of waves due to ship motion causing resistance increase due to drift, yaw, sway and roll and also rudder movement. Maruo et al. (1978) have given a comprehensive account of forces and moments acting on a ship due to pitching and heaving motion.
3. Reduced propeller efficiency caused by
    a. Increased propeller loading
    b. Propeller racing and air drawing and
    c. Reduced hull efficiency

It has been stated that the propulsive efficiency factors in waves remain more or less same as in calm water. But propeller operation thrust, torque and RPM vary in waves. It is very difficult and rather inaccurate to estimate increase in power from its various components. It is possible to estimate increases in thrust, torque or RPM in waves by response analysis, i.e. by computing the RAO and then the spectrum to obtain the mean increases.

Voluntary speed loss is mainly due to the action of the ship's navigating officer leading to reduction in speed. The master of the ship could order speed reduction due to ship oscillations for safety of life and ship. Also, the master could ballast the ship to reduce its motions, thereby increasing the drag of the ship due to higher displacement.

7.3.2.2 Experimental Prediction

The usefulness of seakeeping experiments is same as in any engineering science, i.e. to determine the effectiveness of theoretical prediction methods, to experimentally determine design parameters (in this case, hull form) and to investigate any other effects likely to occur. The accuracy of experimental prediction depends on proper scaling to full scale, which must be verified by full-scale trial.

For seakeeping experiments, one needs to scale only the inertial and gravity forces and hence scaling is done as per Froude's law. Compressibility, viscosity and surface tension have little effect on motion and they are ignored. Care should be taken to ensure proper scaling of weight and weight distribution such that transverse and longitudinal radii of gyration are scaled properly, which are important parameters in roll and pitch motion prediction, respectively.

Generation of regular waves of a given frequency and height in a towing tank can be done by means of flaps, wedges or similar devices. The problem is that at low frequency, secondary waves develop in the tank since the beaches are normally not effective at such frequencies. At high frequencies, on the other hand, the peak of large amplitude wave is likely to break or noise may be added to the frequency required. Thus, one has to be careful about the actual wave generated.

An irregular seaway can be generated by superimposing a number of regular waves in a tank. So if an electronic circuitry is available to do the analysis of a given spectrum, select a few major frequency components and control the wave maker in such a way that the harmonic waves due to these frequency components superimpose on one another; then one could get an irregular wave pattern of a given spectrum. It must, however, be understood that such wave generation is repetitive and hence unlike an actual seaway, it is not random but only pseudo random.

Offline analysis of recorded data is rather tedious in the case of seakeeping tests due to the magnitude of data involved, and hence most of the modern tanks have installed digital computing systems on the carriage itself to analyse the data online and, if necessary, to control the experiments also. While doing tests in regular waves, it is possible that some nonlinearity may affect the measurement slightly. It is, therefore, necessary to filter the data to get a reliable RAO. Filtering can be done through electronic circuitry or by software using an FFT analysis to determine the most dominant frequency. This analysis also gives the quantum of nonlinearity present in the measurement. To get the complete response spectrum, one has to run the test in regular waves of various frequencies and amplitudes. On the other hand, only one run in irregular waves (known spectrum) is enough to give the response spectrum and hence the RAO. Also this serves as a check on tests in regular waves. Full-scale trials are also necessary to verify the accuracy of experimental predictions. But they are expensive and are only occasionally conducted.

7.3.2.3 Statistical Prediction

Apart from experimental and theoretical techniques for accurate prediction of seakeeping characteristics, it is possible to develop a database of seakeeping characteristics of various vessels either from (1) numerous theoretical calculations by systematic variation of parameters, (2) collection of experimental results from towing tank experiments of different types of vessels or (3) theoretical/experimental results of a series of forms by systematically varying the main parameters. Regression analysis of these data can yield usable formulations of various seakeeping characteristics as a function of known parameters. Alternatively, suitable interpolation techniques can be applied to obtain an estimate of the seakeeping behaviour of a vessel from the available data.

Moor et al. (1968, 1970) have suggested regression equations for heave amplitude, pitch amplitude, thrust, torque, RPM and power increase in head seas based on multiple regression analysis of experimental data as a function of waterplane area coefficient, block coefficient, length–breadth ratio, length–draught ratio, radius of gyration and speed–length ratio. The regression coefficients are given by Bhattacharya (1978) also for various Beaufort numbers. It must be remembered that these coefficients are given based on BTTP spectrum. Hence, the results obtained from these data will be valid for another standard spectrum (like ITTC) if the average period associated with significant wave height is same as that of the BTTP spectrum. Similarly, Moor (1967) has also given regression equations and corresponding coefficients for the estimation of bending moment in a seaway. These data give the maximum bending moment as well as its position of occurrence.

Various other regression analysis results are available in publications. Townsin and Kwon (1982) have suggested an empirical method for estimating speed loss due to winds and waves. St. Denis (1983) has analysed the past seakeeping series results and the suggested empirical formulae for estimating the qualitative seakeeping behaviour of ships and their design applications.

Extensive theoretical calculations using strip theory for series 60 models have been presented in the form of average value of heave, pitch, bending moment, added resistance, relative motion, relative velocity and acceleration at various stations as functions of ship particulars (block coefficient, length–breadth ratio and breadth–draught ratio), Froude number and significant wave height. Suitable interpolation can yield results for any new design. The tabular data as well as the parameter ranges for which data are valid are presented by Bhattacharya (1978). It must be mentioned that experimental results agree well with this method of calculation based on statistics.

The use of statistical results must be limited only to preliminary design or to get only an indication of motion characteristics. This is because ship behaviour in a seaway is dependent on the ship's main parameters as well as the local hull characteristics. The statistical data is analysed only on the basis of the main parameters. So for accurate prediction, one has to do either theoretical calculations or model experiments. One can interpolate to get nearly accurate results only if the new vessel has nearly the same hull form as the ones for which data are available.

7.3.3 Effect of Ship Parameters on Seakeeping

The design for good seakeeping behaviour has become important recently due to keen competition as well as because of the various complex operations the ship is required to function. The overall seakeeping design can be divided into three main areas:

1. *Habitability*: Crew efficiency depends on the comfort conditions on board involving motion amplitude and acceleration magnitude. Multidirectional large acceleration causes seasickness. Similarly, vibration and noise reduce work efficiency. The demand on efficiency also varies depending on the ship operation. Habitability is more important in passenger ships than on cargo ships. It is possible that crew efficiency is not likely to suffer very much in adverse conditions if it is for short duration. Therefore, vessels with longer run periods require being better for habitability.

2. *Operability*: The efficiency of various machinery and equipment may be affected due to adverse seakeeping behaviour. The safety of cargo may be affected due to large motion and acceleration. Speed loss may lead to loss of trade and revenue. Also, course keeping and manoeuvrability requirements may demand voluntary speed loss or charge of course due to adverse behaviour.

3. *Survivability*: The ship and lives on board must survive in extreme motion conditions. The dynamic stability in waves must be large to upright the ship during extreme roll motion. In general, the response during synchronous oscillation should be limited and the ship should be controllable.

To satisfy these objectives, it is necessary to first identify the sea conditions in which the ship is likely to operate. The spectrum of the sea should be obtained, and depending on wind speed distribution, the probability of occurrence of different sea states (i.e. significant wave height) during the life of the ship should be established. If the exact sea spectrum of the route is not known, ITTC spectrum can serve as a standard.

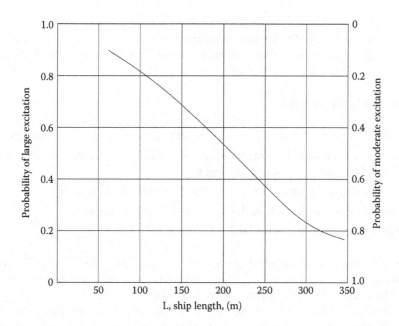

FIGURE 7.21
Probability of large pitch and heave excitation as a function of ship length.

Ship length is one of the main parameters which affect ship behaviour with regard to longitudinal motion such as pitching and heaving. The longer the ship length, the lower the probability of meeting a wave of length equal to ship length. Wave slope is similarly lower for long waves and, therefore, excitation with long waves is less than shorter waves. For waves, still longer waves than ship length, ship simply follows the wave contour. Lewis (1989c) has reproduced the relationship between ship length and the probability of large excitation, taken from Sellars, and this diagram is reproduced with slight modification in Figure 7.21.

A general rule of good seakeeping behaviour is that the motion period (significant) should not coincide with the natural period to avoid synchronization. Even if this occurs, damping should be large to limit the amplitude in this condition. So the initial step in sea-keeping design is to estimate natural periods of heave, pitch and roll. Broad guidelines for natural period of pitch (T_5), heave (T_3) and roll (T_4) are as follows:

$$T_5 = 2\pi * \sqrt{\frac{(1/g)}{L} * \frac{T}{L} * C_{VP} * \frac{C_K^2 + A_5}{C_A^2}}$$

Another useful formula is given by

$$T_5 = \frac{2\pi}{\sqrt{g}} * \frac{k_{yy}}{\sqrt{BM_L}}$$

$$T_3 = 2\pi \sqrt{\frac{(1/g)}{L} * \frac{T}{L} * C_{VP} * (1 + A_3)}$$

$$T_4 = \frac{c * B}{\sqrt{GM_T}}$$

where

g, L, T, BM_L and GM_T have their usual meaning

C_{VP} is the vertical prismatic coefficient and is equal to the ratio of the block coefficient and the waterplane area coefficient

The other constants are as follows:

$C_A = k_a/L$ where $I_L = k_a^2 * A_{WP}$; thus k_a is the waterplane area radius of gyration

$C_K = k_{yy}/L$ where k_{yy} is the mass radius of gyration about the transverse axis and can be taken as 30% of length at initial stages

A_5 = Pitch-added mass moment of inertia/$(\Delta * L^2)$ dependent on speed

A_3 = Heave-added mass/Δ and can be taken as 1 at initial stages

c = constant $* B$ as given in Section 6.2

It can be observed that by altering parameters such as draught–length ratio, length of ship, vertical prismatic coefficient, or waterplane area coefficient and longitudinal mass radius of gyration, the natural periods of pitch and heave can be changed.

Tuning factor Λ is defined as ω_e/ω_n, the ratio of encounter frequency and natural frequency, where $n = 3$ indicates heave natural frequency and $n = 5$ indicates pitch natural frequency. Synchronization occurs at $\Lambda = 1$, and with forward motion with speed (ω_e increases with speed), the maximum excitation can occur at slightly higher Λ, say 1.1 or 1.2. But with damping factor increasing with speed, maximum excitation can also occur at a tuning factor less than 1, say 0.9. The deck wetness criterion may be divided into three zones: spray, wet and very wet. The very wet condition may lead to damage of deck fittings, which should be avoided. This may, therefore, lead to voluntary speed reduction. The limit between spray and wet conditions represents shipping of green seas of about 2 times/100 pitch oscillations, which can be a good guide for design. One to 2 slams/100 pitch oscillations should be enough for slamming. Sometimes with proper forward stiffening, 3–4 slams may not cause any damage. Propeller racing becomes critical if propeller emerges leading to 25% torque reduction about 25 times/100 oscillations. Figure 7.22, taken from Lewis (1989c) with slight modification, shows a division of region of severe pitching and heaving in stormy seas and zone of acceptable longitudinal motion with forward speed as a function of length–displacement ratio, draught–length ratio and block coefficient. The dividing range is for $\Lambda = 0.9$ and 1.0. This diagram can be used as a design guideline at the concept design stage.

For roll, on the other hand, the important parameters controlling the motion are the beam of the ship, transverse metacentric height and the transverse mass radius of gyration. In 1861, in his seminal paper on the rolling of ships, Froude showed that it was the wave steepness (slope) that excited the rolling motion of a ship and not the height of the waves. He further commented that since short waves appeared to be steeper than long waves, there was, then, no advantage in trying to reduce the natural roll period of the vessel. Instead, this period should be extended as much as possible so as to avoid synchronization with the wave excitation frequency. This can be achieved only by increasing the mass transverse moment of inertia and reducing transverse metacentric height. The mass moment of inertia entirely depends on the distribution of the weight of the ship, whereas the metacentric height depends largely on the shape of the hull. With regard to the hull shape and damping, significant increase in damping can be achieved by designing hulls with a small bilge radius and with the appendages located as far as possible from the roll axis.

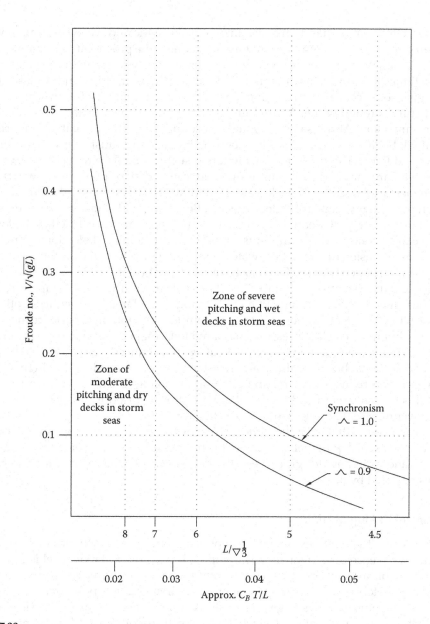

FIGURE 7.22
Trends of speeds for synchronous pitching in irregular head seas.

All three motions – pitch, heave and roll motions – set a limit on human endurance. It is extremely difficult to quantify human endurance since it is a very subjective quantity. In general, therefore, one should try to achieve good seakeeping behaviour for better human endurance, as shown in Figure 7.20.

Beam does not have a marked effect on seakeeping behaviour. Again, low $\Delta/(L/100)^3$, low L/T and high L/B seem to have favourable effect on motion, i.e. narrower, deeper and finer ships are better from the motion point of view. Higher block coefficient generally increases wave-bending moment, speed loss and resonant motion amplitude in heave. A full water-plane area at ends increases motion damping and reduces resonant pitch and heave amplitudes. On the other hand, this also increases the static and wave-bending moments on the hull girder. A minimum draught forward reduces the probability of slamming and, therefore, classification societies require a minimum forward draught. Normally, $T_F \geq$ /0.045L to limit slamming. This draught is difficult to achieve in ships, and so classification societies give some concession of reduction in forward draught as $T_F \geq 0.025L$ with forward structural strengthening. A minimum aft draught is to be maintained to ensure full propeller immersion. At the same time, to avoid unnatural ship orientation, the trim has to be limited. Ballast arrangement has to be suitably made to satisfy these requirements. Freeboard and flare have little effect on motion as such. Good flare reduces deck wetness, but excessive flare may increase slamming incidence and slam pressure. Bow freeboard should be increased to reduced deck wetness, and upper deck forward should be clear so that green water may escape from deck quickly. Bow flare angle may be contained within 20°–25°. Flare may be provided with a knuckle at the upper deck level so that the forward deck area can be contained within limits simultaneously helping deflection of green seas from the bow. Occasionally, to deflect water from bow, spray rails have been provided on the bow at the upper deck level. The forwardmost flat of bottom should be as far aft of FP as possible to reduce slamming. But this is not possible in a bulbous bow ship. Normally, a forward bulb means less pitching and more heaving. But a large bulb reduces both pitching and heaving. V-shaped sections exert large damping on motion, experience more number of slams, but reduced slam pressure compared to U-shaped sections. Thus, a ship with V-shaped section experiences less speed loss at sea but reduced slam pressure compared to U-shaped sections. But the calm water resistance of a V-shaped ship is higher. Lower radius of gyration reduces heaving and pitching motion and hence results in less speed loss. But this means larger bending. Similarly, larger roll period means less roll acceleration but lower metacentric height and hence stability.

Thus, it can be seen that even the indicative guides may be exactly opposite for two different aspects of seakeeping behaviour. So it is not possible to have set rules for seakeeping design, and as in any design process, the design has to evolve satisfying the various objectives required by it.

7.3.4 Control of Ship Motion

It has been stated that large excitations or motions occur at the resonance of ship natural period and wave encounter frequency. If damping is high or excitation level is low due to some opposing or cancellation forces or moments, resonance effects reduce. By manipulating hull form parameters supporting the operational requirements, it is possible to have reasonably good seakeeping qualities as given by Faltinsen (1990). For a semi-submersible, TLP or similar floating or moored structure which is used mostly for oil exploration and production, it is necessary that heave, pitch and roll should not cause displacement of the body beyond a specific operational requirement. Semi-submersibles have very low waterplane area compared to ships which have large waterplane. Reduced waterplane area increases the pitch and heave periods, and these are of the order of 20 s or more, which is much higher than that of incident waves. Ships, on the other hand, have a much lower natural frequency of oscillation, and the wave excitation frequency has a wide range depending on forward speed. The roll motion natural frequency of a semi-submersible is of the same order as that for pitch motion since dimensions in longitudinal and transverse directions are of the same order.

For a planing craft, motion can become severe leading to voluntary speed reduction. Making extreme V-shaped sections forward can produce a wave piercing effect and the craft can maintain some forward speed in rough weather. A catamaran can be designed with reduced waterplane area, higher length–breadth ratio (of each demi hull) and higher radius of gyration so that its natural period of oscillation can exceed the wave excitation frequency in some operating range. This has been further exploited in SWATH ships in which the waterplane area is very low.

Since surge, sway or yaw motion does not have a restoring force or moment, there is no oscillation in these modes. But once the structure is moored (oil exploration and production structures), there is a natural frequency in minutes, which is again much higher than the wave excitation frequency. Thus, there is no danger of resonance. Often, for operational reasons, it may be required to keep a particular course or maintain position, which means that there must be surge, sway and yaw compensation devices. This is easily achieved by providing multiple thrusters (of comparatively low power) in different directions and controls for the thrusters in a centralized control station. Such systems include bow thrusters in ships, rotating water jets and dynamic positioning systems in offshore platforms.

The motion of a ship is related to the body shape and the sea condition or environment. Any device used in a ship would necessarily impart force or moment in the opposite direction to those experienced by the ship by the external sources. Since the forces and moments and, therefore, the motions instantaneously vary, any mechanism must respond instantaneously to control/reduce motion. Of all types of ship motions, roll is the most critical for survival and safety, having large amplitude and so devices are generally designed to reduce roll. It is impossible to eliminate roll completely, and so the approach generally adopted is to understand motion characteristics in the frequency domain and deploy a device which would be effective in the prominent frequency ranges. The devices could be passive or active, i.e. controllable, and these could be also internally installed inside the ship or external to the ship body. If a ship has to operate in heavy seas in rough weather, it may also be necessary to control pitch and heave by controllable mechanisms. Smith and Thomas (1990) have given a survey of various motion control devices and their effectiveness. Some of these devices are mentioned as follows.

7.3.4.1 Bilge Keel

A bilge keel is a long strip of metal of 0.3 m to about 1.2 m width, either a flat plate or having a 'V' section, welded along the length of the ship at the turn of the bilge. Bilge keels are employed in pairs, one on each side of the ship. Each bilge keel could be in single piece or in multiple pieces, which is rare. Bilge keels increase hydrodynamic damping when a vessel rolls, thus reducing roll motion. Bilge keels are effective at all ship speeds, particularly at zero speed.

7.3.4.2 Outriggers or Removable Stabilizers

Outriggers may be employed on certain vessels to reduce rolling. Rolling is reduced by introducing opposing moment required to submerge buoyant floats foils. In some cases, these outriggers may be of sufficient size and removable or fixed so that the vessel can be termed a trimaran. Generally, on vessels they may simply be referred to as stabilizers.

7.3.4.3 Antiroll Tanks

Passive antiroll tanks are normally two tanks with a water tube at the bottom connecting the two tanks and an air tube on top fitted with baffles intended to slow the rate of water

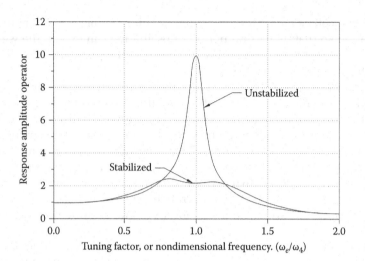

FIGURE 7.23
Roll RAO as a function of tuning factor with antiroll tanks.

transfer from the port side of the tank to the starboard side as the vessel rolls. The tank is designed such that a larger amount of water is trapped on the higher side of the vessel as it rolls providing antiroll moment. The baffles can be adjusted manually or mechanically to be effective at different excitation frequencies. This is intended to have an effect different from that of the free surface effect. The excitation at resonance is reduced to a large extent, as shown in Figure 7.23. Passive roll stabilization tanks can have different geometrical configurations known as U-tubes or antiroll tanks. Instead of being connected by a water tube, these tanks can be open to sea being free-flooding also.

7.3.4.4 Active Antiroll Tanks

The antiroll tank system could be provided with a pump, hydraulic piston or electric actuator to move water from one side tank to the other based in the phase of the roll motion sensed by the requisite sensor. The advantage of this active system is that this system can be effective over a large frequency range.

7.3.4.5 Stabilizer Fins

Fins having an aerofoil section attached to the ship body in port and starboard sides, at the bottom of the ship in the front or aft or at the bottom in between the two hulls of a catamaran or SWATH vessel, can generate lift in the transverse as well as longitudinal directions providing opposing force or moment to the excitation. Fins can be passive where these are fixed as permanent structures providing requisite service at limited excitation frequency range. Fins can also be retractable and can be lowered only in rough weather. Such devices are normally fitted in small vessels such as planing vessels. Fins can also be active where the angle of inclination of the fins can be controlled from within the ship so that their orientation can be optimal in any wave condition. The control can be achieved by hydraulic or electric mechanisms operated automatically by sensing the wave situation. A number of fins and their orientation can be designed and installed such that either roll or pitch or both roll and pitch can be controlled. There have been a number of variations to the fin stabilizers carrying different commercial names such as T-foils, interceptor, trim tabs, spanning foils, lifting foils, etc. Fixed foils or fins can be provided with controlled ventilation from within the ship to produce lift in the desired direction.

7.3.4.6 Translating Solid Weight

A spring-mounted weight sliding over a pre-laid path can be controlled so that it moves in a way which opposes the roll exciting moment. This is like the antiroll tank where solid moves instead of water. The control can be designed such that the weight movement can be effective at all frequencies.

7.3.4.7 Gyroscopic Stabilizers

Gyroscopic stabilizers consist of a spinning flywheel, and gyroscopic precision imposes boat-righting torque on the hull structure. When the boat rolls, the rotation acts as an input to the gyroscope, causing the gyro to generate rotation around its output axis such that the spin axis rotates to align itself with the input axis. This output rotation is called precession, and, in the boat case, the gyro rotates fore and aft about the output or gimbald axis. This device, oriented in a different way, can also be used for pitch control.

7.3.4.8 Rudder Roll Stabilization

Rudder roll stabilization is an attractive method for reducing ship rolling motion produced by waves. When the rudder is rotated off the centre line, a lift force is generated, which generally acts at a point below the roll centre of the ship. This side force, therefore, causes both yaw and roll moment. The roll moment generated by a rudder is small in magnitude but opposes the roll excitation.

7.3.4.9 Maglift Stabilizers

A very interesting solution is using rotary stabilizers which can make use of the so-called magnus effect in the effective reduction of roll stabilization at low speeds. These are perfect for slow-moving trawlers, yachts and at zero-speed condition. These are retractable and easy to maintain. The advantages of these stabilizers are less weight and small size, efficient at low speeds and could be retractable.

7.4 Manoeuvrability

A marine vehicle must be able to steer properly so as to maintain a prescribed path, which means it should

- Have turning ability so as to change direction
- Maintain a constant direction or have *course stability*
- Start and stop changing direction quickly, or have *controllability*
- Start, accelerate, decelerate, stop and reverse, or have *speed changing ability*

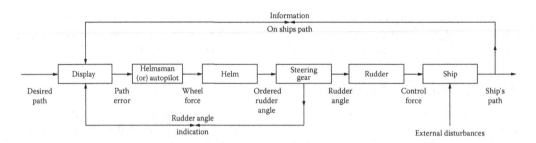

FIGURE 7.24
Direction control system of a ship.

For a surface ship, manoeuvrability means control in the horizontal plane or on the surface of water. For an underwater vehicle such as a submarine or submersible of any type, this should include controllability in the vertical plane also, which means maintaining constant depth as well as changing depth in a specified manner in trimmed condition.

In a very small ore-propelled boat, steering may be done by controlling the ores on port and starboard sides in the absence of a rudder. Conventionally, steering of all types of marine vehicles is done by the rudder fitted below the stern of the ship operated by the steering system. In a manually operated boat such as the non-propelled wooden boat, there is a wooden tiller fitted on top of the wooden rudder stock at the stern operated by the helmsmen sitting at the stern. In bigger vessels, the torque required to turn the rudder increases, and hence such simple manual steering is not possible. Also for better steering control, the helmsmen are required to be located such that they can have an overall view of all around so as to steer the vessel without any accidents. For this purpose, the steering wheel is normally located on the navigation bridge deck, which is the highest deck in the ship's superstructure or deck house longitudinally located between the midship and aft end of the ship. The steering gear which turns the rudder is located above the rudder stock in the aft compartment of the ship. The steering gear is operated by the helmsmen from the top. This operation can be done (1) mechanically through a sprocket and chain arrangement or (2) electrically by servo-controlled electrical motor or (3) by electro-hydraulic steering gear where the rudder is turned by an electric motor controlled remotely by the helmsmen by a hydraulic steering system, which is the most commonly used system today. There are many variations to the basic system such as vane-type or ram-type steering gear or link-type steering gear operating two rudder, port and starboard, synchronously in a twin-screw ship. The system to control the direction of motion of a ship can be represented by a *closed-loop system*, as shown in Figure 7.24.

Starting from the left, there is a desired path or trajectory set by the ship's *conning officer*, which is shown on the display. The actual path being followed by the ship is also displayed. These two paths may not coincide, which may be aggravated by external forces such as wind, waves and current. Corrective action is taken by the helmsmen by changing the helm such that the path error is corrected. This generally involves activating the steering gear to turn the rudder. The rudder then produces a control force which acts on the ship and tends to correct its path. The automatic control of the steering system follows the same principle incorporating a continuous feedback system.

7.4.1 Manoeuvring Trials

The officer on watch at the bridge must know the manoeuvring behaviour of the ship to be able to steer the ship safely avoiding collisions in its path and in the shortest route to its destination. The steering characteristics of the ship are established during the manoeuvring trials, which are performed as a part of sea trial prior to the delivery of the ship (Lewis 1989c, Brix 1993).

7.4.1.1 Turning Circle Manoeuvre

While moving at constant speed in a straight line, the ship's rudder is turned to some (generally maximum) angle δ to starboard and held there. The ship first acquires a linear acceleration to port. This is the 'first phase' of the turning motion, which is very short and may be over even before the rudder has reached its full angle. In the 'second phase' of turning, the drift angle increases and the ship starts moving in a curved path of increasing curvature to starboard. The drift angle β at any point along the length of the ship is the angle between the tangent to the path of that point (usually the centre of gravity) and the centre line of the ship. δ is the rudder angle. When the ship moves on a curved path, it has an acceleration directed towards the centre of curvature, which gives rise to an inertial reaction ('centrifugal force') in the opposite direction, which is inversely proportional to the radius of curvature. The radius of curvature of the path of the centre of gravity keeps on decreasing as long as the inwardly directed component of the hydrodynamic forces is greater than the outwardly directed inertial force. Eventually, equilibrium is reached among all the forces in the subsequent 'third phase'. The ship moves on a circular path at constant speed. Figure 7.25 shows the turning circle test path of the ship and the various terms used to describe test output parameters. These output parameters describe the turning ability of the ship.

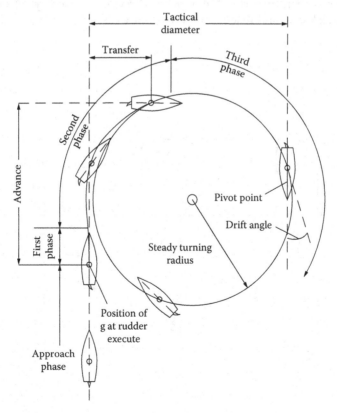

FIGURE 7.25
Turning circle manoeuvre.

7.4.1.2 *Zig-Zag Manoeuvre*

The zig-zag manoeuvre, also known as the Z manoeuvre or the Kempf overshoot manoeuvre, gives an indication of the ability of the rudder to control the turning of a ship. A typical procedure for conducting this manoeuvre is as follows:

1. The ship is held on a straight course at a steady speed for some time, and the rudder is then turned to starboard at its maximum rate to a specified angle $-\delta = 20°$ (could be 5° or 10° also) and held there until the ship's heading changes by a specified angle $\psi = 20°$ to starboard where ψ is the angle between the ship's centre line and the initial straight course.

2. At this moment, the rudder is turned at its maximum rate to the same angle on the opposite side (port), i.e. $\delta = 20°$, and held there until the ship's heading changes to the specified angle to port, i.e. $\psi = -20°$, with respect to the initial straight course.

3. This sequence is repeated through three or more cycles.

4. A record of the rudder angles δ and the heading angles ψ as a function of time t is maintained.

The main numerical measures of controllability which are obtained from the zig-zag manoeuvre are the times of the various rudder movements (first, second, … executes), and the overshoot angles (i.e. the amount by which the ship's heading continues to increase after the rudder has been turned in the opposite direction in each cycle). Other measures are the time to check yaw and reach, shown in Figure 7.26.

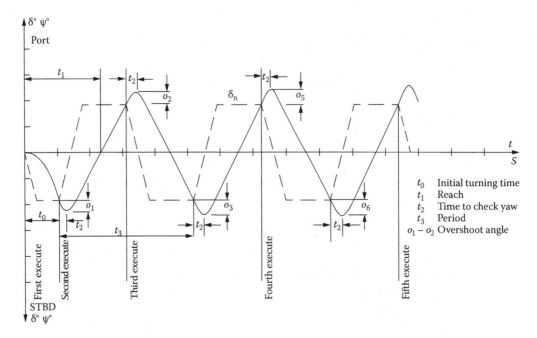

FIGURE 7.26
Zig-zag manoeuvre.

7.4.1.3 Manoeuvres to Determine Course Stability

Three manoeuvres may be carried out to determine the course stability of a ship: the Dieudonne spiral, the Bech (reverse) spiral and the pull-out manoeuvre. In the Dieudonne spiral, the ship is steadied on a course and the rudder turned to (say) 15° to starboard and held there until the rate of turn r becomes constant. The rudder angle δ is then reduced to 10° and held until the rate of turn again becomes constant. This is repeated for different rudder angles from a large angle (15°) to starboard to a large angle to port and back again to starboard in steps of 5°, and the steady rate of turn (yaw rate) as a function of the rudder angle determined. If there is a single ψ–δ curve for the rudder angle being changed from starboard to port and from port to starboard, the ship has course stability. ψ is the yaw angle and r is the rate of turn or $\dot{\psi}$. However, if there is a 'hysteresis' loop in the curve, i.e. if the ship does not follow the same path in the reverse direction, the ship does not have course stability, the width and height of the loop being a measure of the instability, i.e. at small rudder angles, the ship can turn against the rudder. This is demonstrated in Figure 7.27.

In the Bech reverse spiral, which requires less space and time to carry out, the ship is steered until it has a constant rate of turn in one direction and the rate of turn and the corresponding mean rudder angle are noted. This is done for varying rates of turn from a moderate value (corresponding to 15° rudder) to starboard to a moderate value to port and back. A ship with course stability will have an r–δ curve without large curvature, whereas a ship which does not have course stability will have an r–δ curve with a pronounced S shape near the origin. Figure 7.28 shows this diagram for an unstable ship.

In the pull-out manoeuvre, the ship is made to turn in one direction with a rudder angle of say 20°, and the rudder is then returned to zero until the ship reaches a zero or low rate of turn. The procedure is repeated with the rudder to 20° in the opposite direction. A ship with course stability will have the same residual rate of turn for initial rudder angles in either direction, while for a ship without course stability, the residual rate of turn for an initial rudder angle to starboard will be different from that for an initial rudder angle to port. In Figure 7.29, the diagrams for a directionally stable and unstable ship are shown.

7.4.1.4 Stopping Manoeuvres

The stopping characteristics of a ship in an emergency are determined by 'crash stop' manoeuvres. With the ship going ahead at full speed, the order is given to stop and go full speed astern. The time it takes for the ship to stop and the path it follows after the order is given are determined. The distance the ship travels in its original direction of motion before coming to stop is its head reach; the distance travelled by the ship perpendicular to its original path is the side reach and the length of the actual path travelled is the track reach. This is the crash stop astern manoeuvre. A similar manoeuvre carried out with the ship going astern is the crash stop ahead manoeuvre to determine the stern reach.

7.4.1.5 Other Effects during Turn

During turn, the vessel experiences heel because of the heeling moment caused due to the centrifugal force acting with a vertical lever from the centre of water pressure of the underwater hull. Another effect is the speed loss due to drag being increased due to a drift angle experienced by the ship.

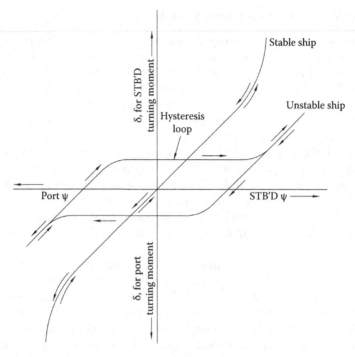

FIGURE 7.27
Dieudonne spiral manoeuvre.

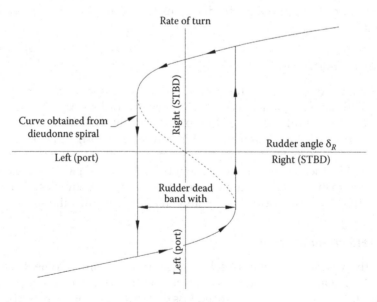

FIGURE 7.28
Bech reverse spiral manoeuvre.

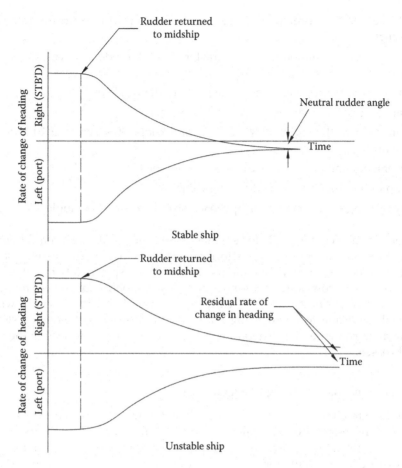

FIGURE 7.29
Pull-out manoeuvre.

7.4.2 Manoeuvring Standards

To ensure that a merchant ship has good manoeuvring characteristics such that it can manoeuvre itself safely avoiding collisions and without causing hazards to itself and other ships in the vicinity, the International Maritime Organization has set standards for ships more than 100 m in length, which are as follows:

- *Turning circle*: Advance not more than 4.5L.
- Tactical diameter not more than 5L.
- Distance travelled before the heading changes by 10° following a 10° rudder angle not more than 2.5L.
- *Zig-zag 10–10 test*: First overshoot not more than 10° if L/V is less than 10 s, not more than 20° if L/V is equal to or more than 30 s and not more than (5 + (1/2) (L/V)) if L/V is 10 s or more but less than 30 s.
- The second overshoot should not be more than 15° greater than the first overshoot.
- *Stopping*: The track reach should not be more than 15L.

- *Spiral*: There should preferably be no hysteresis loop in the turning rate-rudder angle curve.
- Typical values of manoeuvring characteristics of modern naval ships are as follows:
  - Tactical diameter–length ratio 3.25 for a rudder angle of 35°.
  - Rate of turn 3°/s maximum, 0.5°–1.0°/s typical.
  - Time to turn through 20°: 80 seconds at 6 knots, 30 seconds at 20 knots for a 150 m ship.
  - Overshoot angles: 5.5° at 8 knots, 8.5° at 16 knots.
- No loop in the turning rate-rudder angle curve.
- Steady turning speed 60% of the approach speed with rudder angle 35°.

Though there are no set standards for smaller vessels, generally care should be taken to see that the ship can manoeuvre satisfactorily using its own devices. Vessels operating in ports and harbours must have better standards than those specified before. Vessels operating in rivers and narrow waterways must be easily manoeuvrable where manoeuvring characteristics change due to shallow depth of water and restricted width, which gets further complicated due to meandering nature of the waterway. Vessels at slow speed respond slowly to rudder action. Therefore, care must be taken in designing the rudder of slow-speed vessels.

7.4.3 Estimation of Manoeuvring Characteristics

It is not possible to change the manoeuvring behaviour of the ship after construction. Therefore, it is necessary to estimate the manoeuvring characteristics of the ship at the design stage itself. There are many empirical and statistical formulations to estimate the turning circle diameter and such other parameters in literature (Comstock 1967). But with the availability of numerical techniques, computational facilities and model experimental facilities, a better estimate of manoeuvring characteristics can be achieved.

When a ship takes a turn in the horizontal plane, the forces and moments acting on the body are surge force X in longitudinal or x-direction, the sway force Y in y-direction and the yawing moment N about the z-axis in the ship fixed axis system. The forces X, Y and moment N are based on (1) the hydrodynamic action on the hull based on the body geometry, its mass and its mass moment of inertia about the z-axis and (2) the rudder and its angle δ.

If u and v are the velocities along the x- and y-axes, $r = \dot{\psi}$ the angular velocity about the z-axis, $\dot{u}$, $\dot{v}$ and $\dot{r}$ being the corresponding accelerations, the hydrodynamic forces X and Y and moment N are functions of the velocity components u, v and r and the accelerations $\dot{u}$, $\dot{v}$ and $\dot{r}$ of the hull at the centre of gravity, as well as the rudder angle δ and the rate at which the rudder is turned $\dot{\delta}$, i.e.

$$\begin{Bmatrix} X \\ Y \\ N \end{Bmatrix} = f(u,v,r,\dot{u},\dot{v},\dot{r},\delta,\dot{\delta})$$

The equations of motion can be written and be linearized by ignoring the higher-order terms. The coefficients in these equations such as $X'_u = \partial X'/\partial u'$ or the terms $\partial X/\partial u, \partial X/\partial \dot{u}, \partial X/\partial v,\ldots$ are called *hydrodynamic derivatives* and are written using the following notation:

$$\frac{\partial X}{\partial u} = X_u, \frac{\partial X}{\partial \dot{u}} = X_{\dot{u}}, \frac{\partial X}{\partial v} = X_v, \frac{\partial X}{\partial \dot{v}} = X_{\dot{v}},$$

$$\frac{\partial X}{\partial r} = X_r, \frac{\partial X}{\partial \dot{r}} = X_{\dot{r}}, \frac{\partial X}{\partial \delta} = X_\delta, \frac{\partial X}{\partial \dot{\delta}} = X_{\dot{\delta}}$$

and similarly for Y and N. The hydrodynamic derivatives are constant for a given ship form and do not change with time, whereas the variables $u,v,r,\delta,\dot{u},\dot{v},\dot{r},\dot{\delta}$ are functions of time. The forces, moments and velocities can be nondimensionalised (denoted with corresponding primed symbols) by dividing with $(\rho/2)L^2V^2$, $(\rho/2)L^3V^2$ and V respectively where V is the original free stream ship speed.

The linearized equations of motion can only be used for small changes from a steady state, e.g. when considering small deviations from a straight line course at steady speed. For large deviations, it is necessary to use the nonlinear equations taking into account higher-order terms.

Once the hydrodynamic derivatives are obtained, it is possible to do a time step integration with the help of the linearized equations of motion to simulate the path due to a particular rudder movement. In this simulation, it is also possible to include propeller-generated thrust and torque. It is also possible to estimate heel during turn. The hydrodynamic derivatives can be obtained either by numerical calculations or by model experiments.

7.4.3.1 Free Running Model Experiments

A free running model is a self-propelled model fitted with all appendages and systems to control the propeller and the rudder. There are arrangements to measure the position of the model and its orientation as a function of time, as well as the propeller RPM and the rudder angle.

Free running models can be made to carry out all types of manoeuvres – turning circle, zig-zag, reverse spiral, etc. – and the results are compared with established criteria. A large manoeuvring basin is, however, necessary.

The difference in the Reynolds numbers of the ship and the model creates a problem since the model operates at its self-propulsion point instead of the ship self-propulsion point on the model. This problem is overcome by assisting the propulsion of the model through an air propeller so that the model propeller has the same slip as the ship propeller. It is also important to scale correctly the mass moment of inertia and the rate at which the rudder is turned. For proper simulation of speed loss during manoeuvres, the characteristics of the model propeller drive system must be made similar to the characteristics of the ship propulsion plant.

Free running models provide only the overall manoeuvring characteristics without identifying the individual factors that affect the manoeuvring qualities of a ship. Free running models are also used to determine manoeuvring in shallow water and in canals and with hydraulic models of harbours and waterways.

7.4.3.2 Captive Model Experiments

In captive model tests, the model is forced to follow a prescribed path and the forces X and Y and the moment N are measured. By varying the model trajectory and orientation (drift angle), the various hydrodynamic derivatives can be obtained and used with the equations of motion to simulate various manoeuvres.

In straight line tests, the model is towed at constant speed in a towing tank for different values of the drift angle. From the measured values of X, Y and N, the hydrodynamic derivatives with respect to the transverse velocity component v can be obtained. The propeller has an important influence on the measured values, and it is necessary to tow the model with its propeller operating at the ship self-propulsion point. The rudder is kept at zero angle. If the rudder angle is varied in the tests, the derivatives with respect to rudder angle can also be obtained.

In rotating arm tests, the ship model is attached to a horizontal arm rotating about a vertical axis and made to move on a circular path during which X, Y and N are measured. By changing the drift angle, the hydrodynamic derivatives with respect to v can be obtained, while varying the radius allows the derivatives with respect to the angular velocity r to be obtained. Varying the rudder angle allows the derivatives with respect to δ to be determined. Acceleration derivatives cannot normally be obtained in rotating arm tests. All the readings must be completed in the first revolution of the rotating arm.

In model experiments using a planar motion mechanism, the model is attached to a device which oscillates the model transversely at two points along its length while the model is being towed along the length of the tank, and the force exerted at each point to force it to oscillate sinusoidally is measured. By adjusting the phase difference between the two oscillators, the model can be made to execute a pure sway motion or a pure yaw motion. The various hydrodynamic derivatives, including acceleration derivatives, can be obtained.

7.4.3.3 Numerical Simulation

With the availability of high-speed computing and commercially available CFD calculation packages, it is possible to simulate rotating arm basin as well as planar motion procedure numerically and estimate hydrodynamic derivatives. The CFD calculations for the static drift condition and pure yaw condition have been done and compared with experimental data obtained from experiments conducted at Insean, Italy. The agreement with regard to nondimensional surge and sway forces and yaw moment is excellent as discussed by Roychoudhury et al. (2014) for deep-water condition and by Sha et al. (2014) for shallow-water condition. Taking the same data, Figures 7.30a and 7.31a show this comparison for deep water for steady drift and pure yaw conditions, respectively, and Figures 7.30b and 7.31b show the corresponding comparison for shallow-water condition, which demonstrate the excellent agreement.

7.4.3.4 Statistical Analysis

If access to a large number of experimental data and trial data is available, one could do statistical analysis and get the required results after doing a regression analysis. Lewis (1989c) gives such equations for ready use taken from multiple regression analysis done by Clarke using data available. These data are the first-order hull derivatives of velocity and acceleration. It may be noted that these do not contain derivatives of velocity and acceleration in the x-direction and also second-order and cross-coupled derivatives. So their use is limited.

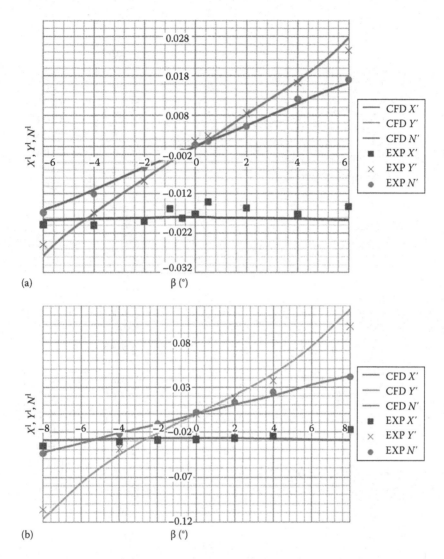

FIGURE 7.30
Sway force and yaw moment for steady drift motion: (a) deep water and (b) shallow water.

7.4.3.5 System Identification–Based Prediction

The inputs to the ship manoeuvring system are the ship hull characteristics such as parameters and hydrodynamic derivatives, the rudder characteristics, its deflection and rate of turn, etc. (Lewis 1989c). The output is the surge, sway and yaw motion of the ship obtained from the noted speed, heading angle and drift angle. The system equation for this process can be developed mathematically. Knowing the output parameters from sea trial and operational data and ship and rudder geometry, it is possible to calculate the system parameters, in this case, the hydrodynamic derivatives. If such derivatives can be calculated for a number of ships of similar type and size, it is possible to presume that the same will hold for the design ship and can be used for manoeuvring behaviour prediction.

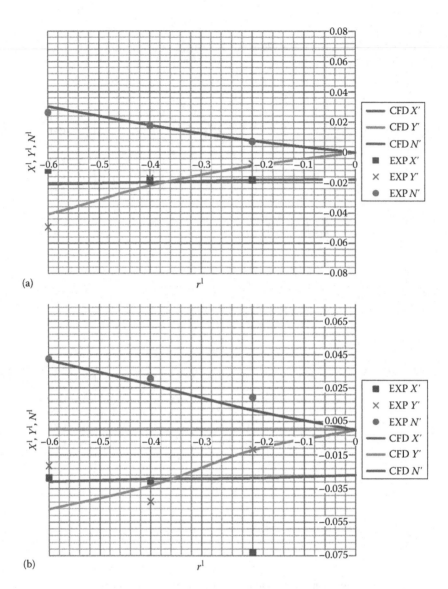

FIGURE 7.31
Sway force and yaw moment for pure yaw motion: (a) deep water and (b) shallow water.

7.4.3.6 Manoeuvring Devices

Manoeuvring devices convert energy to control forces and moments which control the ship as required. This energy can be fed to the device from the ship, in which case these are known as active control devices. On the other hand, if the energy is generated by the hydrodynamic flow around the ship, rudder(s) or fin(s), these devices are known as passive control devices. The effectiveness of passive devices increases with speed, but these devices are practically useless at low or zero speed. At these speeds, active stabilizers become effective and their effectiveness reduces with increase in speed. This is diagrammatically shown in Figure 7.32 (Brix 1993).

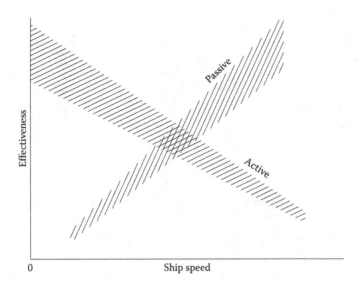

FIGURE 7.32
Effectiveness of active and passive control devices with speed.

Rudder is a passive control device, which is most commonly used in ships from small boats to very large ocean-going vessels. The rudder is moved by turning the rudder stock with the help of the steering gear. The size of the steering gear is determined by the torque demand on it to move the stock. The torque is the moment caused by the rudder lift force acting at the centre of pressure of the rudder multiplied with the lever arm, which is the distance between this centre of pressure and the stock centre line. Based on the lever distance, rudders can be of three types:

1. *Balanced rudder*: The centre of pressure and rudder stock centre line are very near each other such that the torque required for turning the rudder is very small. This also means that the vessel responds with very small effort, which may not be good in certain circumstances.

2. *Unbalanced rudder*: In this case, the rudder stock is at the forward end at leading edge of the rudder since the rudder centre of pressure is approximately one-third chord length from the leading edge.

3. *Semi-balanced rudder*: The rudder stock is somewhere in between the rudder centre of pressure and the leading edge.

Also based on the rudder-fixing arrangement with the vessel, rudders could be bottom supported (older ships) or hanging like a horn or sped rudder. Figure 7.33 shows different propeller–rudder configurations. Rudder cross section could be a straight line if the rudder is a plate rudder like in small ships. For bigger ships requiring large transverse force and turning moment, rudder cross section is generally of the aerofoil type. Rudder could be all movable type when the entire rudder is supported by a single stock. But here the lift generated is by a foil without camber. If it was cambered, the lift generated would be higher. For this purpose, a fixed structure as part of the stern frame could be placed aft of the propeller and the movable part of the rudder could be attached to the aft of it. The fixed part should be so designed that the fixed and movable parts together give an aerofoil section, and when the rudder is moved, the combination of fixed and movable parts gives the

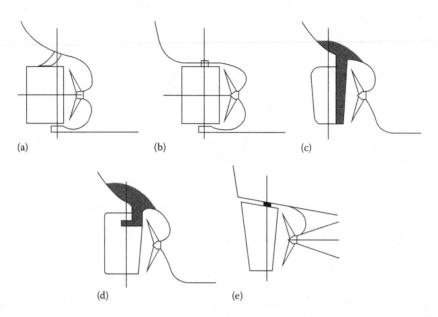

FIGURE 7.33
Different rudder propeller configurations. (a) Unbalanced rudder with rudder post, (b) Fully balanced bottom supported rudder, (c) Balanced or semi balanced with fixed structure, (d) Under hung balanced or semi balanced horn rudder, and (e) Spade rudder-transom stern.

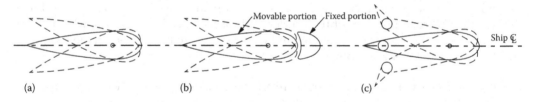

FIGURE 7.34
Different rudder cross sections. (a) All movable rudder – no camber, (b) Movable rudder behind fixed structure – simulated camber, and (c) All movable rudder with flap – simulated camber.

section of a cambered aerofoil. Alternatively, an all movable rudder could have a tail flap connected to the main rudder in such a manner that when the rudder is moved through a certain angle, the flap moves through a larger angle. This also gives a cambered aerofoil shape generating a larger lift than the corresponding fixed foil. These three types of rudder cross sections are shown in Figure 7.34.

Propellers are the simplest form of passive control devices. In a single-screw ship, due to the circumferential water movement in opposite directions at the top and bottom of the propeller and also due to nonuniform wake, there is a small transverse force causing the vessel to turn slightly in control-fixed condition. But this turning is unintentional and can be controlled in straight line manoeuvres by moving the rudder. But it is not possible to control the ship in turning using a single propeller. A CPP has the same nature. In the case of a tug, trawler or supply vessel, a tow force can be generated at low speed, which can be used for other pulling activities. In a twin-screw ship, by controlling the RPM of the propellers or by operating only one propeller or altering the pitch of the propellers in CPP at constant RPM, it is possible to generate different thrust forces on either propeller and an effective turning moment is generated, which turns the ship. The effectiveness of this manoeuvre depends on the distance between the two screws and whether these are CPP or FPP.

The thrust generated by propellers can be used to turn or control a ship by locating such thrusters(s) at the desired locations. If the thrust generated is used for controlling the ship, these thrusters become active control devices. A bow thruster unit is a thruster located at the forward end of the ship located inside a tunnel in the forward body of the ship. The thrust generated is in the transverse direction at a distance from the centroid of the underwater body. Sometimes depending on the thrust requirement and space available, there could be two thrusters in the forward end and/or a stern thruster also. In such cases, the vessel can turn on its own axis. A floating platform, which is required to maintain its horizontal position more or less stationary in spite of sea and wind conditions, may require a number of thrusters located at different positions in different orientations so that the thrust can be directed in any direction to maintain the platform position. In this case, manual control is difficult and automatic control of the thrusters must be provided. Such a system is known as the dynamic positioning system.

A nozzle propeller is one in which the propeller is placed inside a nozzle to generate larger thrust in low-speed conditions, particularly useful in tugs and trawlers. If the nozzle can be provided with azimuth rotation, it can provide thrust in any direction avoiding deployment of a separate rudder. This is known as azimuth propulsion. A steering rudder is one where a rudder flap is attached at the aft end of the nozzle housing the propeller. Such a nozzle can be fixed or steerable with the rudder also being controllable. A podded propeller can also provide the desired control if the pod is steerable. A Voith Schneide propeller is a vertical axis propeller which can provide thrust in any direction by controlling the pitch of each blade separately.

7.4.4 Design Considerations for Controllability

To manoeuvre a vessel successfully, one has to provide devices which should provide both fixed directional stability and good turnability with controls working. For this the designer has to understand the external environment in which the ship is expected to operate and the internal environment which must be provided for the operator to control the ship properly.

7.4.4.1 Environment

The external environment includes the vessel's route, traffic lane and passages it must pass through, the traffic congestion in these areas and the docks and ports it is likely to call at and operational requirements. The environmental conditions in these areas such as water depth, channel or seaway width, current and wind, wave conditions, etc. must be noted. Then it is possible to draw up the operational map of the vessel. For example, a container ship or a RORO ship or a passenger ship may require a quicker port turnaround time than a bulk carrier or tanker. A tug may require to provide services as a harbour tug to a vessel in a port or as an escort tug in a channel.

The internal environment includes the facilities, the people and the work environment. The vessel must be provided with adequate facilities for doing effective manoeuvres such as echo sounder for depth measurement, GPS for location and speed indication, radar and its display for location of other vessels in the vicinity, rudder angle and rate of turn indicator and ship's heading indicator to know the drift angle.

The manoeuvring devices may require specialized skills, which the vessel-operating personnel must possess and based on the ease with such training can be imparted, devices must be chosen. Also bridge layout plays an important role in the effective operation of manoeuvring devices. All-round visibility, comfortable ergonomic environment and accessibility to various displays play an important role in successful control of the ship.

7.4.4.2 Effect on Hull Parameters

In the recent past, manoeuvrability was studied only at a later stage of design after almost all design parameters were finalized, the purpose being to ensure acceptance of steering capability. But recent trends indicate that manoeuvring characteristics should be studied from the concept stage itself and updated through different stages of design. At the concept stage, first-order hydrodynamic derivatives and performance can be estimated using statistical data and later using CFD methods. Normally, hydrodynamic derivatives can be expressed as a function of L, B, T and C_B. It is possible to do a quick parametric study to determine the set of parameters suitable for controllability.

The controllability of a ship depends on the rate of response of a ship to a rudder command. Dynamic stability, on the other hand, depends on the vessel's inherent course stability independent of the rudder. Therefore, if the response of a ship to a turning command is sluggish, its controllability would be poor. Additionally, if the turning ability is poor, directional stability for course keeping would be better. Generally, large full form ships are sluggish to respond to rudder movement but have better turning ability with poor dynamic stability. Fine form high-speed vessels, on the other hand, respond to turning command more quickly and have better dynamic stability. Small vessels, in general, have better controllability than large vessels at good speed. Broader vessels have better directional stability compared to leaner vessels.

Trim by stern and deadwood at stern give better directional stability and are poor in turning. Fine sections at stern improve flow onto the rudder and its effectiveness in turning. Cutaway deadwood at stern improved turning ability, and sometimes it is necessary to improve dynamic stability of such vessels by fitting a skeg, thus increasing the lateral area. Tugs, trawlers and twin-screw vessels with cutaway sterns are generally provided with skegs. The position of the centroid of the lateral area affects the turning moment and, therefore, the turning ability. Trim by bow is not desirable for controllability. Fine sections at forward end improve turning ability slightly.

The above-water portion is important for wind forces acting on the ship in the form of sway force and yaw moment. Trim affects the wind action on the ship. Also single lean deck house structure implies less wind force compared to bigger and numerous deck houses. The wheel house or the navigation bridge ideally should be located at amidships on the top deck of the deck houses so that the forward and aft can be visible adequately. If this is not possible, the navigation bridge should be located in the front portion of the deck houses at a height such that forward is visible adequately leaving a blind zone less than 1.25 times the length of the vessel. The bridge deck should be so arranged that aft of the ship is also adequately visible for better control of the ship.

8

Hull Form Design

Fixed offshore platforms, semi-submersibles, compliant platforms and tension leg platforms are geometric bodies and their form can be accurately defined by mathematical equations which can be used for generating various design drawings as well as for hydrodynamic and structural analysis. But for a floating and mobile body like a ship, the shape of the body or the hull form is long and thin or slender and highly three dimensional having curvatures in all directions. In this chapter, ship hull form design is discussed. The hull form is the starting point for various calculations such as hydrostatics and stability, hydrodynamic behaviour, structural information and production information generation. Production information pertaining to ship structural elements like shape, area, cutting length and weight can only be obtained if the hull surface is defined accurately and in a form suitable for the purpose. Thus, there are different stages of ship hull form design, which can be broadly defined in three stages – the concept design stage, the final design stage and the manufacturing information generation stage.

At the concept design stage, the principal design parameters and design guidelines are used to generate curves and surfaces and a preliminary shape is given to the hull form. The hull shape at the concept design stage can be generated by using the deck outline, stern and stem profiles and the main design particulars and carrying out a systematic interpolation from an existing set of faired hull form data (Kuo 1977, Munchmeyer et al. 1979). A hull form can also be developed by using the main particulars and other design parameters which are fundamental to the shape of the hull. This method of hull form development is known as the form parameter approach. Finally, the hull form can also be developed by systematic distortion of the hull form of an existing faired ship (Soding and Rabien 1977). In the latter case, however, the desired shape should be close to the basic faired shape that is being distorted. At this stage, the geometry of the hull form is normally generated only in the form of an offset table and preliminary lines plan drawing where x is the longitudinal direction, y is the offset in the transverse direction and z denotes the vertical direction. The origin of this Cartesian coordinate system is generally the intersection of the centre line longitudinal plane, the transverse plane at midship and the horizontal plane at the base line. An offset table contains the half-breadth y (offset) of the ship surface at discrete water planes z and stations x, and also the defining end contours such as stem and stern profiles and deck at side for 10 or 20 equidistant stations between after perpendicular (AP) and forward perpendicular (FP). As an example, an offset table is shown in Table 8.1. More closely spaced offset table defines the ship more accurately. In the final stage of numerical definition of ship hull form, the offset table based on longitudinal locations defined by frame numbers may also be generated. A lines plan drawing consists of three views: (1) sheer plan, which is the cross-sectional view of the ship surface in the x–z plane parallel to the centre plane at different distances, commonly known as buttock lines, (2) half-breadth plan, which is the cross section of the ship surface with x–y plane at different distances, commonly known as water lines and (3) the body plan, which is the transverse sectional view (on y–z plane) of the ship at various longitudinal positions (stations) as shown in Figure 8.1.

TABLE 8.1

Typical Offset Table

Stations	0 – AP	1	2	3	4	5	6	7	8	9	10 – FP	Stern Contour	Stem Contour
Distance from AP (m)	0	11	22	33	44	55	66	77	88	99	110	Contour from AP (m)	Contour from FP (m)
Water Lines							Half Breadth						
0	0.00	0.00	0.00	3.73	5.59	5.94	5.55	4.16	2.28	0.18	0.08	26.18	0.33
1	0.00	0.00	6.37	7.95	8.52	8.62	8.22	7.14	4.63	1.35	1.13	16.52	3.45
2	0.00	2.90	8.02	8.72	8.89	8.95	8.74	7.88	5.42	2.14	1.66	10.20	4.11
3	0.00	7.23	8.61	8.90	8.98	9.00	8.91	8.21	5.97	2.74	1.70	3.46	4.03
4	6.23	8.23	8.83	8.96	9.00	9.00	8.97	8.44	6.45	3.24	0.00	−4.39	0.00
5	7.86	8.65	8.92	8.99	9.00	9.00	9.00	8.60	6.89	3.73	0.19	−4.48	−0.35
6	8.45	8.84	8.96	9.00	9.00	9.00	9.00	8.72	7.30	4.28	0.49	−4.56	−1.16
7	8.75	8.92	9.00	9.00	9.00	9.00	9.00	8.83	7.68	4.90	0.93	−4.64	−2.00
8	8.90	8.97	9.00	9.00	9.00	9.00	9.00	8.92	8.04	5.58	1.56	−4.73	−2.84
9	8.97	9.00	9.00	9.00	9.00	9.00	9.00	8.98	8.36	6.30	2.35	−4.81	−3.68
10	8.99	9.00	9.00	9.00	9.00	9.00	9.00	9.00	8.65	7.06	3.31	−4.90	−4.52
11	8.99	9.00	9.00	9.00	9.00	9.00	9.00	9.00	8.92	7.81	4.47	−4.98	−5.36
Deck at side	8.99	9.00	9.00	9.00	9.00	9.00	9.00	9.00	8.92	7.81	4.47		
	Aft end −4.89										Fore end 115.36		

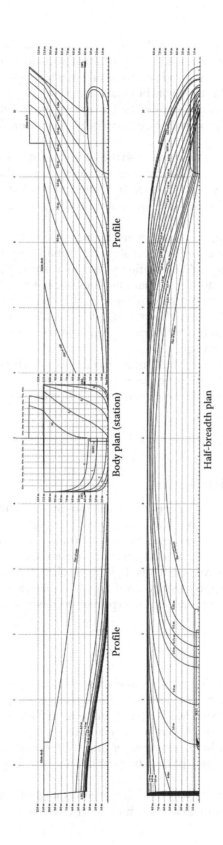

FIGURE 8.1
A typical ship lines plan.

A faired ship hull form consists of faired curves and surfaces. The definition of a curve as fair is highly subjective: a curve is said to be fair if it is pleasing to the eye and is smooth and continuous along its length. A faired surface is one in which all curves are smooth and continuous and the surface is bound by its boundary curves. None of the methods used at the concept design stage guarantees a faired hull form and the resulting preliminary hull shape must undergo a fairing process before it can be used for detailed design or production purposes.

In the final design stage, the preliminary shape definition obtained from the concept design stage is further refined by an elaborate fairing process, and the hull form definition obtained at the end of this stage can be used for different fabrication purposes. Various computer-aided design (CAD) techniques are used for 2D curve fairing and 3D surface fairing (Nowacki 1980).

With the introduction of computer-aided manufacturing (CAM), the hull form information in numerical form is to be generated and stored in a manner such that it can be used for detailed design work using CAD techniques and also for CAM activity. This is an important aspect of hull form design so that one does not have to prepare separate hull form data as input to different programs while using computers. The different hull design stages are shown in Figure 8.2. The hydrodynamic design of hull form shown in the flow chart of Figure 7.8 is an input to the hull form design at the concept design stage and, to a lesser extent, in the final design stage.

The next section discusses general hull form characteristics which can serve as guidelines for preliminary hull form design. Subsequent sections deal with different hull design methodologies, which are followed by computer applications in hull form design.

8.1 Hull Form Characteristics

Boats and ships encompass a large variety of vessels moving in inland and coastal waters as well as offshore and deep oceans. The vessels may be single-man boats or passenger ferries carrying a large number of passengers over short distances or high-speed planing boats or large ocean-going vessels. Similarly, the vessel characteristics also vary depending on widely different service requirements. Hull form plays the most significant role in combining the operational requirements with the hydrodynamic behaviour of the craft. Therefore, hull form design is undertaken by experienced designers supported by model test results.

8.1.1 River Vessels

River vessels have limited carrying capacity and these vessels move at low speed and are affected by river geometry and river currents. At low speeds corresponding to low Froude number, say around 0.2, the resistance is mainly viscous and wave making is negligible, as discussed in Section 7.1.1. The viscous resistance consists of the frictional component, which is mainly a function of Reynolds number and wetted surface, and the separation drag which depends largely on sharp curvatures of the hull shape, particularly in the forward and after ends.

For cargo barges in inland waterways moving at low speed or vessels ferrying a large number of passengers over short distances, hull form cross section can be rectangular.

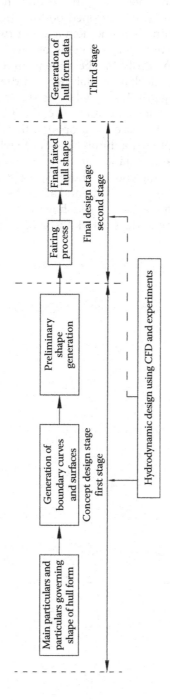

FIGURE 8.2
Different stages of hull form design.

However, to reduce separation drag, the forward and after ends may be shaped in the profile. The resistance characteristics of barges or vessels with completely rectangular cross section may be further improved by modifying the section shape. For ease of construction, the form may be single chine having a sloped bottom towards the ends, or double chine having a shaped bilge. A chine is a line demarcating two different surfaces joining each other along this line. Further improvements of this shape at the bilges lead to round-bilge forms (Latorre and Ashcroft 1989). The dimensions of such vessels are decided from operational requirements like deck area, stability and draught available. Passenger boats are normally broad with large deck area whereas shallow river vessels may be of low draught–depth ratios with fine forward and aft ends. These vessels could be self-propelled and sometimes, to provide space for propeller, there could be a tunnel at the stern. Figure 8.3a shows the lines plan of a dumb river barge where the fore end is raised for better flow at the forward end and rectangular cross section with a double chine bilge. Mark the nearly complete straight line construction. Figure 8.3b shows a self-propelled barge with the fore end shaped for improved flow and aft end shaped to house a propeller. This form also has a double chine bilge. Figure 8.3c shows a self-propelled river passenger boat carrying passengers on the upper deck. Mark the shaped cross section compared to wall-sided cargo barges.

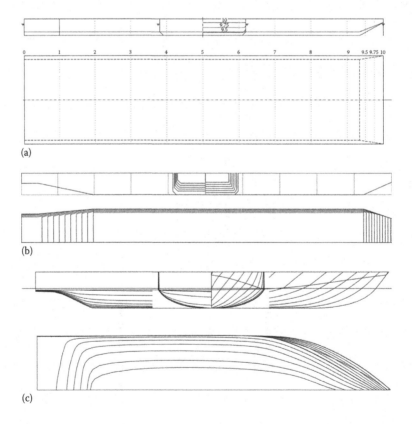

FIGURE 8.3
Hull forms of river vessels. (a) Dumb cargo barge. (b) Self-propelled cargo barge. (c) Self-propelled river passenger boat.

8.1.2 Yachts

Another category of relatively low-speed vessels, which has attracted much attention in recent years, is the sailing yacht. In an open expanse of water, the yacht picks up speed driven by wind, the resultant of which acts at a considerable height above the water, based on sail geometry. This category of vessels has picked up much interest from pleasure and sports aspects in recent years. The racing yacht is generally made of light material like fibre-reinforced plastics and has a very smooth narrow hull form so that there is absolutely no separation drag. Wetted surface is kept at a minimum and the vessel causes minimum free surface disturbance. The course stability and the vessel's stability are achieved by fitting additional appendages called fin keels and large rudder. Design of yacht hulls is discussed in Larsson and Eliasso (2000). Performance of yacht hulls with sails is discussed in Marchaj (1996). A typical yacht hull lines plan is shown in Figure 8.4. Gerritsma et al. (1978, 1991) give the form and resistance characteristics of a systematic series of yacht hull tested at the Delft Ship Hydrodynamics Laboratory.

As speed increases, apart from viscous resistance, wave-making resistance becomes prominent. The vessel hull is shaped in the forward end to reduce the wave-making resistance. To reduce separation drag, the bilges are rounded. The hull is tapered at the stern and curved upwards towards the water line. Such sterns are normally known as *Canoe sterns* or *Counter sterns*. When buoyancy forces alone support the total displacement, its shape tends to be the same as the conventional ship shape.

8.1.3 Semi-Planing and Planing Vessels

At a Froude number of about 0.40, the ship creates waves having length equal to the length of the vessel on water line. At this speed, wave making becomes a virtual barrier to any increase in speed for true displacement forms. This is because the increased local velocities caused by the rounded hull form result in negative pressures which cause the vessel to settle deeply and trim down by the stern. To cross the speed barrier, it is necessary to generate dynamic lift which should support a part of the weight so that the vessel starts planing or generating vertical lift by hydrodynamic interaction between the bottom surface and the water beneath. Semi-planing and planing vessel characteristics are given in Saunders (1957), Savitsky (1985) and Gonzalez (1989).

As the speed increases further, it is necessary to depart from the small sterns of the low-speed vessels and to make transom sterns that allow for flatter buttock lines. These flat buttock lines avoid negative pressures and cause the flow to separate cleanly at the stern,

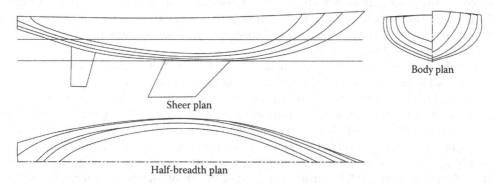

Sheer plan

Body plan

Half-breadth plan

FIGURE 8.4
Typical sailing yacht hull form.

thus keeping the separation drag to a minimum. For higher speeds, even straighter buttock lines are required and the transom must be broader and more completely immersed. For $F_n > 0.9$, viscous resistance is dominant and it may be beneficial to provide large flat bottom surface to generate lift though there may be some induced drag. Running trim, which is a function of longitudinal centre of gravity (LCG) position and centres of action of lift force and buoyancy, determines the angle of inclination of the bottom surface, and care should be taken to see that it is maintained within 4°–6° to get adequate lift.

Round-bilge forms are recommended for semi-displacement high-speed ships operating at Froude number between 0.4 and about 1.3. These hulls have fore bodies of fine entrance, fine and straightforward water lines with low entrance angles dictated by the maximum beam and relatively high values of deadrise which is the inclination of boat bottom to horizontal in the transverse plane. Often, the round-bilge sections incorporate knuckles and/or spray rails that help throw out some of the higher waves encountered in rough waters as well as the spray that results due to the convex shape of sections and the curvature of the fore-body buttock lines. A knuckle is a line in the longitudinal direction which provides discontinuity between the bottom surface and the side of the boat, whereas a spray rail is an additional longitudinal strip attached to the boat surface approximately at the running water level. Knuckles and spray rails also help in increasing roll stiffness at high speeds. A wide transom stern, with flat buttock lines rising gently aft, is introduced in the after-body round-bilge forms in order to avoid stern squat, which means an increased drag. But normally, the transom section is not made too flat at the underside to avoid slamming in following seas. As the lateral projected area of transom stern, round-bilge ships is reduced, a centre line skeg is provided in the aft to give some directional stability by increasing the lateral submerged area.

Round-bilge hull performance can be spectacularly refined by adding dynamic trim control devices. The control of LCG along with the incorporation of fixed transom wedges or movable transom flaps (provided at the transom) should provide the optimum, or near optimum, dynamic trim angle for different speeds and load conditions of the vessel. The cheaper fixed transom wedges work well for a small range of speeds with a small resistance penalty at other speeds while the adjustable transom flaps work well for a large speed range.

For vessels moving at fully planing speeds, hard chine forms are recommended, as they are highly desirable to keep the wetted surface to a minimum. A chine is a knuckle line which sharply divides the flat bottom surface or lifting surface from the nearly flat sides of the vessel. The lift generated by the hard chine hull causes it to rise bodily above its at-rest position and to trim by stern, thereby reducing wetted surface considerably. An important part of the total resistance at high speeds is frictional and hence, a reduction in wetted surface means a noticeable reduction in total resistance at planing speed. Furthermore, a hard chine hull provides more roll damping than similar round-bilge hull. This roll damping decreases as speed increases and could be an important factor while considering transverse stability at high speeds.

The main characteristic of the planing hull is its effective flow separation not only at the transom, as happens with the high-speed displacement ships, but also at the sides. The hard chine hull form prevents the formation of negative pressure areas in the bottom of the hull, and full flow separation at transom occurs. By proper selection of section shape and placement of spray rails, it is also possible to achieve effective flow separation at the sides. Sometimes, the same effect is achieved by providing a step along the chine.

Deadrise is a very important factor in designing fast motorboats. Though a flat bottom on the aft side is desirable for planing, it would suffer from high impact accelerations and bad steering and manoeuvring characteristics. Normally, deadrise is low, say to 4° at the transom gradually increasing to 14°–18° at midship. The section shape should be such that

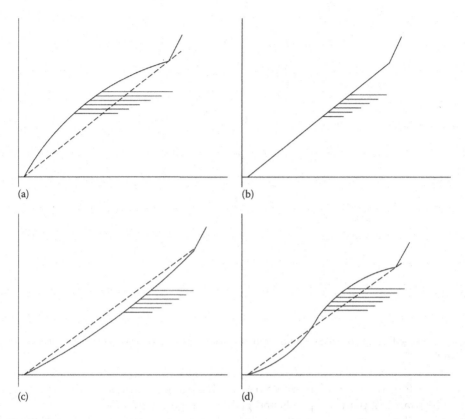

FIGURE 8.5
Four types of fore-body section shapes of planing craft. (a) Concave. (b) Straight or pure 'V'. (c) Convex. (d) Inverted bell or yoke.

the buttock shape at the mid position of the half breadth in the aft body is nearly flat. Good planing can then be achieved with trim by aft. Low deadrise in the aft is required for good planing whereas high deadrise in the forward is required for good motion characteristics in waves and also for good steering and manoeuvring. While designing the lines plan, these conflicting requirements are to be kept in mind.

Figure 8.5 shows four basic shapes of fore-body sections used for planing boat design. The concave section with a relatively low chine and small rise of floor gives excellent smooth water performance, good flow separation at sides, full planing behaviour at comparatively low speeds and low power requirement. However, these sections are unsuitable for wave going since the incidence of pounding and slamming is large resulting in speed loss and bottom damage. This section shape throws the water out at the sides. By raising the chine forward and with a fine entrance, this type of section can keep the spray out, but there is no improvement in the incidence of slamming and porpoising. Porpoising is the motion instability of planing boats at high speed due to the periodic coupled roll and pitch motions. Convex sections, on the other hand, give more internal volume in the forward, but flow separation at the sides is not as easy as in concave sections. Therefore, there is higher wetted surface, higher resistance and planing at a higher speed. There is also considerable deck wetness due to spray. But these sections are suitable for wave conditions. Straight sections combine the disadvantages of both the aforementioned types of sections without any specific advantage. The yoke or inverted bell section provides a good

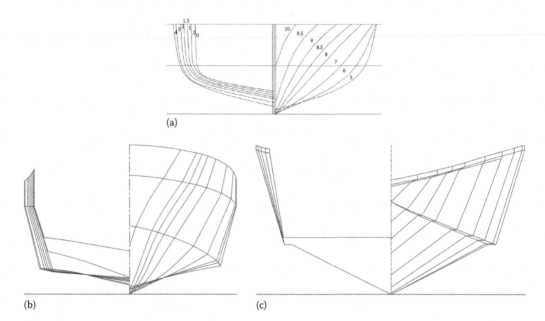

FIGURE 8.6
Different high-speed craft body plans. (a) Round bottom semi-planing. (b) Hard chine semi-planing. (c) Hard chine planing.

compromise. It provides good internal volume, sharp angle at the chine to throw the water out, low resistance at planing speeds and good wave-going qualities.

Figure 8.6 shows three different sectional shapes of semi-planing and planing hull forms. Figure 8.6a is the sectional drawing of a round bottom semi-planing boat hull form. Figure 8.6b shows a warped bottom hard chine hull with deadrise increasing from stern to stem. Figure 8.6c is a deep V hull with high deadrise angle throughout. This is a hard chine hull with a step at the chine which acts like a spray strip deflecting water from the vessel side. Sometimes a number of spray strips are added to the vessel side for deflecting water from the sides.

8.1.4 Catamaran Vessels

Catamarans or twin hull vessels are used as research boats, crew boats, excursion boats, passenger ferries, survey boats, police/rescue boats and patrol/reconnaissance boats. For all these types of craft, the payload is a small fraction of the total displacement. A planing boat is severely restrained in its use by sea conditions. Catamarans can be designed to behave better in such seas and also provide large deck area for use. For catamarans at low speed, the emphasis is on deck area whereas at high speed, hydrodynamics plays an important role. Catamarans also have an advantage over mono-hull vessels as sail boats. Due to the spread of hulls, catamarans have good stability and hence, ballasting is not required to counter the heeling moment due to sail. Therefore, effective weight to be moved in water is relatively less compared to mono-hull vessels.

Catamarans have a potential for large deck area per tonne of displacement and there is the possibility of achieving higher speed. Stability without weight is the main thing that makes multi-hull vessels more attractive than mono-hull vessels. By dividing the displacement of the vessel between two hulls, the displacement length ratio is lowered for each hull and the hulls may be designed for minimum resistance at high speed with no regard to stability of each. The basic advantages over mono-hull vessels can be listed as follows (Gonzalez 1989):

- The transverse separation between hulls gives large moment of inertia of the water plane and hence, the catamaran has high transverse stability.
- At displacement speeds, the separation distance between the two hulls can be adjusted such that the interference between the waves of the inner sides of both hulls is favourable and wave resistance is reduced. At planing speeds, since there is no wave resistance, separation distance is immaterial.
- The useful deck area is greater than that of a mono-hull of equal length.
- The turning ability is excellent because of the wide separation of its propulsors.
- The freedom of designing the layout is greatly enhanced due to large deck area and stability.

The principal problems of catamarans are as follows:

- If not designed carefully, there is a large total resistance at low speeds due to increased wetted surface.
- The cross-structure between the demi-hulls has to be properly strengthened and hence, payload to structural weight ratio becomes low and less competitive.
- There is severe wave impact on the bottom of the cross-structure while moving in a seaway. To keep this bottom of the cross-structure high above the water surface, large variations of draught are not permissible. Therefore, catamarans have a heavy restriction on payload. Normally, payload should not exceed about 10% of the total displacement.

Catamaran hulls can be classified broadly into three types: symmetric, asymmetric and split hulls based on the demi-hull geometry. If the volume of the demi-hull is distributed equally on its centre line, it is symmetric, and if distributed unequally, it is asymmetric (Bhattacharya 1978, Couser et al. 1997, Bruzzone et al. 1999). If the weight of the demi-hull is distributed on only one side of the centreline, i.e. if a mono-hull has been split into two along its centre line, it is called split hull. It may be pointed out that the hull form and separation distances are important parameters at displacement speeds. For $F_n < 0.7$, symmetric hull forms are normally preferred. For $F_n \geq 0.7$, asymmetric or split hull forms are preferred. At planing speeds, separation distance is immaterial to resistance and is dictated by considerations of adequate deck area and stability. At this speed, each demi-hull is designed as a planing hull. Fast catamaran boats are extremely weight sensitive. Both steel and fibreglass are too heavy for such boats. Perhaps aluminium is the most suitable material. Figure 8.7a shows the symmetric demi-hull geometry, Figure 8.7b shows a chine section shape suitable for high speed operation and Figure 8.7c gives the aft and forward section views of an unsymmetrical demi-hull of a catamaran vessel.

8.1.5 SWATH Vessels

The small waterplane area twin (or triple) hull (SWATH) vessel (Bhattacharya 1978) is a displacement ship having two demi-hulls. Each demi hull is made up of a semi-submerged hull resembling a body of revolution and a strut which pierces the water surface. The separation

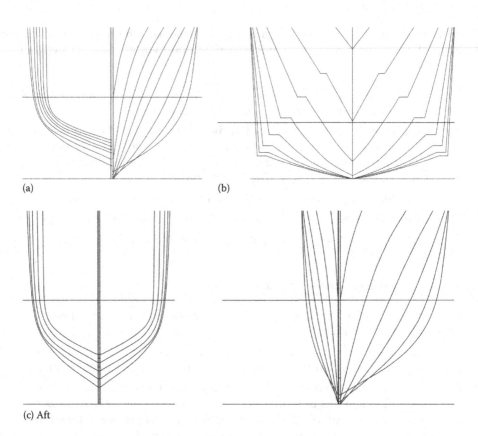

FIGURE 8.7
Different catamaran demi-hull body plans. (a) Round bottom. (b) Chine high speed. (c) Unsymmetrical.

between the demi-hulls is bridged by a box cross-structure. Thus, the SWATH ship combines the favourable features of catamarans and semi-submersibles. This enables the SWATH ship to retain the essential advantage of a large deck area combined with controlled motion and reduced wave making drag by placing its displacement well below the water surface.

The features of a SWATH vessel are the same as the catamaran vessel at displacement speeds. The essentially different features from those of catamarans are as follows:

- A SWATH ship has a larger wetted surface and, therefore, has a large skin friction drag.

- A significant reduction in wave drag can be obtained at higher displacement speeds because of the placement of displacement well below the water surface, thus reducing waves produced by the struts and the submerged hulls, because of favourable mutual interference of the waves generated by the strut and the hull of each demi-hull and because of favourable mutual interference of the two demi-hulls.

- SWATH ships have very low water plane area and hence, their response to sea, particularly pitching and heaving, is very low.

- SWATH ships are not designed for planing speeds.

- Since the water plane is small, these vessels are extremely weight (payload) sensitive. A small increase in weight means a large increase in draught which may cause wave impact on the bottom of the cross-structure.

8.1.6 Seagoing Vessels

Seagoing vessels are normally mono-hull displacement type of craft. For such vessels, lines plan is developed for optimum hydrodynamic performance satisfying the geometrical constraints and constraints imposed by ship general arrangement (double bottom width to be a required minimum at the ford end of ford hold to accommodate containers, etc.) and ship production requirement (straight line for single curvature forms – cylindrical bow instead of normal/bulbous bow). Optimum hydrodynamic performance normally means

- Optimum calm water resistance (until recently, this was the main consideration)
- Good seakeeping characteristics
- Good manoeuvring performance
- Proper flow into the propeller

The hull form development from resistance point of view is discussed in Section 7.1.7 and shown in Figure 7.8. By following the guidelines mentioned in Section 7.3.3 for seakeeping and Section 7.4.3 for manoeuvring, the hull form can be developed and estimation of seakeeping and manoeuvring behaviour can be done, and the hull form can be changed, if necessary, and finalised.

However, until a final tank tested lines plan is available, a realistic hull shape is necessary to proceed with the design activity. Discussed in the following are some guidelines which may be used for preliminary hull form development once the main dimensions and the block coefficient have been established. These guidelines have been compiled over a period of time from various sources including Saunders (1957), Comstock (1967), Schneekluth and Bertram (1998) and Watson (2009).

8.1.6.1 Midship Section Design

The fullness coefficient of the midship section area, or C_M, is rarely known in advance by the designer. While deciding on the midship area coefficient, the designer must have the following considerations – favourable resistance, plate curvature in the bilge area, stowage space for cargo (e.g. containers) and roll damping. Ships with small midship coefficient tend to experience large rolling motion in heavy seas when compared with those with higher coefficient. The simplest way of providing roll damping is to provide bilge keel. The length of bilge keel on a full ship is approximately 30% of L_{BP}.

Various formula for midship area coefficient calculation are given in Section 5.4. With a flat of keel of width K and a rise of floor F at $B/2$, the relationship between bilge radius r and midship area coefficient can be stated as follows:

$$C_X = 1 - \frac{\{F[((B/2)-(K/2))-r^2/((B/2)-(K/2))] + 0.4292r^2\}}{BT}$$

It can be observed that if there is no rise of floor,

$$C_X = \frac{1 - 0.4292 \cdot r^2}{BT} \quad \text{or} \quad r = \sqrt{\left[\frac{BT(1-C_X)}{0.4292}\right]}.$$

However, from producibility considerations, many times the bilge radius is taken equal to or slightly less than the double bottom height.

8.1.6.2 Bow Profile and Forward Section Shape

Normal bows mean non-bulbous conventional bows. The vertical or straight stem was first used in 1840 in the United States. This shape was also very popular until the 1930s when it became more raked both above and below the water line. The 'deadwood' cut below the water line reduces the resistance. Maier first introduced this form in 1930 in conjunction with V sections to reduce frictional resistance.

Forward section shape could be U of V or in between the two extreme shapes. It is diffi-cult to numerically rate these sections. An extreme U section and extreme V section having the same sectional area below the water line (i.e. they satisfy the same sectional area curve [SAC]), the same depth up to the deck at side and the same angle of flare at deck level, as shown in Figure 8.8, are compared in the following giving the features of V shaped sections.

- Greater volume of top sides.
- Greater local breadth of the design water line: associated with this is a greater moment of inertia of the waterplane and a higher centre of buoyancy. Both effects increase the value of KM.
- Small wetted surface, lower steel weight.
- Less curved surface, cheaper outer shell construction.
- Better seakeeping ability caused by higher reserve buoyancy and reduced slam-ming effects (though slam frequency may be higher).
- Greater deck area – particularly important for the breadth of the forward hatch on container ships.
- In ballast condition of a given displacement, the wedge form provides a greater draught and hence a decrease in block coefficient. At a smaller draught, the decreased block coefficient leads to lower resistance than in the case with equiva-lent U-form. Also, less ballast is needed to achieve the desired immersion.

The disadvantage of the V section form (advantages of the U section form) is that V sections in the fore body have a higher wave-making resistance (due to higher angle of entrance)

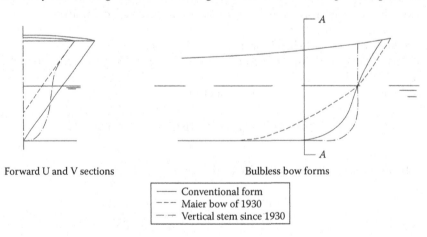

Forward U and V sections   Bulbless bow forms

> —— Conventional form
> − − − Maier bow of 1930
> — — Vertical stem since 1930

FIGURE 8.8
Different non-bulbous bow profiles and sections.

with lower frictional resistance. This leads to higher overall resistance than U sections in the range $0.18 < F_n < 0.25$ (depending on other influencing effects of form).

The following conclusions can be arrived at for U and V forward section shapes. In the Froude number range, when V section has an optimum effect on resistance, extreme V section should be used, since all the advantages here are in V section form. In the range when U section has an optimum effect on resistance, the advantages and disadvantages of this form must be assessed. At points of transition between the ranges, a near U section form is used. At the range when $F_n \approx 0.27$ where U sections are hydrodynamically most advantageous, an extreme U section is suitable.

Bows with vertical stems or box forms have been developed for very full ships with block coefficients above $C_B = 0.8$ and Froude number below $F_n = 0.18$. These bows are used in tankers and bulkers, and also on less full ships with high B/T ratios. These box forms have elliptical water lines with minor axis of the ellipse equal to the ship's breadth. To improve water flow, the profile has been given a rounded form between keel and stem. The effect of these bows is to create a relatively large displacement in the vicinity of the perpendicular and less sharp shoulders positioned further back in comparison with alternative designs with sharp or inclined stems. Parabolic bows can also be fitted with bulbs, for which cylindrical bulb forms are usually employed. Comparative experiments using models of bulk carriers have demonstrated the superiority of parabolic bows with fullness of over 0.8 and low L/B ratio over the speed range investigated between $F_n = 0.11$–0.18. With increase in speed, the stem is raked with either V or U sections at the forward part of the ship. A development from this form is the removal of deadwood at the forward end giving V-shaped forward sections. Figure 8.8 shows three non-bulbous bow profiles and the corresponding sections just aft of the FP showing the U and V sections at the forward part of the ship.

8.1.6.3 Bulbous Bow

It is commonly believed that a bulb reduces wave-making resistance. So for ships that experience a large percentage of wave-making resistance, one can profitably design a bulb to reduce effective power. But the designer has to be careful about bulb design so that the bulb reduces and does not increase wave-making resistance instead. Furthermore, the bulb must also be effective in full-load as well as ballast conditions. One also has to be careful that the bulb is fully or largely immersed in ballast condition so that it does not cause avoidable slamming. The classification societies have minimum draught requirement for this purpose. In full-form ships like tankers and bulk carriers, wave-making resistance is less, only about 5%–10% of the total resistance. The rest of it is viscous resistance. It is believed that most of this is composed of wave-breaking resistance due to high wave slope and also to shedding of bilge vortices near the forward turn of the bilge. It is possible to smoothen the free surface flow as well as bilge vortex shedding by carefully designing a bulb. So, a well-designed bulbous bow can also reduce 10%–15% resistance for a full-form ship.

Bulbous bows are defined using the following form characteristics:

- Shape of section
- Shape of bulb in the profile
- Length of projection beyond perpendicular

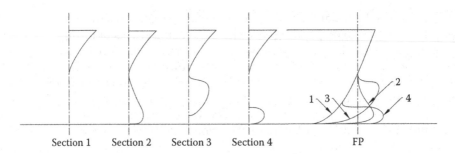

Section 1   Section 2   Section 3   Section 4   FP

FIGURE 8.9
Different bulbous bow profiles and sections.

- Height at maximum breadth of bulb at FP
- Area ratio (transverse sectional area of bulb at FP/midship area)
- Transition to hull

Bulb forms can be of various types. Figure 8.9 shows three different bulbs and a normal raked bow for comparison. Sections at FP show the shape in transverse plane. Some designers have a preference for bulbous bow that tapers sharply underneath, as this offers greater protection against slamming. The lower water planes also taper sharply, so that when the vessel is ballasted, the bulb has the same effect as a normal bow lengthened. This has the advantage of avoiding a build-up of additional resistance and spray formation created by the partially submerged bulb. These bulbs have V-shaped transverse sections. The teardrop-shaped bulb has volume moved downwards so that the half angle of entrance at load water line (LWL) is reduced and sections are not V-shaped but rather more extreme U-shaped. These are effective in reducing wave-making resistance and the frequency of slam. Cylindrical or ram-type bulbs are preferred where a simple building procedure is required and the potential danger of slamming effects can be avoided.

8.1.6.4 Forward Section Flare above Water

Stem more or less raked above water is a common designer's practice. In cargo ships, the forecastle sides can be flared to a maximum angle of about 40°. The advantage of a raked stem above water is that it deflects water easily, increases reserve buoyancy by providing extra length above water, gives better protection to the ship against collision and can be more attractive aesthetically.

Shipping company requirements often lead to a pronounced forward section flare above water so as to provide extra forward deck space to carry cargo, particularly containers. The deck area, having increased due to flare, can have portal crane tracks laid right up to the forward hatch. Flare also helps car and train ferries where there must be minimum entry breadth near the design water line within a limited distance from the stem.

Increased forward section flare helps deflect green seas and increases local reserve buoyancy, therefore reducing pitching amplitude and increasing the righting arm over a longer range of stability. The disadvantages of large flare may be that in a rough sea condition, the flared section produces spray. A flared above water fore body may lead to large pitching acceleration and slamming or impact loading of the fore body. Such a fore body also requires more material and more work to be produced.

8.1.6.5 Inverted Bow or X-Bow

The Ulstein Group of Industries developed the inverted bow concept in 2005 where the bow turns inwards (instead of being raked outwards) as it goes above water towards the top deck. So there is a bulbous projection at the water line level with the water lines being full as these move from the bottom to the water line level and reduce again as these move towards the top. So there is no outward flare towards the top. It is claimed that with this bow shape, the vessel behaves better in rough sea conditions and has a reduced speed loss compared to a normal ship form. Thus, in a voyage, the vessel reaches an average speed higher than the corresponding normal ship. The first vessel, *M/V Bourbon Orca*, won the Ship of the Year Award in Norway in 2006, reported in Ulstein Brochure (2005–2006). Figure 8.10 shows three different ships fitted with X-bow or inverted bow. Figure 8.10a shows a photograph of a vessel built with Ulstein X-bow. Rolls-Royce has a slight variation of this bow known as the wave piercing bow. Figure 8.10b shows such a built and operating vessel. An innovative application of this inverted bow concept was done by Tarjan and later by Morelli and Melvin in designing catamaran yachts (Tarjan 2007, Morelli & Melvin 2012). A photograph of such a catamaran vessel is shown in Figure 8.10c.

8.1.6.6 Sectional Area Curve

An important aspect of hydrodynamic design is the distribution of sectional area up to the load water line over the length of the ship. The geometrical significance of the curve is that the area under it represents the volume of displacement, the maximum ordinate of this curve is the midship area and thus one can obtain the block coefficient, the midship area coefficient and the prismatic coefficient of the vessel. The longitudinal centre of this curve

FIGURE 8.10
Inverted bow or X-bow shapes on different ships. (a) Ulstein X-bow. (b) Rolls-Royce inverted bow. (c) Inverted bow in a yacht hull form.

gives the position of the longitudinal centre of buoyancy. Figure 8.11a shows the various features of the SAC such as length of entrance, parallel middle body (PMB) and length of run. The curvature at the joining of the entrance and PMB represents the smoothness of the forward shoulder and similarly, the joining of PMB and run represents the smoothness of the aft shoulder. These shoulders dictate the formation of shoulder waves and their interference with other wave systems. An increased midship area would lead to a SAC with lesser length of PMB and smoother forward and aft shoulders. A reduced midship area would similarly show just the reverse effects.

8.1.6.7 Load Water Line

Another important factor in designing the forward lines is the shape of the design water plane. The area of the water plane gives the tonnes per centimetre immersion (TPcm) and the water plane area coefficient. The centroid of this water plane is the longitudinal centre of floatation (LCF). Figure 8.11b shows the various features of a typical load water plane. This area is representative of the transverse form stability aspect of the ship and also of the moment to change trim by 1 cm. One important characteristic property is represented by the half-angle of entrance i_e which is related to the section shape, SAC and ship's breadth. Table 8.2 shows the half angle of entrance as a function of C_P. These values of i_e are valid for $L/B = 7$. For different values of L/B, i_e has to be multiplied by the factor $7/(L/B)$. These recommendations are primarily applicable to ships without bulbous bow. Various resistance calculation methods also give the optimum angles of entrance.

8.1.6.8 Stern Forms

The choice of stern form should be based on criteria such as low resistance, high propulsive efficiency with uniform flow of water to propeller and good hull efficiency coefficient and avoidance of propeller-induced vibration.

In discussing stern forms, distinction should be made between the form characteristics of underwater part and the top side part of the vessel.

The above water portion of the stern form for merchant vessels could be elliptical or cruiser stern and transom stern. Cruiser stern is rarely used nowadays. The guidelines for transom stern design are presented in Table 8.3. In slow-speed operation, transom stern gives noticeably higher resistance than cruiser stern. But as speed increases, this difference reduces and perhaps gives less resistance than cruiser stern.

Separation at the stern is a function of ship's form and propeller influence. The suction effect of the single screw propeller causes the flow lines to converge. This suppresses separation. The effect of propellers on twin screw ships is conducive to separation. Separation is influenced by the radius of curvature of the outer shell in the direction of flow, and by the inclination of flow relative to the ship's forward motion. To limit separation, sharp shoulders at the stern and lines exceeding a critical angle (15°–20°) of flow relative to the direction of motion should be avoided.

The stern water lines above the propeller should be straight (hollows are to be avoided) in order to keep water line angles as small as possible. When adherence to the critical angle is impossible, greatly exceeding the angle over a short distance is usually preferred to marginally exceeding it over a longer distance. This restricts the unavoidable separation zone to a small one.

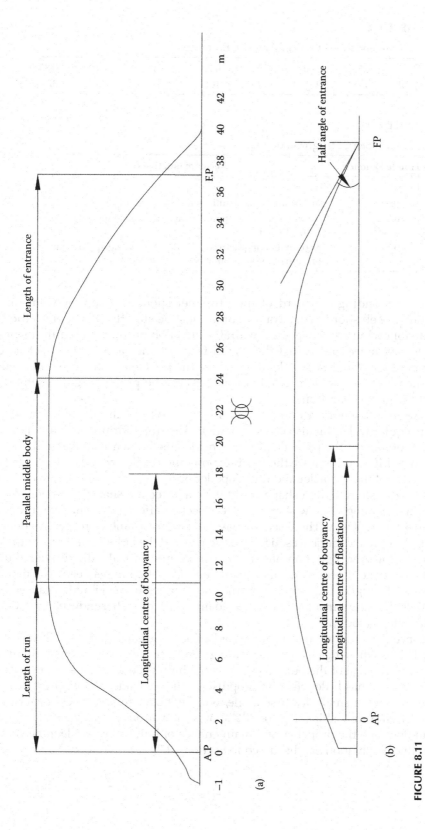

FIGURE 8.11
Sectional area and load water line. (a) Sectional area curve. (b) Load water line.

TABLE 8.2

Recommendations for Half Angle of Entrance

C_p	0.55	0.60	0.65	0.70	0.75	0.80	0.85
i_e (°)	8	9	9–10	10–14	21–23	33	37

TABLE 8.3

Transom Stern Design Guidelines

Froude Number	Recommendations
$F_n < 0.3$	Stern above L_{WL}.
$F_n \approx 0.3$	Small stern – only slightly below LWL.
$F_n = 0.5$	Deeper submerging stern with average wedge submergence $\approx$10%–15% of T.
$F_n > 0.5$	Deep submerging stern with wedge, possessing approximately breadth of the ship. Submergence $\approx$15%–20% of T.

The water line endings forward of the propeller should be kept as sharp as possible. The outer shell should run straight, or at most be slightly curved, into the stern. This has the following advantages – favourable effect on propulsion requirements and reduction of resistance and thrust deduction fraction and favourable effect on quiet propeller operation. The lines in the area where the flow enters the propeller must be designed such that the suction remains small. Here, the propeller gains some of the energy lost through separation.

A nonuniform inflow reduces propulsion efficiency. While diminishing propeller efficiency, an irregular wake can also cause vibration. The ship's form, especially in the area immediately forward of the propeller, has a considerable influence on wake distribution. Of particular significance here are the stern sections and the horizontal clearance between the leading edge of the propeller and the propeller post.

The underwater stern section shape could have a V section shape, U section shape or bulbous stern. On a single-screw ship, each stern section affects resistance and propulsion efficiency in a different way. The V section has the lower resistance, irrespective of Froude number. The U section experiences higher resistance and the bulbous stern form the highest resistance. However, V section has the most nonuniform wake distribution, bulbous stern form the most uniform wake distribution and U section in between. The more uniform the wake, the higher the propulsion efficiency and lesser the vibration caused by the propeller. Bulbous sterns, installed primarily to minimise propeller-induced vibration, are of particular interest today.

The influence of stern form on propulsive efficiency is greater than its influence on the resistance. This is why single-screw ships are given U or bulbous sections rather than V form. The disadvantage of the bulbous stern is the high production cost. The stern form of twin-screw ships has little effect on propulsive efficiency and vibration. Hence, the V form aft sections with small deadrise angle, with its better resistance characteristics, is preferred on twin-screw ships. Figure 8.12 shows the various stern types. Stern design for uniform flow into the propeller and maintenance of hull propeller clearance to avoid propeller-induced vibration are discussed in Section 7.2.3.

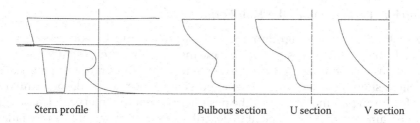

FIGURE 8.12
Stern profile with different types of sections.

8.2 Geometrical Design

The design of a ship hull is based on a specific geometrical definition of the hull. This shape largely influences the general arrangement, hydrostatics and stability, dynamics, strength and, last but not least, the aesthetics of the vessel. In designing, a naval architect applies his engineering knowledge and draftsmanship to arrive at the optimum hull form.

The unique geometrical characteristics of a ship hull form begin traditionally with an empirical shape rather than a simple geometric form like a sphere, cylinder or any other mathematically defined shape. This has led to rather specialised hull form design practices when applied to lines creation and fairing. An important aspect of preliminary design is to create a set of faired ship lines. At this initial stage of design, the naval architect has to develop a set of ship lines of sufficient accuracy to be used for later design calculations from relatively sparse information. The resulting lines plan must not violate any of the implicit, and sometimes intuitive, rules of hull form design with respect to fairness and practicability.

Today, through the use of computers, analytical methods for ship lines creation have been extended well beyond systematic series development and hull form research, and have become accepted as practical tools in everyday design work. Mathematical methods for ship form design, especially when supported by computer systems with interactive graphics capabilities and skillfully applied by experienced designers, have become competitive to conventional graphical techniques with respect to time and cost. This new orientation in ship lines creation has somewhat shifted from the systematic to the automatic, but fundamentally the problem remains the same, and many established mathematical procedures can be adapted to computer use. Other procedures from computer graphics techniques and related industries have also been used. In a ship design office, the approach for the development of ship lines for a new design may be one of the following:

- *Form parameter approach*: from geometrical hull form parameters of the new design
- *Lines distortion approach*: from a single successful parent design
- *Standard series approach*: from a series of parent designs

8.2.1 Principal Parameters of the Hull Form

To apply mathematical or diagrammatic techniques to ship lines creation, a system of descriptive form parameters is needed by means of which the designer can express, as uniquely as possible, the desired properties of any design curve. For practical reasons, such a system should be kept as simple as possible; hence, it should be limited to those relatively few form parameters which the designer can readily interpret. The smallest possible set of parameters that is sufficient to describe a desired range of form variations is the best for this purpose. The following set of hull form parameters, discussed earlier, is fairly universally accepted in ship design practice:

$$L_{WL}, L_{BP}, B, T, D, \nabla, C_B, A_X, C_X, C_P, LCB,$$

SAC including lengths of entrance, PMB and length of run and load water line including length on water line and half angle of entrance.

In this context, it must be recognised that the set of form parameters discussed, or any other finite set of parameters, can by no means be sufficient to specify an arbitrary form completely in all local details. With all the given form parameters fixed, a curve can still vary locally, giving an experienced designer some personal freedom over the nature of the shape of the desired curve. In a similar manner, it must be understood that some subtle and complex changes in the hull form can also be affected by simple form parameter changes. The general purpose of mathematical lines creation procedures is to reduce the design of complex shapes to a sequence of simple steps based on form parameter design.

8.2.2 Form Parameter Approach

In this approach, lines are created according to specified values of the parameters that define the significant curves of the hull form. Traditionally, the design of a ship surface by naval architects has been reduced to a sequence of curve design problems. Therefore, in initial hull form design, a system of curves is first developed, e.g. sectional area curve and design water line curve. Each basic curve is developed from its own form parameter inputs.

Prior to starting further design activity, it is necessary to determine the types of fore body and aft body one prefers to have. Correspondingly, the profile outline can be drawn as the first step. To draw a preliminary SAC, one has to determine the midship sectional area first. This is done by choosing a suitable midship area coefficient, as discussed earlier. Then, $C_P = C_B/C_M$ is determined. The next step is to determine a suitable longitudinal centre of buoyancy (LCB) position from a hydrodynamic point of view. Lewis (1989a) gives an indication of likely LCB position. One way to draw a SAC is to draw a trapezium and then, smoothen it to get the desired shape, as shown in Figure 8.13. In this diagram, suffices E, M and R represent entrance, PMB and run, respectively and X is the location of the corresponding longitudinal centroid position. Alternatively, one can draw an approximate SAC from published data like the one shown in Lewis (1989a).

In a manner similar to drawing the SAC, the half-breadth plan of the LWL can be drawn to satisfy water plane area A_{WL} and LCF requirement. It must be remembered that the lengths of entrance and run on L_{WL} are less than the L_E and L_R on SAC. The other important

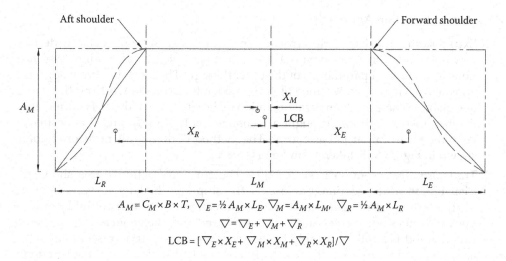

$$A_M = C_M \times B \times T, \quad \nabla_E = \tfrac{1}{2} A_M \times L_E, \quad \nabla_M = A_M \times L_M, \quad \nabla_R = \tfrac{1}{2} A_M \times L_R$$

$$\nabla = \nabla_E + \nabla_M + \nabla_R$$

$$\text{LCB} = [\nabla_E \times X_E + \nabla_M \times X_M + \nabla_R \times X_R]/\nabla$$

FIGURE 8.13
Geometric generation of sectional area curve.

criterion for LWL is the half angle of entrance based on C_P, as given in Table 8.1. A deck outline plan can then be drawn, the main consideration being the desired deck area and the type of flare above L_{WL}.

The next step is to draw the section shapes. Figure 8.14 demonstrates how to draw any section satisfying area and L_{WL} requirements. Once all sections are generated, half-breadth plan and buttock lines are drawn and by an iterative process, surface fairing is done to generate a three-dimensionally smooth surface.

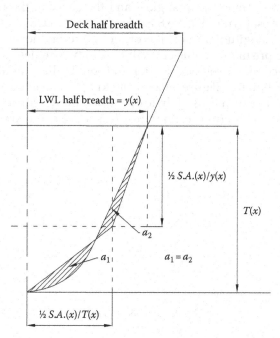

FIGURE 8.14
Geometric section shape generation.

8.2.3 Lines Distortion Approach

1. Extrapolation from basic ship – In a design office, it is a common practice to derive a new lines plan from a given parent design by deliberate but often minor modifications in some form parameters of the parent design. The lines distortion approach aims at arriving at a new lines plan by moderate extrapolation from the parent design by suitable mathematical operations. The simplest method of extrapolation is when the breadth and draught are changed only slightly, and it is desired to keep the form similar to basic form. The method of doing this transformation is shown in Figure 8.15, taken from Lewis (1989a).

2. Transformation from basic ship SAC – The curve of sectional area shows the distribution of the sectional area over the ship length. This curve is the base for the set-up of the lines drawing. If the position of the longitudinal centre of buoyancy does not correspond with the desired one, the general known correction method is applied, namely transformation of the curve of sectional areas, as shown in Figure 8.16. The dotted line is the transformed curve of sectional area. For small angles α, the area under the curve is almost unchanged and corresponds with the volume of the ship below the load water line, i.e. the area between this curve and the axis is unchanged, and also the prismatic coefficient is not changed. The new sectional area of the design ordinates can be read-off out of this figure. Instead of calculating the new sectional areas of the design ordinates, the distances over which the original design ordinates are moved (a_j) are calculated.

3. If a new ship form has to be designed, starting from a basic ship form, in many cases not only the position of the centre of buoyancy will change but also the prismatic coefficient. To make a ship fuller, the curve of sectional areas of the afterbody is transformed by an angle α and the curve of the fore body with an angle α ($\alpha > 0$), as shown in Figure 8.16a for change of LCB. A combination of the changing of the longitudinal position of the centre of buoyancy and increasing or reducing the prismatic coefficient will lead to two angles, α_a and α_f, by which the curve of sectional areas of the after and fore bodies are to be transformed, respectively, as shown in Figure 8.16b. By making both half curves dimensionless in x and y directions by means of dividing by half the length between the perpendiculars and the main cross section area, respectively, the area of each of these

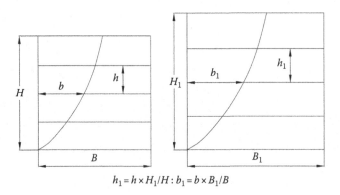

$$h_1 = h \times H_1/H : b_1 = b \times B_1/B$$

FIGURE 8.15
Change of section shape by linear proportions.

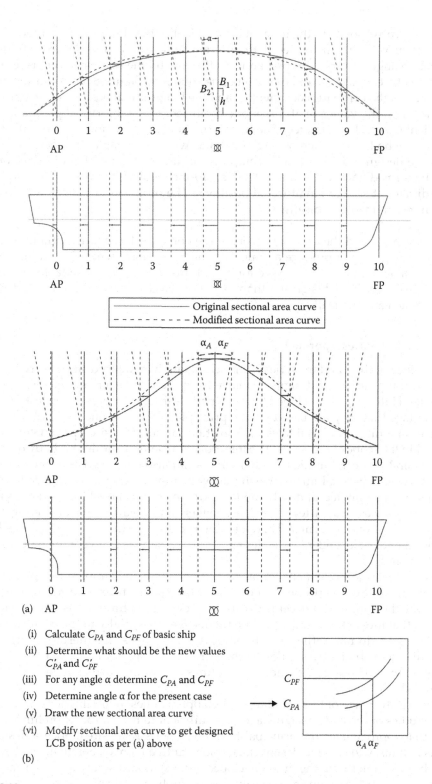

(i) Calculate C_{PA} and C_{PF} of basic ship

(ii) Determine what should be the new values C'_{PA} and C'_{PF}

(iii) For any angle α determine C_{PA} and C_{PF}

(iv) Determine angle α for the present case

(v) Draw the new sectional area curve

(vi) Modify sectional area curve to get designed LCB position as per (a) above

(b)

FIGURE 8.16

Transformation of sectional area curve. (a) Change of LCB. (b) Change of CP and LCB.

curves corresponds to the prismatic coefficient of the ship part concerned. When these curves are transformed with different angles α, a relation between transformation angle α_a versus aft prismatic coefficient and the relation α_f versus forward prismatic coefficient can be obtained, as shown in Figure 8.16b. With the given values of the prismatic coefficient C_P and the longitudinal position of the centre of buoyancy, the values of aft prismatic coefficient C_{PA} and forward prismatic coefficient C_{PF} can be estimated. The transformation angles α_a and α_f for the after and the fore bodies can then be computed and with these angles, the new positions of the design ordinates can be computed. In this way, a number of points of the transformed SAC can be found. The transformed SAC will have ordinates equal to ordinates at shifted longitudinal positions. The ordinates at equidistant stations can be obtained graphically.

The transformation of the curve of sectional area of a fine basic form without a cylindrical midship region will produce a knuckle at half the length when C_P is transformed to a smaller value than that of the basic form, and this method is not suitable for this case. If the prismatic coefficient is greater than that of the basic form, a cylindrical midship part will be created automatically.

8.2.4 Standard Series Approach

The basic idea of the standard series approach is simply to interpolate a desired new hull form within the variety of designs available from the rich experience of systematic hull form series. Hull designs developed for systematic resistance or propulsion test series offer the advantage that they are based on the principle of varying one significant hull from parameter at a time. This facilitates interpolation for newly desired parameter combinations within the scope of the series. The interpolation method within this body of information is generally linear, but higher-order curves can also be used. The range of variation in the series is, however, limited, which means that the forms that can be deduced from the series are also limited and only modest extrapolation is possible. Besides, in most of the standard series, only a few form parameters can be varied independently, the rest being dependent on the former. Therefore, interpolation of the new hull form is limited to the number of independent variables and the designer has to accept the outcome of the dependent ones.

Standard series approach encompasses only some of the simpler variations in hull form with respect to the proportion of main dimensions, fullness and, in some cases, location of the longitudinal centre of buoyancy. Other limitations of the standard series are that most of the series confine themselves essentially to the underwater part of the hull and do not define the shape of the main deck. Furthermore, the parents are not always sufficiently similar or the variations may not be close enough to rely entirely on the fairness of the interpolated lines, and some corrective fairing may still be required.

The main advantage of this standard series approach lies in its simplicity. In many feasibility studies of the early design stage, it is sufficient to have a rough preliminary lines design that approximates the principal form characteristics desired. Such a design may serve as a dummy for several design calculations until a more elaborate lines plan is produced. In this connection, the lines developed from standard series approach can, therefore, be very useful. Also one can use any successful lines distortion technique on parent design from standard series.

8.3 Computer-Aided Design of Hull Form

As has been discussed, generation of preliminary hull shape is a complex procedure involving decision making almost at every step of form generation starting from choosing the parent form and guidelines until the local curvatures. It also involves preliminary fairing of curves in two dimensions and broad matching of water lines, buttocks and sections. Also the shape should satisfy the functional requirements like displacement and LCB position. If done manually, this process becomes very time-consuming and it excludes the possibility of studying alternate hull shapes. Also only 2D fairing can be done by going from one view to another on a reduced scale. By this process, the full-scale fairing is done in a mould loft at a later stage. Even then, generation of form data for CAM application is not possible.

A number of CAD packages including graphics are available for use by hull form designers either for preliminary development of lines plan or for complete 3D surface fairing of rough lines plan and generation of required graphical documents (lines plan drawing) and numerical output for CAM applications. Such a flow chart of lines plan development is given in Figure 8.17, showing the different stages of lines plan development presented in Figure 8.1. Figure 8.18a shows the lines plan (profile and body plan only) in two dimensions, and Figure 8.18b shows the wire mesh diagram of the same vessel in three dimensions generated using Cad software tools. Figure 8.19 shows the wire mesh diagram of a bulbous ship surface which has no knuckle line and has been designed as one smooth surface.

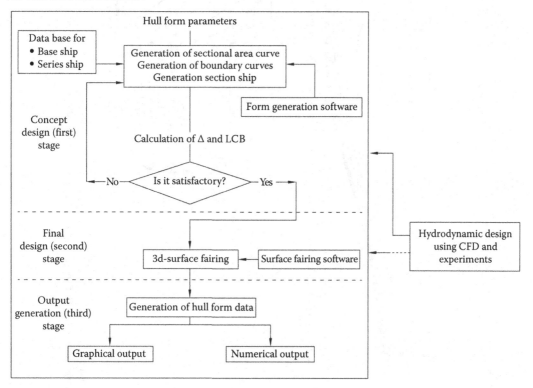

FIGURE 8.17
Flow chart for hull form information generation.

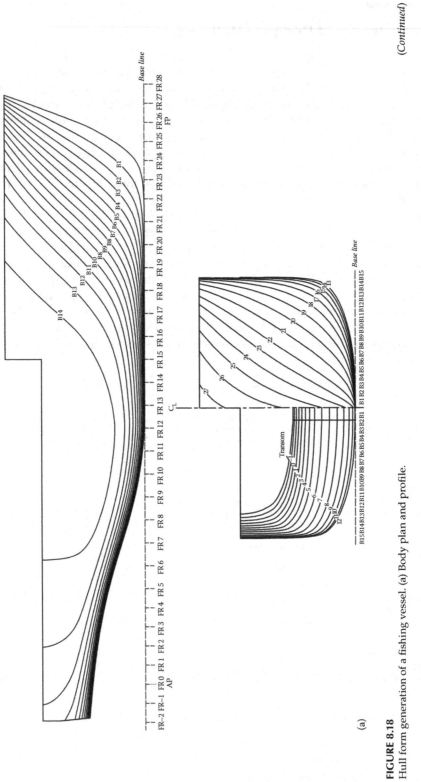

(a)

FIGURE 8.18

Hull form generation of a fishing vessel. (a) Body plan and profile.

(Continued)

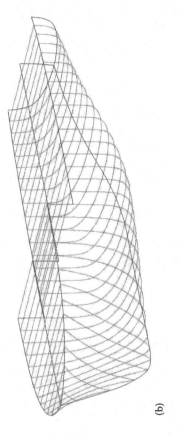

(b)

FIGURE 8.18 (*Continued*)
Hull form generation of a fishing vessel. (b) Wire mesh diagram.

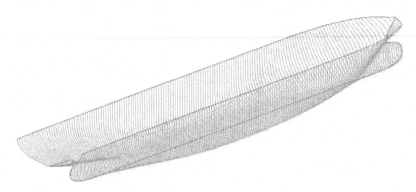

FIGURE 8.19
Wire mesh diagram of a merchant vessel hull form.

9

Machinery System

A marine platform, may it be a vehicle or structure, requires a main machinery system to perform its main operational task at sea. It also requires a lot of auxiliary machinery to support the work of the main machinery as well as to do other auxiliary services required for operation, maintenance and emergency services of the platform. These main and auxiliary machinery and equipment form small subsystems of their own which must be connected by pipelines and other means to form bigger subsystems and finally become an integral part of the overall platform system working efficiently and effectively together. The machinery and equipment and the pipeline details have been described in marine engineering books such as Harrington (1992), Taylor (1990) and Rowen et al. (2005). The main focus of this chapter is the integration of the main and auxiliary equipment with the platform and its operation.

Normally, marine platforms use fossil fuels as the main source of energy. But recently, concerns for the environment have changed the focus from the use of fuel oil to the use of alternative fuels. This will be discussed in Chapter 13. In this chapter, the discussion is based on the use of oil fuels. Marine vehicles, particularly ships, use the main machinery for propulsion purposes and this is commonly known as propulsion machinery. In the present chapter, the discussion is based on ship propulsion and its auxiliary and support systems. The discussion can be easily extrapolated to stationary or temporarily stationary marine platforms.

9.1 Main and Auxiliary Machinery and Equipment

Marine diesel engines normally use heavy fuel oil (HFO) which is the residual oil left after certain distillates have been extracted from crude oil. The impurities and chemical composition of crude oil vary based on location of extraction of oil from the ground. Some diesel engines such as medium or high speed ones use light diesel oil of different grades. The impurities in HFO include water and sediments, ash, fibres and oxidation products such as gums and varnishes, apart from alumina salts, silica, hydrogen sulphide and elements such as sulphur, sodium and vanadium. The impurities reduce in quantity as the fuel used changes to light diesel oil. The density of HFO is about 991 g/m^3 at 15°C. The flash point of marine diesel oil (MDO) is normally taken as 16.7°C (62°F) which makes it a dangerous liquid to carry in bulk. The viscosity of marine fuel oil is high ranging from 180 to 700 centistokes at 50°C, increasing with impurities content. Thus, HFO can have a viscosity of 700 cst. The pour point of fuel oil is 30°C, thus requiring heating when transported from one location to another inside or outside the ship. A ship's fuel oil is normally stored in steel tanks inside the ship (until recently, in double bottom tanks) which require to be heated by passing steam through the tubes laid in the tank (electrical heating is not encouraged) for transportation to the oil treatment system before going to the day tank for feeding to the main engine. The oil treatment system consists of filters, centrifuges (purifiers) and settling tanks. The cleaned oil is stored in the day tank to be fed to the engine as and when required. This treatment system generates a dirty oil or oily water mixture and sludge which must be stored, disposed or treated as required. Starting the diesel engine is affected by the supply of compressed air and so the ship must carry air compressors and compressed air bottles including emergency facilities. There is also a closed loop lubricating oil system for lubrication of engine parts which must be cooled and recirculated. Similarly, the engine is cooled by fresh water which must, in turn, be cooled by sea water. A part of the energy contained in the exhaust gas could be reused by utilising it to run an exhaust boiler running a turbine to generate electrical power. Thus, it can be observed that a diesel engine installation requires a large number of auxiliary machinery which is run by electricity. Thus, a part of the electrical power generated by the diesel generator (DG) sets is utilised for the auxiliaries running the main engine and the diesel engines of the DG sets. Figure 9.1 shows a typical line diagram showing the connection between various components required for running a ship's diesel propulsion plant. The exhaust gas has an exit temperature of 300°C–490°C and has a lot of energy content. The engine cylinders and the fuel nozzle get heated and require to be cooled by fresh water. The fresh water, on exit, after cooling the engine components and similarly other components of auxiliary machinery, requires to be cooled, and this is done through a heat exchanger by circulating sea water around the fresh water tubes. An engine designer or manufacturer is faced with a dilemma as to how much the engine will be cooled by fresh water. If the engine is cooled more, more heat must be generated inside the engine for the same power; on the other hand, if it is not cooled, then the engine is likely to overheat. Normally, the inlet of fresh water to the engine is about 70°C and the outlet after cooling is 85°C. Fresh water, being rare at sea, must be recirculated. Fresh water also has other uses, particularly for domestic purposes such as cooking and sanitary. Fresh water is also used as a boiler feed water for the generation of steam. It is not possible to carry all the fresh water required for the entire duration of the journey including in emergency situations. Therefore, ships generally carry a fresh water generator to provide fresh water from sea water on a daily basis. Similarly, sea water has multiple uses at sea. Sea water is taken from the sea through a sea chest and is used for cooling the fresh water and normally discharged to the sea. This heated sea water contains a lot of waste heat, the temperature being around 45°C which goes out to the sea. Sea water is also used directly for cooling some auxiliary parts and bearings. Sea water is used in many ship systems such as provision for fire fighting by providing deck lines on port and starboard sides on the main deck as well as on all accommodation decks and also for domestic purposes. This line could also be used for general purposes including cleaning and deck wash. Since sea water and fresh water are used for domestic purposes, this supply is made through hydrophore tanks to ensure constant flow at all times. There is an entirely separate pump and pipe line system of sea water for filling and emptying ballast tanks at ports as per journey requirement. This system is also used for ballast water management at sea as discussed in Chapter 13. A line diagram of sea water and fresh water system is shown in Figure 9.2.

Since main and auxiliary engines run on diesel oil, starting of such engines must be done by providing compressed air at 30 bar pressure to a few cylinders of the engine such that the corresponding pistons are initially turned by the compressed air and combustion starts. Also while running, it is necessary to provide compressed air at 7 bar pressure for control of the combustion process. Thus, the ship has to carry air compressors and air bottles of sufficient capacity for effective engine operation. She must also carry emergency compressor and air bottles. Normally, classification societies stipulate the requirement of compressed air for main engine operation including emergency operation. Compressed

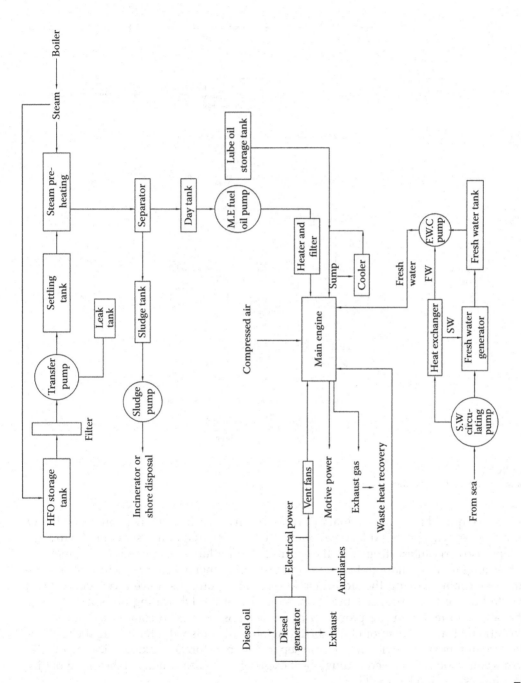

FIGURE 9.1
Components of a typical marine propulsion plant.

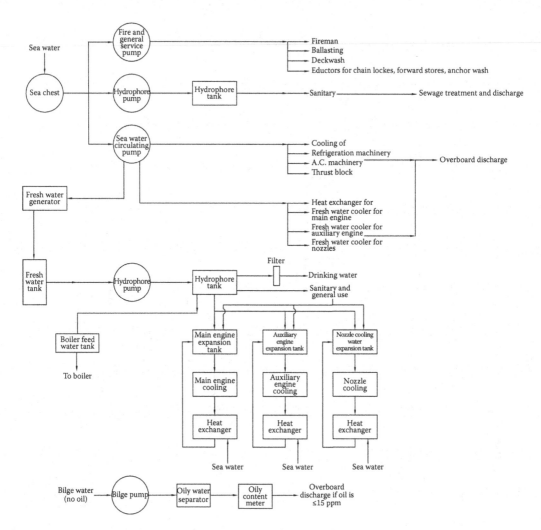

FIGURE 9.2
Sea water and fresh water lines.

air is also required for some auxiliary purposes such as use in workshop, provision in CO_2 room and for hydrophore tanks. The compressed air line diagram is shown in Figure 9.3.

Ships also carry lubricating oil in the engine sump for lubrication of engine components such as engine crank shaft and piston cam shafts and other auxiliary machinery and systems, one example being the stern gland. Lubricating oil must have a certain viscosity level to be able to do proper lubrication. Viscosity of the lubricating oil is lost through lubrication. To maintain the proper viscosity level as well as to compensate for leaked lubrication oil, a certain amount of lubricating oil is replenished after testing the used oil. This amount must be carried in the ship separately in a storage tank. Lubricating oil, like fresh water, should be cooled through heat exchangers by sea water. A lubricating oil line diagram is shown in Figure 9.4.

It can be observed that the diesel engine system has to carry a lot of auxiliary equipment to function properly. Furthermore, it may be noted that slow-speed engines (up to 400 RPM) can run on HFO, whereas medium-speed engines (up to 1500 RPM) require lower viscosity

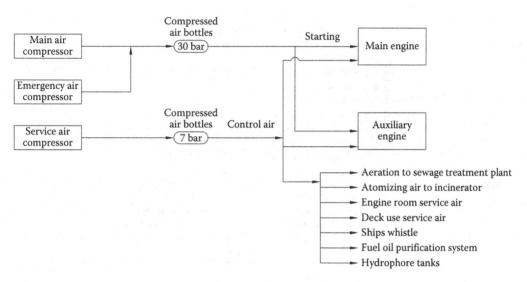

FIGURE 9.3
Compressed air line.

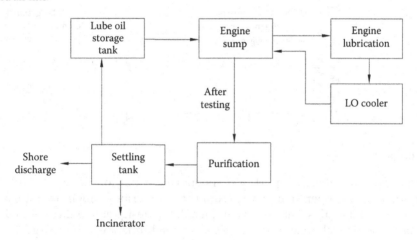

FIGURE 9.4
Lubricating oil line.

MDO for their operation. Since diesel engines required for electrical power generation are medium-speed engines, they require MDO. Thus, marine vehicles work on a two-fuel system which increases the number of auxiliary equipment. This type of operation is expensive since two oils are required to be carried in separate tanks. Efforts have been made to make ships work on one fuel by redesigning medium-speed diesel engines to run on HFO. Other alternatives include the use of a shaft generator by taking off power from the main propulsion engine to generate electricity to cater to the needs of the sea load which may reduce (if not eliminate) carriage of MDO. Another alternative, considered actively by many stake holders, is to use lighter oil or oil of lower viscosity for both main propulsion and auxiliary power which reduces SO_x emission. Of course, if diesel–electric propulsion is installed, then automatically the two-fuel system is eliminated.

A proper load analysis must be carried out to determine the electrical power required for the platform in question. In the case of a ship, electrical load varies depending on the

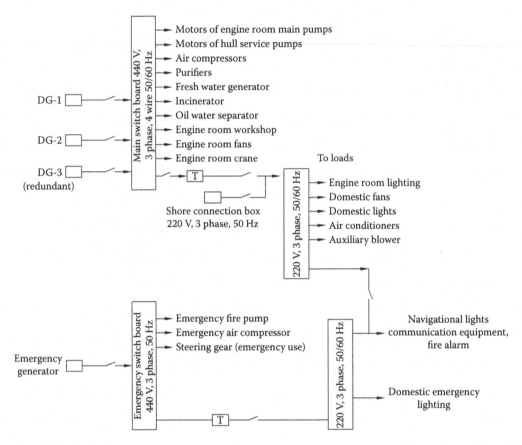

FIGURE 9.5
Electrical distribution system.

operating condition – at sea, in port or manoeuvring conditions. Power could also vary based on night or day operation. Harrington (1992) describes a method of load analysis for ships. Once load analysis has been completed, the designer should able to determine the number of electrical generator sets including redundancy and emergency requirement. Normally, if refrigerated cargo is carried, such as cargo in a refrigerated hold or in refrigerated containers, one could provide a separate electrical power supply system for cargo alone with the arrangement of redundancy. Figure 9.5 shows a typical electrical power distribution system with the necessary redundancy and emergency requirement. In actual installations, the diagrams change as per design requirement and component availability.

9.2 Energy Consumption Pattern

A ship's energy consumption pattern is quite complex. The primary objective of a merchant ship is to carry a certain amount of cargo from point A to point B. A ship is, therefore, a housing for the cargo and the auxiliaries and personnel required for this transportation. Assuming the ship form to be optimal, the minimum energy requirement is, therefore,

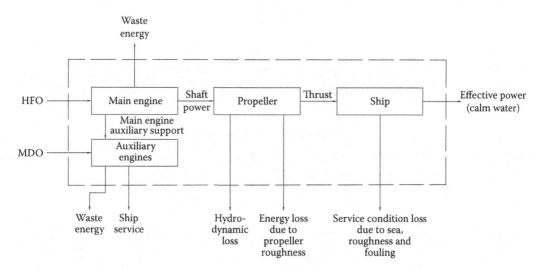

FIGURE 9.6
Energy flow in the overall ship system.

the effective power of a clean new freshly painted ship as predicted from tank tests. The energy input to a ship is in the form of stored chemical energy in fuel (generally, HFO) for the main propulsion plant and MDO for the auxiliary systems such as electrical power plants, and steam and fresh water generating systems. The input energy is first converted to useful energy or effective power through a hydrodynamic device such as a propeller. This is shown diagrammatically in Figure 9.6.

The energy input to the main engine is from the diesel oil and also from the electrical energy provided to various auxiliary systems for running the main engine. The useful output energy is in the form of motive power in the shafting system. The losses are mostly in the form of heat energy in exhaust gases, scavenge air, cooling water, lubricating oil and radiation losses. An energy balance can be carried out to ensure that the input energy is equal to the output energy and then the thermal efficiency of the main engine can be calculated as useful output/input energy. This is normally between 49% and 51% in a modern diesel engine. The thermal efficiency of a DG set is slightly less, about 45%, due to losses in the generator set. The energy efficiency of the total propulsion engine system including auxiliaries is about 48%–50%. Normally, the shafting efficiency is between 95% and 98% based on the number of bearings and reduction gear box, and the propeller efficiency is between 55% and 60%. Assuming a clean newly painted ship, the energy efficiency of the total ship system is between 25% and 30%. Figure 9.7 shows the energy flow in diesel propulsion machinery and Figure 9.8 shows the energy flow in a DG set. Table 9.1 shows the energy efficiency of the ship machinery system. As the ship goes into service, the effective power changes due to fouling and corrosion, the propeller efficiency reduces due to propeller roughness and the engine may also deteriorate reducing its thermal efficiency. There may also be a loss of speed due to rough sea conditions. Thus, the energy efficiency continuously reduces as the ship goes into service until drydocking in which the efficiency goes up but never to the original new ship level.

Steam turbines for marine use require that a boiler burns fuel oil which generates steam and, in turn, runs a steam turbine for a geared propulsion shafting system. Similarly, one could also use a gas turbine run on an air boiler in burning fossil fuels. A turbine system is simpler than a diesel engine system since it has less number of auxiliary

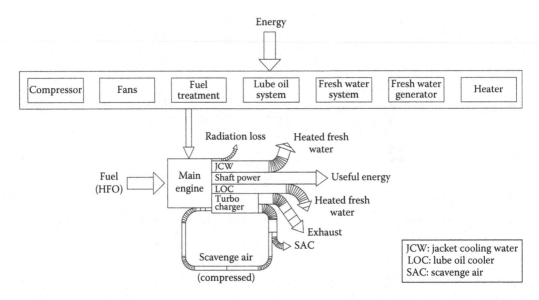

FIGURE 9.7
Energy flow in a diesel main propulsion system.

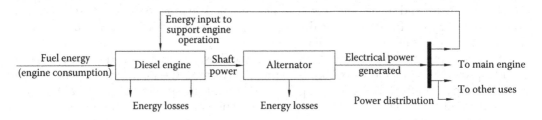

FIGURE 9.8
Energy flow in a diesel generator system.

TABLE 9.1

Overall Energy Efficiency of Diesel Engine–Driven Ships

Item	Efficiency (%)	Comment
Marine diesel engine	49–51	Thermal efficiency
DG set	45.00	Diesel engine and AC generator
Main diesel engine with support system	48–50	Support is the electrical power to auxiliary systems
Shafting system	95–98	Based on the number of bearings and gears
Propeller	55–65	Depends on wake distribution and propeller loading and design
Hull	100.00	Optimised best hull form having a minimum effective power in trial condition
Overall energy efficiency with diesel engine in trial condition	25–30	Effective power (trial)/chemical power input to ship through oil

TABLE 9.2

Thermal Efficiency Range of Different Engine Types

Type of Engine	Capacity Range (MW)	Thermal Efficiency Range (%)
Slow-speed diesel engine	2–52	46–55
Medium-speed diesel engine	1–10	40–50
Gas turbine	2.5–100	22–38
Steam turbine	15–500	27–40
Combined cycle gas turbine	15–500	35–50

components consisting of feed pump for water (compressed air supply system in the case of a gas turbine), boiler with safety valve, multistage turbines and condenser. Also, it is easier to combine a number of turbines to a single system. Steam and gas turbines are normally used in naval vessels to satisfy the large power requirement even if the thermal efficiency of these engines is lower. Table 9.2 shows the comparative thermal efficiencies of various engines with varying capacities in megawatts (MAN B&W May 2009). This is

TABLE 9.3

Conventional Propulsion Machinery Alternatives

Type of Propulsion Mechanism	Shafting System	Application
Diesel engine – slow speed	Direct drive	Most commonly used for merchant vessels
Diesel engine – medium speed	Geared drive	Most commonly used for merchant vessels
Diesel engine – high speed	Geared drive	Small craft
Boiler and steam turbine	Geared drive	Merchant and naval vessels
Air boiler and gas turbine	Geared drive	Naval vessels and merchant vessel application is being explored
Diesel engine and electrical motor	Geared/direct drive	Merchant and naval vessels
Electrical motors from electrical power generation	Geared/direct drive	Merchant and naval vessels and small craft
Electrical motors with batteries	Geared drive	Submarines and underwater vehicles
Multi-diesel engine or electrical motor or steam turbine/gas turbine systems	One/two or more engines connected to one or two or three shaftings through gear mechanism	Merchant and naval vessels, submarines, small craft
COGAG, COGOG	Combined gas and/or gas turbines – gas and gas turbines work singly or together depending on load demand	Naval surface vessels
COGAS	Combined gas and steam turbines – gas turbines and steam turbines work singly or together depending on load demand	Naval surface vessels
CODAG, CODOG	Combined diesel engines and/or gas turbines – diesel engine and gas turbines work singly or together depending on load demand	Naval surface vessels
Main propulsion supported by auxiliary power	Main propulsion plant provides thrust reduced by that provided by auxiliary power source	Merchant vessels and small craft

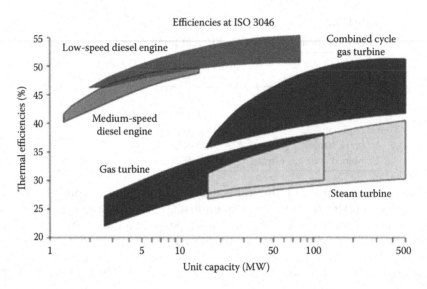

FIGURE 9.9
Thermal efficiency of different engine types.

also shown in Figure 9.9. As can be seen, diesel engines have the highest thermal efficiency, and therefore, in spite of its many disadvantages, it is the preferred machinery in ships due to its cost implications. Naval vessels also operate on multiple screws to cater to different load requirements such as normal operating conditions at cruising speed and high-speed conditions during offence or defence. Often it may be required to combine different types of power plants in the same vessel to satisfy the operating requirement. Table 9.3 presents the different conventional propulsion systems for merchant and naval vessels which is shown diagrammatically in Figure 9.9.

There is a continuous attempt to improve the thermal efficiency of the main propulsion and auxiliary systems by various methods. There have been improvements in diesel engine design and manufacture such that the fuel injection is controlled and optimised based on the actual load on the engine. Furthermore, for variable load conditions, some consideration may be given to alternative diesel–electric propulsion system. In that case, a part of the electrical power generated could be routed through the main switch board for electrical power supply for uses other than propulsion. Even in a conventional diesel drive system, power could be taken off from the main engine to run a shaft generator to provide electrical power even if there is no propulsion power requirement. Partial utilisation of waste heat in the exhaust gas can be done by passing it through an exhaust gas boiler which, in turn, could run a multistage steam turbine running an alternator feeding to the main power supply system of the ship. A part of the exhaust gas could also run a multistage power (gas) turbine which could also run the alternator generating electrical power. Taking reference of MAN B&W&Turbo (August 2009b), a block diagram showing the use of both steam turbines and gas turbines running on exhaust gas to generate electrical power is shown in Figure 9.10.

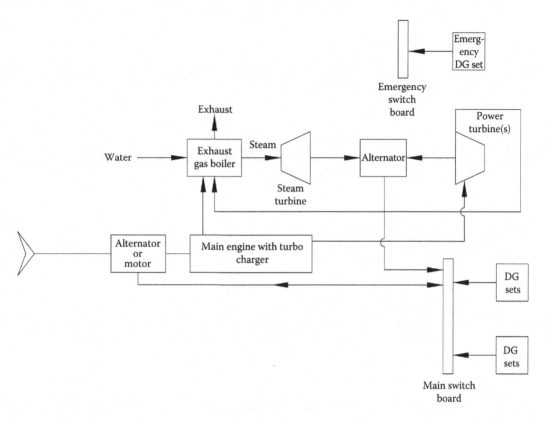

FIGURE 9.10
Suggested system for optimum energy uses of a diesel engine.

10

Structural Design

Structural design of a marine structure or vehicle, floating, submerged or fixed, includes the structural layout and determination of geometrical properties, commonly known as scantlings, of each structural element, each element being interconnected with other elements. The basis of this design is the mechanical properties of the material used such as yield strength, ultimate strength, compressive strength, elongation characteristics and plastic deformation, fracture and fatigue characteristics and other properties. The mechanical properties of various materials are discussed in the next section. The loads that come on the ship or marine structure are varied in nature and this is discussed subsequently. Next we discuss the design principles of structural layout followed by the determination of scantlings.

All non-naval marine structures and vehicles confirm to codes and practices set by various classification societies. Military vessels follow similar codes set up by national naval codes or by approved classification societies. Marine platforms, being high-value items, require to be insured by an insurance company against damage or loss and one of the main reasons for the same is structural failure at sea. Codes are necessary to ensure that the vehicles and structures follow independent structural standards so that the insurer can insure the product without seeking to justify structural adequacy. The classification societies certify the structural design if the codes are adhered to; they also certify the construction of the vessel if it is done under their survey. Classification societies also register vessels during their life time if these are maintained as per their guidelines and remain under their survey. The international classification societies include the Lloyd's Register of Shipping (LRS), American Bureau of Shipping (ABS), Bureau Veritas (BV), China Classification Society (CCS), Croatian Register of Shipping (CRS), Det Norske Veritas – Germanischer Lloyds (DNV-GL), Indian Register of Shipping (IRS), Korean Register of Shipping (KR), Nippon KaijiKyokai (Class NK), Polish Register of Shipping (PRS), Italian Classification Society (RINA) and Russian Maritime Register of Shipping (RS). All classification societies have their own codes and standards which differ from each other based on their experience and knowledge base. The International Association of Classification Societies (IACS) is aware of this fact and it has been making efforts to standardise codes for all societies. The IACS has now been able to develop common structural rules (CSRs) for bulk carriers and tankers which have been incorporated into the codes of all its members. The effort towards unifying these codes for all other vehicles and structures is in progress. Since these code books give detailed guidelines on material properties, layout design as well as scantling calculation, the structural design aspects here deal mainly with the main properties and design principles and details can be found in classification code books or structural design texts such as Muckle (1967), Lewis (1989a), Hughes and Pike (2010) and Mansoor et al. (2008).

Any marine structure is designed such that the material resists the environmental load without failure. But both load and resistance of the material are uncertain and deterministic approach of structural analysis may not be safe always. Therefore, reliability-based design has evolved. If load is greater than resistance, failure occurs and the structure should be designed for the limit state of zero failure probability. Ships have traditionally been designed based on a standard safety factor with deterministic loads and resistance, the preliminary (only for bulk carriers and tankers) reliability-based design being introduced only recently. But offshore structures which stand in the marine environment for a long period and are subject to large environmental loads are being designed for limit state conditions for a long time. In this chapter, discussion is focussed on safety factor-based design of ship structures and reliability-based limit state design concepts are introduced later which may be utilised for offshore structure design.

10.1 Marine Structural Material

Selection of construction material for marine fixed, floating platforms, ships and submarines is an important aspect of structural design. The selection process depends on a number of parameters such as

- Strength-to-weight ratio – Marine structures and vehicles are generally weight sensitive. The displacement for a vessel being constant, higher weight of material means lower payload. Submersibles and submarines are particularly weight sensitive. In vehicles and structures where strength of the platform is a major consideration, such as deep-diving submarine or decks of large ships, higher strength-to-weight (or density) ratio means lower material weight.

- Fracture toughness – This property is known as notch toughness. This is a measure of material's ability to absorb energy before plastic deformation leading to fracturing. This also means ability of a material to resist brittle fracture in the presence of a notch.

- Fatigue strength – Fatigue failure can occur in a material due to low-cycle high-stress fatigue, high-cycle low-stress fatigue or corrosion fatigue leading to stress corrosion cracking. The mechanism of fatigue is complex, failure being initiated due to generation of a small crack which grows to a major failure under repeated cyclic loading.

- Ease of fabrication, weldability and ease of maintenance – Welding causes a lot of problems in high-strength steels, aluminium and titanium. An additional problem is hydrogen induction into the metal leading to hydrogen embrittlement.

- Cost and availability – Cost of material depends on unit cost of material and the quantity of material used. Generally, in steels of various types, unit cost is quite high so that a small reduction in weight may not have significant effect on cost. So using higher strength steels is expensive and they can be used only if other technology considerations predominate. Titanium alloy is very strong though very expensive and used in submarine construction. Aluminium alloy are much lighter than steel though unit cost is higher.

Steels with different alloying elements, carbon content and heat treatment have varying strength properties and depending on strength requirement, the type of steel can be chosen. Alternative materials available for use are titanium alloys and aluminium alloys. Fibre-reinforced plastics (FRPs) have varying structural strengths based on the combination of fibres and matrix used. Generally, FRP is used as a structural material in small boats and structures. Though wood and cement are also used as structural material in small boats and fixed offshore platforms, their use is very limited and these are excluded from the present review.

10.1.1 Structural Steel

Structural steel in marine construction is designated as Grade A, B, D or E which have gone through a process of deoxidation by thermomechanically controlled processes (TMCP) or have been cooled by a process of normalisation (N) or have been controlled rolled (CR). Grades D and E have aluminium (Al) as an additive material. High-tensile steels for ship structures are generally specified at three levels of yield strength – 32, 36 and 40 (representing yield strength in kgf/mm²) with four designated grades – AH, DH, EH and FH based on increasing notch toughness. For these steels, fully killed and grain-refined deoxidation process is applied with addition of grain-refining elements such as aluminium (Al), niobium (Nb), vanadium (V) and titanium (Ti) with residual elements such as copper (Cu), chromium (Cr), nickel (Ni) and molybdenum (Mo). Quenched and tempered steels provide still higher strength and there are six different levels of yield strength – 420, 460, 500, 550, 620 and 690 (representing yield strength in N/mm²) with three grades of steel – DH, EH and FH. These steels are used in very large merchant vessels on decks and specific high-stress areas, but mainly in naval vessels requiring higher strength and higher fracture toughness. For use in low-temperature conditions, steel requires to have minimum prescribed notch toughness at low temperature. For this purpose, one uses carbon, manganese and nickel alloy steels. There are three levels of yield strength – 32, 36 and 40 with four grades of steel – LT-AH, LT-DH, LT-EH and LT-FH. Apart from these, higher nickel-content steels give higher notch toughness and there are four levels of Ni content designated as 1.5 Ni, 3.5 Ni, 5 Ni and 9 Ni. Different grades are chosen for different thicknesses of steel plates and sections for required usage, particularly for use in arctic and temperate climates in ships as well as platforms and pipelines. The chemical composition and mechanical properties of all these grades are described in all classification society rule books. Table 10.1 gives a brief description of mechanical properties of different grades of steel. All these steels have a carbon content varying between 0.18% and 0.21%, the FH grade having a slightly lower carbon content of 0.16%.

When high-tensile steel was initially used for naval vessels, three grades of quenched and tempered steel were developed with alloying elements Ni, Cr, Mo and V. Three grades of steel designated as HY-80, HY-100 and HY-130 have been developed. These could be equivalent to the HT steel for ships mentioned earlier. The properties of these three steels are also given in Table 10.1. Subsequently, a low-carbon steel known as high-strength low-alloy (HSLA) steel has been developed whose carbon content is low, of the order of 0.06%, nickel content lower than the HT steel discussed earlier, but has a higher copper precipitation, being about 1.0%–1.75%. Of course, these are quenched and tempered steels. The properties of HSLA steels are given in the same Table 10.1. Depending on carbon content and composition of alloying elements, HSLA steels with required strength can be chosen.

Marine-grade structural steels are either plates of various thicknesses or rolled sections which are used as stiffening members for plates. The most common rolled sections are flats, angles, T sections, I sections and bulb plates. The way these stiffeners are attached to the plates, as shown in Figure 10.1, is such that the moment of inertia (or section modulus) is higher, compared to the construction where the flange is attached to the plate. If rolled sections are not available, stiffeners could be fabricated to the same scantlings. All other materials described in the following sections also use similar plates and sections.

TABLE 10.1

Mechanical Properties of Various Structural Material

	Yield Strength N/mm²	Ultimate Tensile Strength N/mm²	% Elongation
A, B, D, E	235	400–490	22[1]
AH-32, DH-32, EH-32, FH-32	315	440–590	22[1]
AH-36, DH-36, EH-36, FH-36	355	490–620	21[1]
AH-40, DH-40, EH-40, FH-40	390	510–650	20[1]
DH-420, EH-420, FH-420 (Q and T)	420	530–680	18[1]
DH-460, EH-460, FH-460 (Q and T)	460	570–720	17[1]
DH-500, EH-500, FH-500 (Q and T)	500	610–770	16[1]
DH-550, EH-550, FH-550 (Q and T)	550	670–830	16[1]
DH-620, EH-620, FH-620 (Q and T)	620	720–890	15[1]
DH-690, EH-690, FH-690 (Q and T)	690	770–940	14[1]
AH-32, DH-32, EH-32, FH-32 (LT)	315	440–590	20–22
AH-36, DH-36, EH-36, FH-36 (LT)	355	490–620	20–22
AH-40, DH-40, EH-40, FH-40 (LT)	390	510–650	20–22
1.5 Ni (LT)	275	490–640	22
3.5 Ni (LT)	285	450–610	21
5 Ni (LT)	390	540–740	21
9 Ni (LT)	490	640–790	18
HY-80[2]	>552	*	20
HY-100[2]	>689	*	18
HY-130[2]	>896	*	15
HSLA[2] (depends on composition)	>= 552 (or, 80000 KSI*)	*	18
Aluminium*** - 5083 -H116	215	305	10
Aluminium (welded)*** -5083-H116	125	275	14
Aluminium*** - 5456 -H116	255	350	10
Aluminium*** - 5086 -H116	207	290	9
Aluminium*** - 6082 - T6	260	310	8
aluminium (welded)*** - 6082 - T6	115	205	
Aluminium*** - 6005A - T6	225	270	6
Pure titanium[3]	170–480	240–550	As per ASTM
Titanium alloy Ti-7Al-2Cb-1Ta[4]	720	810	As per ASTM
Titanium alloy Ti-5Al-2.5Sn[3]	760	>790	As per ASTM**

[1] Subject to special requirements in the rules based upon the test specimen dimensions
* KiloPound-weight per Square Inch
** American Society for Testing and materials
*** For aluminium alloys yield stress is taken as proof stress which corresponds to 0.2% of strain limit or 70% of breaking strength whichever is lower
(Source: IRS Rules for Steel Ships: Part 2 – Materials and Welding.)
[2] Source: Czyryca,E.J. and Vassilaros, M.G., *Advances in low carbon, high strength ferrow alloys*, Naval surface warfare centre, Report CARDEROCKDIV-SME-92/64, 1993
[3] Source: Donnachie, M.J., Titanium - A technical guide, ASM International, 2000
[4] Source: Hydrofoils materials research program – final report no. 2-53100/5R-2179 prepared by LTV vought Aeronautics division, 1965.)

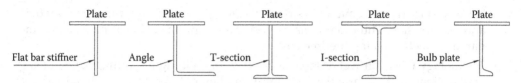

FIGURE 10.1
Different rolled sections used in shipbuilding.

There is a special type of steel plate known as Z plates whose properties perpendicular to the plate surface are better than those properties of the corresponding grade of general structural steel. The letter Z refers particularly to the plate thickness. The through-thickness deformation properties are determined during tensile test using test pieces in through-thickness direction of the plate. The result is the perpendicular reduction of area at fracture, that is, the so-called Z value. These plates may be designated as Z15, Z25 or Z35 based on their reduction of cross section in through-thickness direction. These plates are normally used for higher thickness applications, typically above 15 mm. At places where welding of plates may not give good-quality structure, Z plates are used. Though not commonly used as shipbuilding steel, Z plates are used in vulnerable locations having welding restrictions but requiring higher plate thickness.

10.1.2 Aluminium

In recent years, aluminium alloys have found extensive use in marine applications. Main hulls of small boats such as commercial pleasure craft (planing boats), sailing craft, personnel and work boats, fishing vessels, survey boats and naval boats have been made of aluminium and its use for such purposes is on the rise. Superstructures and deck houses of large commercial vessels and many naval vessels have been made of aluminium alloys. Marine fittings such as hatch covers, ladders, railings, gratings, windows and doors are frequently made of aluminium alloys. Another major use of aluminium is in making sacrificial anodes. This is not a structural item, but is a zinc-enriched alloy attached to the underwater portions of ships and platforms which slowly waste away due to corrosion by galvanic action so that the anode is sacrificed protecting the steel hull. This requires to be replaced regularly as it gets wasted away.

Advantages of using aluminium as the material of construction in marine applications can be listed as follows:

- An aluminium alloy has lower density, which is almost one-third of steel and its strength can be made high with proper alloying material and heat treatment. Thus, aluminium has a strength-to-weight ratio much higher than steel and its use can save weights up to 55%–67% over steel. Used as a hull construction material, for the same payload, one can have a smaller boat resulting in higher speed, fuel saving, higher endurance and better manoeuverability. Alternatively, for the same size of boat, payload carried is more resulting in better use and increased profitability.
- If an aluminium alloy is used in superstructures or deck houses, weight is reduced at the top, lowering the centre of gravity and thus giving improved stability.
- Being non-magnetic, aluminium is the preferred material in naval vessels.
- Aluminium has high thermal and electrical conductivity and it can be used if such is the requirement.
- Being highly corrosion resistant, aluminium is the preferred material in marine environment. An aluminium alloy surface need not be painted, giving aesthetically

better surface which remains as such without degradation with time due to avoidance of corrosion. Thus, maintenance cost of aluminium surface is much less compared to steel, reducing lifetime cost.

- Technology has been well established today for welding aluminium alloy plates and sections without any flaw. Aluminium alloy cutting, forming, bending and machining can be done easily, and aluminium poses no specific problem for production.
- Aluminium alloy plates, rolled sections, extruded products and finished products are available easily and they are easy to procure and install.
- Aluminium is almost completely recyclable and its use does not pose any environmental problem.

Aluminium alloys are manufactured by adding different alloying elements to aluminium. The standard designation of aluminium is by 1000, 2000, ..., 9000 series where each series can be further designated based on the quantity of the alloying material. Thus, the series may be 1*xxx*, 2*xxx*, ..., 9*xxx*. The most commonly used aluminium alloys in marine applications are the 5000 series in which magnesium is the main alloying material and the 6000 series which has magnesium and silicon as the main alloying materials. The alloy strength can be manipulated by proper choice of heat treatment. The common heat treatment used are H*xxx*, which means strain hardening with or without thermal heating and T_x, which means a solution heat treated and artificially aged. The '*x*' can be a set of numbers indicating further definition of the heat treatment process. The types of alloy used in marine applications (5083-H116, 5456-H116, 5086-H116, 6082-T6, 6005A-T6) and their strength properties are shown in Table 10.1 along with other marine construction material.

Aluminium has certain disadvantages over steel as a marine material which must be evaluated properly before it is chosen for construction.

- Aluminium alloys are highly ductile material and, if not properly stiffened, are likely to give large deflections.
- Aluminium alloys have lower fracture toughness making them unsuitable where such requirement is required.
- Aluminium alloy welding has to be done with care. Since strength of welded aluminium is less than non-welded plate, defects in welds cannot be afforded.
- If an aluminium alloy and steel are joined as would be required for aluminium deck house on steel hull, aluminium will waste away due to galvanic action, and so, the aluminium and steel must be separated at the joints by non-metallic separators.
- One of the main disadvantages of use of aluminium as structural material such as partition bulkheads is that aluminium deforms easily and melts when exposed to fire for a long time. Since fire hazard increases with aluminium, it cannot be used in fire-prone areas.

10.1.3 Titanium

Titanium has been used in aerospace engineering applications for some time now. It is now slowly emerging as a potential material in marine applications due to its lightweight, corrosion resistance, potentially high strength and its behaviour over large temperature ranges including cryogenic temperatures. It has been used as the hull material of deep submergence submersibles. Its advantages and uses can be listed as follows:

- High strength-to-weight ratio which is a critical factor in pressure hulls of deep-diving submarines and some areas of high-speed surface vehicles
- High corrosion resistance making it suitable for certain machine components
- High corrosion fatigue strength
- Non-magnetic properties making it suitable for naval applications
- Resistance to higher-than-normal temperature such as hot brine or engine exhaust
- Resistance to cavitation and crevice corrosion due to high-velocity sea water
- No loss of strength in low-temperature environment such as large ocean depths

Titanium in its pure form has relatively low strength and is very ductile. But presence of impurities like oxygen, nitrogen and carbon increases strength and reduces ductility, making it suitable for marine use. Also if alloying elements such as aluminium, iron, tin, chromium, vanadium, molybdenum etc. are added, strength increases. The titanium alloy structures can be subdivided into three categories – alpha, alpha-beta and beta structures depending on the type of alloying material. These structures define the microstructure of the alloy and also define whether these can be heat treated or not. Titanium alloy strength varies with temperature, sometimes increasing with lower temperature, thus making it suitable for deep-diving vehicles. Its notch toughness changes gradually with temperature. This material has a higher fatigue strength than steel and is also stress corrosion resistant since it is generally corrosion resistant. However, one has to take care during its production and welding so that hydrogen does not affect the metal which may cause embrittlement. Table 10.1 shows the properties of titanium alloys.

10.1.4 Fibre-Reinforced Plastics

Fibre-reinforced plastics (or polymers) are normally known as FRP composites where fibres are bonded together with the help of polymers or resins to form a strong structural material. As a structural material, FRP has been long used in bridges, roads, building construction, aircraft and automobile industry and marine industry. FRP material varies in strength and stiffness based on the type of fibre used and the laminate construction. The advantages of FRP over the conventional structural material in marine industry such as steel and aluminium can be listed as follows:

- FRP is much lighter than steel or even aluminium for similar requirements.
- Generally, FRP requires lesser skill levels than metals and it is easy to manufacture, to install and to do in-service repair and maintenance without much down time.
- It is corrosion resistant in harsh marine environment and durability is, therefore, high.
- Being magnetically transparent, it is suitable for many naval applications.
- FRP can be engineered to a particular need with regard to strength and stiffness. FRP is not an isotropic material like a metal. Whereas a metal has the same mechanical properties in all directions, FRP has the mechanical properties based on the fibre orientation and quantity. Though a composite is a heterogeneous mix of fibres and resins, its properties are averaged in any direction considering it

as a homogeneous material. Based on the type of fibre, orientation of fibre and layers of fibre, the strength and stiffness can be manipulated or engineered to get the required values in the structure in a particular direction. If the fibres are uniformly distributed in all directions on a surface, it gives orthotropic mechanical properties, that is same properties in perpendicular x and y directions on the surface plane, but different properties in the thickness or z direction.

- FRP composite manufacture technique allows any three-dimensional shape to be developed and no further effort on shape development is required unlike metals where the sheets require to be prepared, formed and welded to get a shape.

Though FRP has been used in marine industry for a long period, the initial cost is still higher than steel or, even, aluminium. Enough long-term studies have not been conducted to study its effects on fatigue and fracture toughness, a primary requirement for maritime use. Delamination is a major problem with FRP which reduces durability to a large extent. Resins used in composites are generally not fire resistant. By adjusting the chemical composition or adding filler material, composites can be made fire retardant, but their fire-resisting properties are much poorer than metals. A major disadvantage of FRP is that the material is neither recyclable nor bio-degradable with time. So it becomes an environmental hazard in the long run.

In spite of these few disadvantages, use of FRP is on the rise primarily because of it being lighter and giving a long-term cost advantage in terms of reduced maintenance cost.

Naturally occurring fibres such as cotton, jute, wool, hemp or coir can be used as reinforcement, but cannot give adequate strength for structural purposes. Therefore, our focus will be limited to man-made fibres such as glass, carbon, aramid or boron fibres. Man-made fibres are thin filaments produced in a continuous process, the diameter being 2–13 μm except boron, the filament of which has a diameter of 100–200 μm. It is difficult to measure all mechanical properties using a single fibre. It has been possible to measure only two properties, the breaking tensile strength and modulus. That is also dependent on cross-sectional area which is difficult to measure. So there is a lot of variation in the reported properties of fibres. Glass fibres generally have high strength-to-weight ratio comparable to metal. But these have low modulus of elasticity or stiffness. Amongst man-made fibres, these are the cheapest and are most commonly used. There are two verities of glass fibres, E and S glass, which are mostly used in marine industry, E glass being the preferred material of the two. Carbon fibres have the primary advantage over glass fibres that these have a much higher modulus, lower density, better fatigue properties and improved creep rupture resistance. But their impact resistance is low. Carbon fibre varieties include high-strength (HS), high-modulus (HM) and ultra-high-modulus fibres. Two varieties of aramid fibres, Kevlar 29 and Kevlar 49, are mostly used in composites as structural material. These fibres have high strength and high strength-to-weight ratio, but low modulus compared to carbon fibres. On the other hand, it can take much higher strain than carbon fibres at failure. It has excellent fatigue resistance. Though carbon and Kevlar fibres have much better strength characteristics, these are selectively used because of their high cost. Table 10.2 gives approximate comparative characteristics of various fibres.

A common form of E glass available in the market is the chopped strand mat (CSM) which is composed of strands of 25–30 mm length oriented in a random fashion so as to give similar strength properties in two perpendicular directions. It can be laid easily with resin to form a laminate. This is most commonly used glass mainly because it is the cheapest form of glass fibre. Surface tissue is a lightweight mat made from discontinuous glass fibres held together with a binder. It is very thin and mostly used as the first or last layer

TABLE 10.2

Strength Properties (Approximate) of FRP Laminates along with Steel and Aluminium

Type of Material	Density (g/cc)	Young's Modulus (GPa)	Tensile Strength (GPa)	Tensile Strain at Failure (%)	Specific Strength (GPa·cc/g)	Specific Modulus (GPa·cc/g)
Material	ρ	E	σ		σ/ρ	E/ρ
E-Glass	2.6	69–72	1.7–3.5	3	1.18	27.60
S-Glass	2.49	85	4.8	5.3	1.90	34.30
Carbon – high strength	1.8	295	5.6	1.8	3.11	164.00
Carbon – high modulus	1.96	517	1.86	0.38	0.95	264.00
Carbon – ultra-high modulus	2.0–2.1	520–620	1.03–1.31	0.38	0.57	278.05
Kevlar 49	1.45	135	3	8.1	2.10	93.10
Laminate of CSM	1.5	7.5	0.14	2	0.09	5.00
Laminate of CSM and WR	1.6	12	0.18	2	0.11	7.50
Laminate of woven roving	1.7	14	0.21	2	0.12	8.24
Laminate of uni-directional roving	1.7	30	0.15	2.3	0.09	17.65
Laminate of UD Kevlar 49 roving	1.3	50	0.9	2.2	0.69	38.46
Laminate of UD carbon HS roving	1.44	100–120	1.0–1.9	1.5–2.2	1.01	76.39
Laminate of UD carbon HM roving	1.48	140–240	0.8–1.4	0.6–1.4	0.74	128.38
Steel	7.9	200	0.45	20	0.06	25.32
Aluminium	2.7	70	0.26	17	0.10	25.93

of a GRP laminate to give a smooth finish. Fabrics are composed of yarns or woven rovings (WR) running in two directions, the longitudinal ones being known as warp threads and the ones running perpendicularly across being the weft threads. Fabrics are available in various weave patterns such as plain weave, twills and satin weave. Uni-directional (UD) rovings generally have 70% of fibres woven in one direction, and, therefore more of the fabric properties are concentrated in one direction. These are used when the strength required in one direction is much higher than that in the other direction.

Matrices used in FRP composites are generally polymers such as thermosetting resins, thermoplastics and rubber. Of these, for structural FRP laminates, thermosets are used. Thermosets can be polyester resins of three varieties – orthopthalic, isopthalic and bisphenol. Thermosets could also be epoxy resins requiring controlled atmosphere for correct curing. Vinyl ester resins, phenolic resins and polyamides are other types of thermosets. Of all the thermosets, the most commonly used ones are isopthalic resin and epoxy resin.

Many of the FRP materials suffer from the problems of low stiffness, and in many applications such as boat body construction, it is necessary to increase stiffness. One method of increasing stiffness is to adopt a sandwich construction where the faces of the laminate are separated by a core material, each face having a few layers of fibres. The commonly used core materials are balsa or similar wood, PU or polystyrene foam, PVC foam and honeycombs, mostly made of aluminium. The core materials contribute little to the strength of the sandwich panel.

There are many different techniques of FRP composite manufacturing, one of the most common being the contact moulding process which could be semi-automatic using the spray-up process or manual, the commonly used hand lay-up process which is commonly used in boat hull manufacture. The other automatic and semi-automatic processes include matched-dye moulding technique, vacuum bag technique and autoclave moulding. For geometric shapes such as rods, spheres and cylinders, one could draw pultrusions or use filament winding technique.

The mechanical properties of an FRP laminate depend on the amount of matrix and fibre used.

$$\text{Fibre volume fraction: } V_f = \frac{\text{volume of fibre}}{\text{laminate volume}}$$

$$\text{Matrix volume fraction: } V_m = \frac{\text{volume of matrix}}{\text{laminate volume}}$$

$$\text{Density of composit: } \rho_c = V_f \cdot \rho_f + V_m \cdot \rho_m$$

$$\text{Volume fraction of void: } V_v = \frac{(\rho_c - \rho_{exp})}{\rho_c}$$

where
   suffixes f, m, v, c and exp denote fibre, matrix, void, composite and experimental, respectively
   ρ is the density
   V is the volume fraction

Estimation of mechanical properties of FRP is very involved and procedures have been described in text books on FRP. However, for a quick estimate of tensile strength σ_c of composite, a thumb rule can be used as follows:

$$\sigma_c = \sigma_f \left[V_f + \left(\frac{E_m}{E_f} \right) \cdot (1 - V_f) \right]$$

where
   σ is the ultimate tensile strength
   E is the Young's modulus

Table 10.2 shows some approximate values of mechanical properties of laminates along with those for steel and aluminium.

Since use of wood has been restricted all over the world for environmental reasons, FRP has become the replacement material for small boats. For dinghies, canoes, pilot launches, lifeboats and tourist boats and launches, FRP has been the primary structural material for the entire boat. Hulls of high-performance vehicles such as planing boats, planing catamarans, hydrofoil boats, hovercrafts and SES vehicles are being made of FRP. Racing boats and yachts where low weight is a primary requirement, FRP (particularly carbon) has been the primary structural material. To reduce topside weight of ships and offshore structures, the superstructure is made of FRP integrating with steel main hull. Superstructure skin can be made of FRP laminates, whereas the framing can be of steel, thus providing stiffness of steel structure. FRP submarines and submersibles have been manufactured and operated successfully due to their low weight and high strength-to-weight ratio (particularly Kevlar and carbon fibres). FRP has also found wide applications in naval applications. Appendages of pressure hull of submarines such as free-flooding casings, fins and fairings, sonar domes and radomes have been made of FRP. Due to their magnetic transparency, FRP is the favoured material for mine countermeasure vessels. There is a possibility of using high-strength and -stiffness FRP composites for mechanical system components such as ship propellers, propeller shafting etc. to reduce weight and have better durability.

10.2 Loads on Marine Structures and Vehicles

A marine structure or vehicle is subject to different types and quantities of static and dynamic loads during its operational life. Some of these loads are listed in the following sections.

10.2.1 Static Loading and Vertical Bending Moment

Static loading on a marine structure is its own weight distributed based on the weight per unit area over the total area of the structure. In case of a ship which is a long slender structure with length being considerably more than breadth or depth, weight per unit length gives the weight distribution in the longitudinal direction. The weight of a ship consists of the lightship weight which is a constant distribution over the life of the ship and deadweight which is the distribution of payload which may vary from voyage to voyage. This weight is supported by the upward buoyancy force proportional to the immersed sectional area along the length of ship. For equilibrium, the weight and buoyancy distribution curves must be of equal area and have the same longitudinal centroid. Figure 10.2

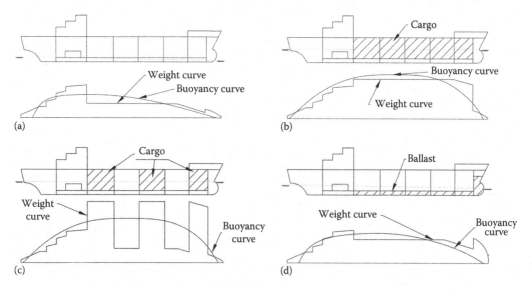

FIGURE 10.2
Still water loading on the ship girder. (a) Lightship condition. (b) Fully loaded with uniformly distributed payload. (c) Fully loaded with specific hold loading. (d) Ballast load condition.

shows four typical loading conditions for a merchant vessel, the lightship condition, fully loaded condition in which cargo is uniformly distributed over the entire cargo space, fully loaded with specific holds only loaded with heavy bulk and ballast condition. Because of the ship's geometry, it can be considered as a longitudinal girder being loaded by an amount of load = weight – buoyancy (per unit length) along the length. This gives rise to a static shear force along the length and vertical bending moment. The ship's structure must withstand the shear stress and flexural or bending stress generated due to this loading.

10.2.2 Wave Bending Moment

As waves pass a ship in the head or following sea condition, the buoyancy curve changes and, accordingly, the shear force and bending moment also change. The maximum bending moment experienced by a ship due to this is when the wave length is equal to ship length. Considering the ship poised on a wave of equal length, it can experience hogging bending moment if the wave crest is at the midship and sagging bending moment if the wave crest is at the ends. This is shown in Figure 10.3a. Needless to say that the extent of bending moment depends on wave amplitude and the total bending moment can be

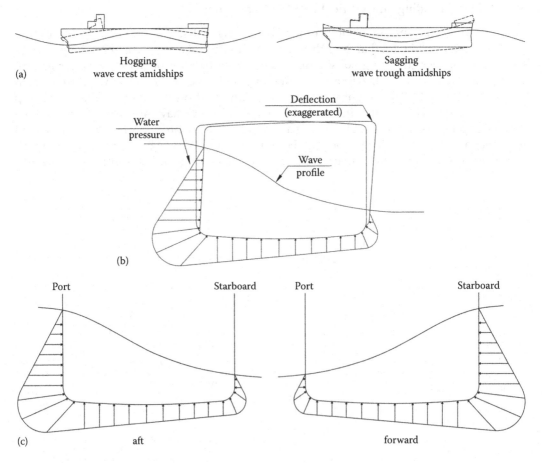

FIGURE 10.3
Ship poised on a wave. (a) Ship poised on a wave of equal length. (b) Beam sea – horizontal bending. (c) Oblique sea–torsion.

considered as the still-water bending moment augmented by the wave bending moment. Wave bending moment depends on the elevation of the actual sea surface which is very rarely a single wave having a length equal to ship length (a swell wave may be an exception) and the ship is never poised statically over the wave. Even then the wave bending moment is estimated as an augmentation over the still-water bending moment which is statistically determined as a ship poised on a static wave.

10.2.3 Horizontal Bending Moment

When a wave passes a ship in the beam sea condition with its length in the direction of ship length, the ship rolls and the hydrostatic pressure due to wave orientation is different on port side from that of starboard side. This difference of pressure causes horizontal bending of the ship girder as shown in Figure 10.3b and is maximum when the ship is heeled towards the wave.

10.2.4 Torsional Moment

In oblique sea condition when a wave passes the ship, hydrostatic pressure varies from port to starboard in different manners from aft to forward causing torsion of the ship girder as shown in Figure 10.3c.

10.2.5 Static External Hydrostatic Load

A floating or submerged structure experiences static lateral loading due to hydrostatic pressure normal to its surface proportional to its submergence from the free surface. Hydrostatic loading is normally compressive in nature. Figure 10.4a through c shows such loading on a floating ship, a semi-submersible and a submarine, respectively.

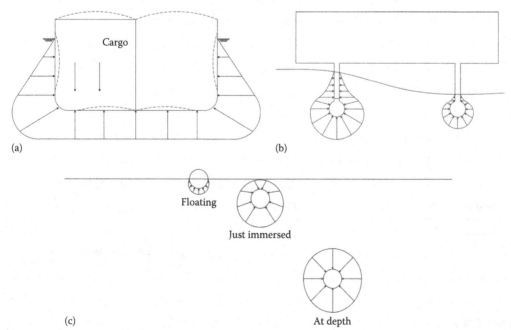

FIGURE 10.4
Lateral hydrostatic loading. (a) Floating ship. (b) Semi-submersible. (c) Submarine.

10.2.6 Static Internal Load

There can be local or regional load which may be concentrated, semi-concentrated or uniform on ship structure and can cause bending, torsion, shear or compression of the local structure. Figure 10.5a through d gives some examples of such loading. Figure 10.5e shows whirling vibration of the shafting system which is an internal load where the loads are oscillatory.

10.2.7 Dynamic External Load due to Waves

As the wave passes a ship in head/beam/oblique direction, the external hydrostatic pressure keeps changing from a maximum to a minimum as shown in Figure 10.6a.

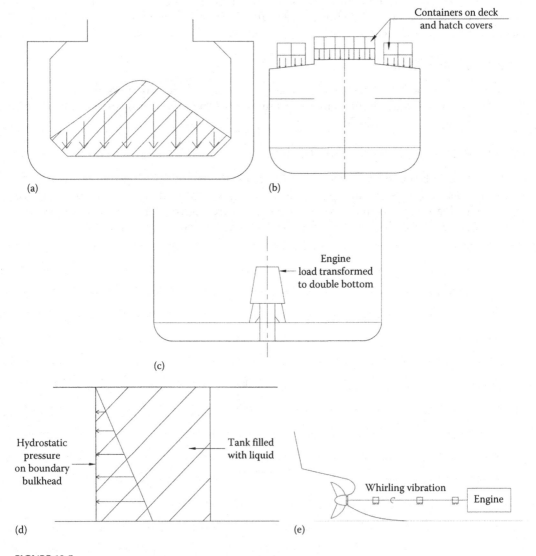

FIGURE 10.5
Internal loads. (a) Cargo on double bottom – distributed loading. (b) Containers on deck – concentrated/semi-concentrated loading. (c) Machinery and equipment on ship structure – concentrated loading. (d) Lateral hydrostatic loading on tank boundaries and bulkheads. (e) Whirling vibration of shafting system.

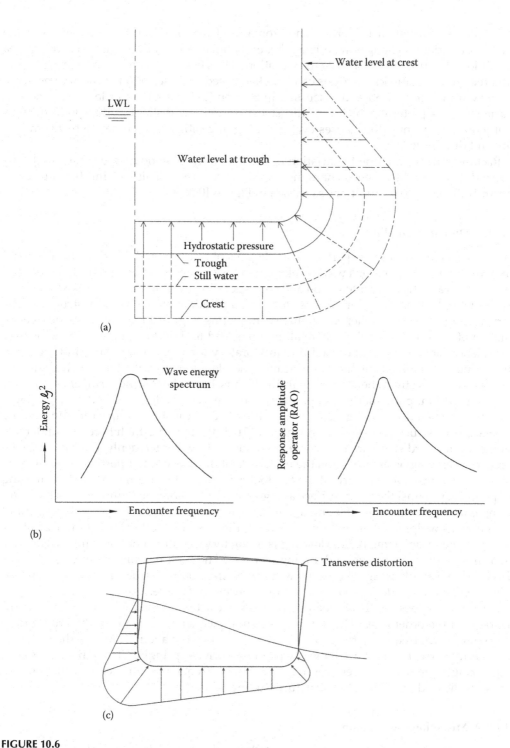

FIGURE 10.6
Dynamic external load due to waves. (a) External lateral pressure on hull side. (b) Distribution of wave bending moment. Variation of wave bending moment due to actual sea condition causing vertical, horizontal and torsional wave b.m. due to head/following, beam or quartering seas. (c) Variation of lateral hydrostatic pressure.

This causes a variation in vertical, horizontal and torsional bending moment and the nature of variation changes with head, beam, oblique sea condition or short crested sea condition. It is difficult to estimate the variation in this bending moment in a particular sea and recourse to statistical analysis can be taken based on description of the sea spectrum as shown in Figure 10.6b and discussed in Section 7.3.1. This kind of loading generates variable stress patterns which the ship structure must withstand at that point of time, but more importantly, this causes cyclic loading and fatigue which has to be taken into account during design.

Racking stresses are due to variation of the lateral hydrostatic pressure causing transverse distortion which keeps changing as the wave passes. This loading takes place in beam and oblique sea conditions as shown in Figure 10.6c.

10.2.8 Dynamic Loading

A floating platform at sea experiences six degrees of freedom and has three translational and three rotational motions with acceleration in each degree. This acceleration is caused due to the wave structure interaction and corresponding loading. Further, this acceleration is oscillatory in many cases, giving rise to vibratory loading on the platform structure. The structure must be strengthened against such loading. Figure 10.7a shows six degrees of motion of a semi-submersible where all motions lead to fluid loading. A ship's ends may experience racking stresses caused due to vibratory force on the forward and aft structures due to combined pitching and heaving acceleration. To withstand such loads, the ship's fore end is strengthened with panting beams, breast hooks and stringers.

As a platform, particularly a ship, pitches and heaves, the forward end may go down in water momentarily taking green sea on board. This causes an impact loading on the forward deck structure as shown in Figure 10.7b. If the bow has high flare, the underside of the top forward structure may come in contact with water suddenly and there may be an impact loading or slamming on the underside of the above-water portion of the stem as shown in Figure 10.7c. Similarly, if the forefoot comes out of water, then there is slamming or pounding against the water surface as it goes down as shown in Figure 10.7d. This slamming is an impact loading affecting the immediate as well as the entire ship structure and is known as whipping. Due to impact loading, there is high vertical wave-induced shear force and bending moment. This loading is oscillatory in nature and has a high frequency and dies down quickly due to damping of water. The high bending moment and shear force cause overall strength concerns with high stress on the structure and, since whipping oscillation dies down soon, this is not a concern of fatigue.

A ship undergoes oscillatory loading due to wave action. If the frequency of such oscillation is of the order of 0.5 Hz or 3 rad/s circular frequency which is of the order of the encounter wave frequency (ship having a forward speed of about 20 knots), then there is possibility of excitation of oscillation which is known as springing. Though stresses during springing may not be very high, the encounter frequency can excite oscillation and cause cyclic loading on the ship structure leading to fatigue.

10.2.9 Miscellaneous Loading

Floating offshore platforms anchored or tethered to the sea bed are often affected by underwater current. The loading on the mooring rope or the tether has an effect on the motion of the top structure and hydrodynamic loading. Figure 10.8a shows such loading. Figure 10.8b shows wind loading on the top structure. Figure 10.8c shows sloshing load

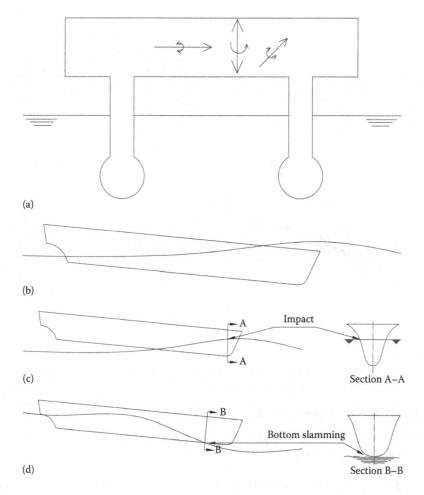

FIGURE 10.7
Dynamic loads. (a) Motion of a semi-submersible in 6°. (b) Shipping of green seas. (c) Impact of bow flare. (d) Slamming.

on a partially filled tank installed on a floating vessel. Figure 10.8d shows hydrodynamic loading due to propeller on the ship's aft structure.

10.2.10 Operational Loads

Due to cargo operations, there can be different loadings on the platform, some of which may be the following:

1. Grab operation on loading (dropping of cargo from a height) and unloading of bulk cargo (abrasion of double bottom).
2. Due to loading and unloading of bulk cargo or oil cargo in a tanker and unloading or loading of ballast water, there is likely to be fluctuations of shear force and bending moment at intermediate stages of operation at a port.
3. Corrosion and degradation of plate material may aggravate stress concentration at sensitive locations and lead to stress corrosion cracking accelerated by fatigue.

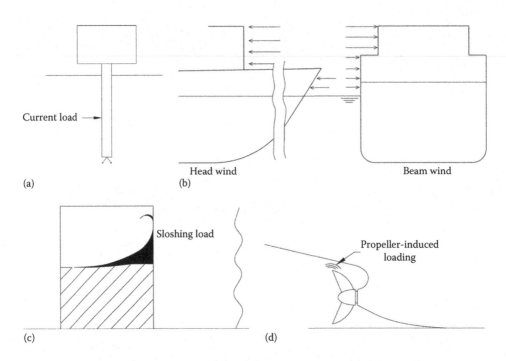

Current load →

(a)

Head wind

(b)

Beam wind

Sloshing load

(c)

Propeller-induced
loading

(d)

FIGURE 10.8
Miscellaneous loading. (a) Current load on a tethered platform. (b) Wind load on top structure. (c) Sloshing inside a tank. (d) Hydrodynamic loading due to propeller.

10.2.11 Loads due to Ship Handling

Different kinds of loads may be experienced by the ship due to handling some of which are mentioned in the following:

1. During launching, the ship undergoes stress changes from concentrated loading at the supports on land to concentrated loading on fore poppet during stern lift or tipping at the way ends as in Figure 10.9a and finally buoyancy support.
2. During drydocking, the ship is supported on blocks inside the drydock and the loading is concentrated on particular locations shown in Figure 10.9b.
3. In the event of grounding, the ship experiences concentrated load at the point of contact and the bending moment of the ship girder changes, leading to different stress distribution shown in Figure 10.9c.
4. Changes in shear force and bending moment take place due to flooding in the event of damage.
5. The ship experiences large impact loading in the bow or on ship side during collision as shown in Figure 10.9d.

Assuming a period of life, say 25 years for marine structures and vehicles, design is generally done for the worst condition. The loads on a particular structure lead to stresses which the structure must withstand. Apart from this, the structure must be fracture resistant and withstand fatigue. Thus, vibration becomes an essential part of structural design. The stresses get compounded due to stress concentration, structural discontinuity and stress corrosion cracking.

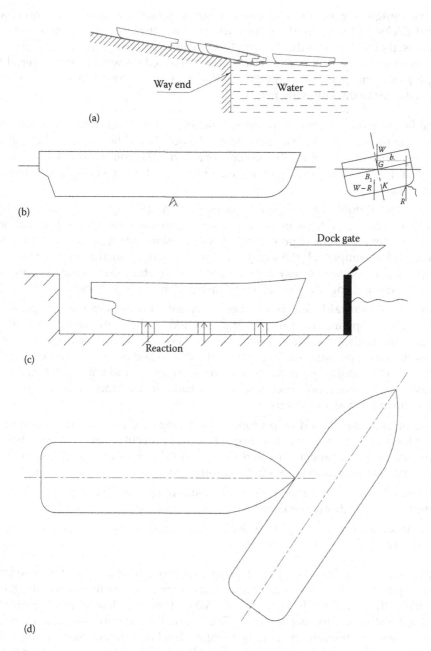

FIGURE 10.9
Loads due to ship handling. (a) Launching. (b) Drydocking. (c) Grounding. (d) Collision.

10.3 Structural Layout

The layout design is generally expressed diagrammatically as the structural elevation, structural decks and structural sections which show the major (and minor) structural components, their extent and their connectivity with other structural components and the particulars of plates and rolled/fabricated sections based on which steel material listing can be prepared and steel mass and its centre of gravity can be estimated. The structural layout must cater to the following needs.

1. The layout must integrate with other functional and operational requirements of the structure or vehicle. For example, the layout should be so made that the specific space requirement for the contents of a particular compartment are well met, for example clean hold without any encumbrances for carriage of cargo and ease of access.

2. The layout should help installation of equipment, piping, ducting and electrical cable for various systems in the ship. This is necessary for ease of installation of all these items as per the production schedule. Normally, ships are outfitted after steel work is complete. If the structural layout is done in such a manner that outfitting could be advanced and done during steel erection itself which is known as advanced outfitting, the time of ship construction can be reduced.

3. Effective structural layout is also necessary for ease of movement of personnel during ship operation for monitoring function of the equipment. For example, engine room girders can be so laid out that these can also be used as engine seatings. Similarly, pipings, ductings and cables can be laid out in the shortest possible route so as to minimise interference with the structure and with each other. There must be adequate clear space free of structural encumbrances for access of the equipment from all directions.

4. Structural layout should help proper monitoring of the structural health of the ship for carrying out surveying, inspection and preventive maintenance. For this purpose, access arrangements to difficult spaces such as double bottom and forward peak tank must be provided and integrated with structural layout.

5. The structural layout should ensure adequate strength in all possible loading conditions which is discussed later.

6. The structural layout should ensure continuity and avoid stress concentration discussed later.

Table 10.3 shows the utility of various structural components and parts from strength and operational points of view. The major structural components are the shell plating, bottom and double bottom plating, decks and bulkheads. These are all steel plates and they are required to be stiffened to bear the load. Two alternative framing systems are normally considered for strengthening the plating – longitudinal and transverse framing. In longitudinal framing system, plates are stiffened by rolled sections laid on the plate in the longitudinal direction and the section properties are calculated based on the span between two major transverse members. Similarly, in transverse framing system, plates are stiffened by rolled sections laid on the plate in the transverse direction and the section properties are calculated based on the span between two major longitudinal members. Traditionally, and as per classification codes, the portion forward of the forward perpendicular (FP) and

TABLE 10.3

Strength and Operational Utility of Various Structural Parts and Components

Item	Function
Strength deck, side shell and bottom plating	Form a box girder resisting bending and other loads.
Freeboard deck, side shell and bottom plating	Function as a watertight envelop providing buoyancy.
Bottom plating	Withstands hydrostatic pressure.
Forward bottom plating	Withstands slamming; plating thickness is increased; intermediate frames are provided.
	Breast hooks and stringers are fitted.
	Minimum forward draught is recommended.
Inner bottom, bottom plating DB floors and girders	Act as a double-plated panel to distribute the secondary bending effects due to hydrostatics loads and cargo loads to main supporting boundaries such as bulkheads and side shell.
	Resist docking loads.
Inner bottom	Acts as tank boundary for bottom tanks and withstands local loading due to cargo.
	Contributes to longitudinal strength.
Strength deck, upper deck	Withstands cargo handling equipment loading and cargo loading in some case as that of the container ship.
	Withstands loading due to shipping of green seas.
Remaining decks	Mainly withstand cargo loading, depending on extent and distance from neutral axis; contribute to longitudinal bending strength.
Side shell	Withstands hydrostatic pressure, dynamic effects due to pitching heaving rolling and wave loads.
Transverse bulkheads	Act as internal stiffening diaphragms for the hull girder and resist in plane torsion.
	Do not contribute to longitudinal strength.
	Generate watertight longitudinal subdivisions.
Longitudinal bulkheads	Contribute to longitudinal strength.
Bulkheads, in general	From tank boundaries support decks and loads generating equipment such as king posts and add rigidity.
	Serve as watertight partitions.
Stiffening of Plates	
Corrugations on bulkheads	Stiffen the bulkheads in place of vertical horizontal stiffeners.
Deck beams	Stiffen the deck.
Deck girders	Support the beams, deck transverses and transfer the load to pillars and bulkheads.
Transverse framing	Stiffens the side shell; supports the longitudinal stiffening.
	Supported in turn, by the decks, stringers and the longitudinal girders.
Longitudinal framing	Stiffens the shell, decks, tank top etc.
	Is supported by the deep transverses.
Side shell framing (general)	The web size is an important factor as regards
	a. Cargo stowage
	b. Panelling and insulation
	c. Running of wiring, vents, piping etc.
Vertical plates in double bottom (floors, side and centre girders)	Stiffen the bottom panel as tank boundaries.

the engine room and the aft body are transversely framed. In the cargo hold portion, the shell, the decks and the double bottom could be separately transversely or longitudinally framed. The decision to select the framing system depends on the span of the stiffener and the spacing between stiffeners which determine the weight and work content (weld length). The connectivity between stiffeners between shell, deck and double bottom has to be considered in the layout to avoid stress concentration and discontinuity.

A marine platform experiences bending or flexural stresses due to differential loading on the structure. It also experiences in-plane axial and transverse stresses, both compressive and tensile, shear stresses, buckling stresses as well as impact loading and increased stresses due to corrosion. In addition, the structure experiences fatigue due to cyclic loading. The following sections introduce the hull girder bending stress, shear stress and buckling stress. It is emphasised that ship structure is very complex and is subject to many different and uncertain loads throughout its life. So, while designing simple structural elements for a pre-determined load inside a ship can be done using simple linear formulations, more complex analysis must be done for loads on the entire ship structure.

10.3.1 Bending Stress on Hull Girder

A ship can be considered as a long and slender structure since its length is large compared to its two other dimensions – breadth and depth. The ship girder should be so designed that it can withstand any global loading – vertical, horizontal or torsional, static or dynamic. One of the fundamental calculations required to be done initially is the ship girder vertical bending moment. The loading on the ship girder is the difference between the weight at any longitudinal location and the upward buoyancy force there. To get the weight distribution curve, one of the fundamental requirements of ship design is the longitudinal distribution of lightweight of the ship. Depending on data available, the longitudinal distribution of weight can be estimated. At concept design stage, the exact weight distribution is not known and approximations have to be made. Figure 10.10a shows weight distribution in a trapezoidal manner such that the weight is equal to the area under the curve with its longitudinal centroid at the centre of gravity (LCG). Another method of weight distribution is that half of the weight is distributed equally along the length and the other half is distributed in a parabolic manner. Figure 10.10b shows how this can be done with alteration of the parabola so that the centroid is at the LCG position. Generally, the lightweight of a ship can be subdivided into three groups – concentrated, semi-concentrated and continuous items. Bulkheads are a typical example of concentrated items. Superstructure, engine room weight, forecastle, etc. are examples of semi-concentrated items. Generally, concentrated items can be merged with continuous items since these are distributed all along the ship length. The semi-concentrated items can be distributed as rectangles/trapeziums such that their individual LCG positions are maintained in the distribution. The final weight distribution can be manipulated geometrically to get the LCG at the correct location (see lightship weight distribution curve in Figure 10.2a). Once the detailed structural design is ready, the exact weight distribution can be obtained. In any load condition, each deadweight item can be superimposed on the lightship curve to obtain the total weight distribution with the correct LCG so that one can obtain the equilibrium floatation condition. For this equilibrium condition, the buoyancy curve can be obtained with the longitudinal centre of buoyancy (LCB) and the LCG coinciding. The load distribution on the ship girder is obtained by plotting the weight curve reduced by the buoyancy such that the area under the load curve is zero. Integration of the load curve gives the shear force and integration of the shear force curve gives the distribution of bending moment. Figure 10.11 shows this static bending moment

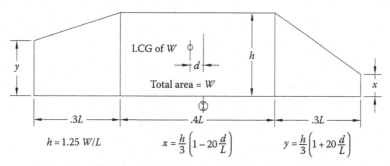

$$h = 1.25 \; W/L \qquad x = \frac{h}{3}\left(1 - 20\frac{d}{L}\right) \qquad y = \frac{h}{3}\left(1 + 20\frac{d}{L}\right)$$

Assumption: 50% of weight W is distributed uniformly over 0.4L amidship
Remaining is distributed in two trapezoids at 30% L forward and 30% L aft giving
an LCG at distance d from amidship

(a)

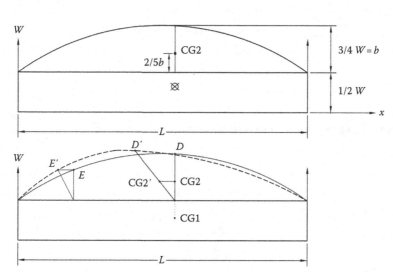

Assumption: 50% weight distribution uniformly with centroid at CG1, 50% weight distributed
parabolically with centroid at CG2 at a vertical height of 2/5 b from its base. The new
Centroid is such that CG2 moves to CG2′

(b)

FIGURE 10.10
Lightship weight distribution. (a) Trapezoidal distribution. (b) Parabolic weight distribution and adjustment of LCG.

distribution for a ship floating at still-water level. This bending moment can be modified to include the wave bending moment by adding a prescriptive amount suggested by the code books of classification societies, by adding an amount determined by the ship poised on a wave giving additional hogging (wave crest in the midship region) or sagging (trough at midship) and/or by adding an amount representing the dynamic wave loading in a seaway based on the response amplitude operator (RAO) of the wave bending moment. The ship girder must withstand this total vertical bending moment $M(x)$ at location x. We know that

$$\frac{E \cdot I \cdot \delta^4 y}{\delta^4 x} = l(x)$$

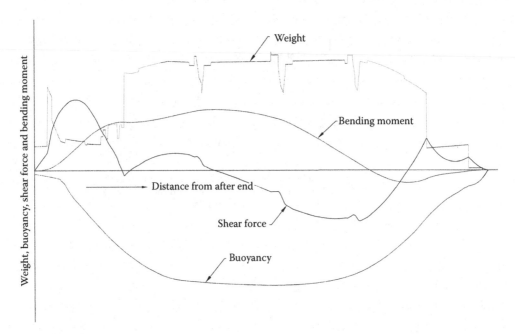

FIGURE 10.11
Shear force and bending moment distribution along the length of ship girder.

and

$$\frac{E \cdot I \cdot \delta^2 y}{\delta^2 x} = M(x)$$

where
    y is the vertical deflection
    x is the longitudinal position along the length
    $l(x)$ is the load at x
    E is the Young's modulus
    I is the mass moment of inertia of the section at x about its neutral axis in y–y direction
        and hence, $I = I_{yy}$

Integration of bending moment curve twice gives the ship girder deflection curve. The bending equation gives bending stress σ as

$$\sigma = \frac{M}{(I/y)}$$

where
    M is the bending moment at that section
    I is the moment of inertia about the neutral axis
    y is the distance of the structural element from the neutral axis
    I/y is known as section modulus

In a ship section, the farthest element is either at deck or keel where bending stress is maximum and is zero at neutral axis. If the ship is subject to horizontal bending moment, bending stress increases as

$$\sigma_l = \frac{M_v}{(I_v/y_v)} + \frac{M_h}{(I_h/y_h)}$$

where suffices l, v and h represent longitudinal, vertical and horizontal quantities, respectively. If there is longitudinal bending, there is a corresponding transverse strain giving rise to transverse stress which can be given in a first approximation as

$$\sigma_t = \nu \cdot \sigma_l$$

where ν is Poisson's ratio and is given as $\nu = |\varepsilon_t/\varepsilon_l|$ where ε is strain. Further ship section material is subject to other stresses also such as shear stress which is a function of shear force and shear flow in the section. This is further aggravated due to dynamic loading due to various causes such as slamming, racking, etc. So the total stress experienced by a ship section is due to compounding of all these stresses known as von Misses equivalent stress given by

$$\sigma_e = \left[\sigma_l^2 + \sigma_t^2 - \sigma_l \cdot \sigma_t + 3\tau^2 \right]^{1/2}$$

where τ is the shear stress. It can be understood that the primary hull girder strength comes from the vertical section modulus of the section. Based on the structural layout, all longitudinal structural elements that are continuous and of large length, typically extending over a length of 0.4 times the length of ship, are taken into account for moment of inertia calculation and determining the position of neutral axis. Therefore, the elements that are considered for moment of inertia calculation include the bottom plating, the side shell plating, the double bottom plating, upper and between deck platings, deck girders, centre girder and side girders in the double bottom, longitudinals on the side shell, double bottom and deck, longitudinal stingers if any and continuous hatch coamings on upper deck. Transverses, deep transverses, deck beams, transverse stiffeners on shell, deck and double bottom, brackets, beam knees and hatch coamings which are not continuous are not included in the moment of inertia calculation. The structural layout plays a major role in determining the transverse moment of inertia I_v. It can be understood that though the physical bending process is in the vertical plane, the stresses generated are the in-plane stresses of tension and compression.

10.3.2 Shear Stress

Shear stress at a radial distance r from a rotating shaft centre line due to torsion is given as

$$\tau(r) = \frac{Q \cdot r}{J}$$

where
  Q is the torque
  r is the radial distance from shaft centre line
  J is the polar mass moment of inertia

But the shear stress caused due to bending of a beam of rectangular cross section is

$$\tau(y) = \frac{V \cdot M(y)}{\left(I \cdot t(y)\right)}$$

where

V is the shear force at any particular longitudinal position

$M(y)$ is the first moment of the area of the section from one end (top) about the neutral axis

I is the transverse moment of inertia about the neutral axis

$t(y)$ is the thickness of the cross section which is constant

$M(y)$ is maximum at the neutral axis and zero at the top and bottom ends

Thus, shear stress is maximum at neutral axis and bending stress is zero. Figure 10.12 shows both the stress distributions in the rectangular beam.

In case of a thin rectangular shell of constant thickness (symmetrical about the vertical centre line), shear stress varies along the shell and y is replaced by s in the earlier formula where s is the distance from one end of the cross section varying as one moves from one end along the shell. The thickness t being constant along the entire shell, one can define a term shear flow $q(s)$ which is

$$q(s) = \tau(s) \cdot t(s) = \frac{V \cdot M(s)}{I}$$

One can observe that shear flow increases as one moves along the shell from one end till neutral axis as shown in Figure 10.13. Ship cross section is much more involved with regard to shear flow since ship cross section is not complete, has a hatch opening; thickness along the deck, shell and bottom may not be constant and there are a number of other longitudinal partitions such as bulkheads, double bottom and sloping tanks. Relevant literature can be studied from Hughes and Pike (2010) and Mansoor et al. (2008).

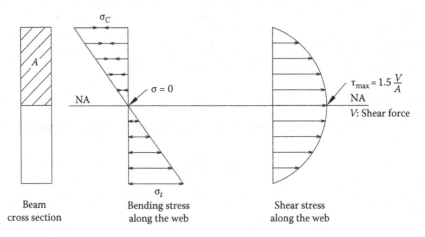

FIGURE 10.12
Bending stress and shear stress distribution along the web of a rectangular beam.

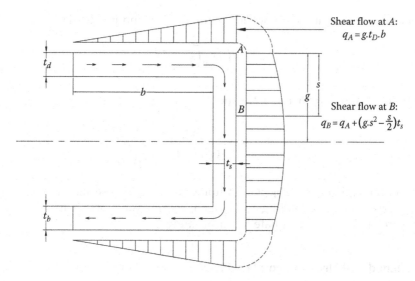

FIGURE 10.13
Shear flow in a rectangular shell.

10.3.3 Buckling Stress

Buckling load is the compressive load applied to a structural member from opposite sides. Normally, pillars are subject to buckling load when load is applied axially. Also in-plane compressive loads on welded structures such as stanchions, stiffeners and plated structures are also subjected to buckling due to in-plane compressive loading. A structure can withstand a certain amount of deformation due to buckling before failure when compressive stress crosses the material yield stress. This load is known as the critical buckling load and is given by

$$P_{cr} = \frac{\pi^2 \cdot E \cdot I}{L_e^2}$$

where
L_e is the effective length $= k \cdot L$ (L is the length of the column, k is the effective length factor)
E is the Young's modulus
I is the moment of inertia of the section about its neutral axis

Values of k can be taken as 1 for both ends pinned so that ends do not deflect but can rotate, and can be taken as 0.5 if both ends are fixed like pillars in ships and land structures. The value of k is 2 if one end is pinned and the other fixed. Very often for welded pillars, stiffeners and stanchions, k is taken as $\sqrt{2} = 0707$. Then critical buckling stress is given as

$$\sigma_{cr} = \frac{P_{cr}}{A} = \frac{\pi^2 \cdot E \cdot I}{\left(A \cdot L_e^2\right)}$$

$$I = A \cdot k_{yy}^2 \ \text{ or } \ k_{yy} = \sqrt{(I/A)}$$

k_{yy} for a hollow circular pipe with outside diameter as r_o and inside diameter as r_i is

$$k_{yy} = 0.5 * \sqrt{\left(r_o^2 + r_i^2\right)}$$

L_e/k_{yy} is known as the slenderness ratio and

$$\sigma_{cr} = \frac{\pi^2 \cdot E}{(L_e/k_{yy})^2}$$

In case of ships, members are subject to complex buckling load situations such as eccentric loading, uncertain in-plane loading and bi-axial compression. Such cases are conventionally taken care of by taking adequate factor of safety.

10.3.4 Stiffened and Unstiffened Plate Panels

The global load described so far works on the entire ship structure and the reaction of the box girder is the primary reaction and the stresses developed due to global loading are the primary stresses. A maritime structure is made up of a number of stiffened panels such as the double bottom panel between two bulkheads, deck panel between bulkheads, bulkhead panel and similar other plate panels stiffened by longitudinal and transverse stiffeners and having an end fixity due to primary structural members at the ends. The loads on such a panel cause reactions which are known as secondary reactions. The reaction of an unstiffened plate panel due to local loading is known as tertiary reaction. Figure 10.14 shows the three types of reactions. The structural layout is the first step to determining the secondary and tertiary responses of the ship structure to sea loads.

10.3.5 Continuity and Structural Alignment

A structure is aligned when the load or stress in a structural member has a direct path to the supporting structure. The structure is continuous when it is capable of transferring the loads on the structure without creating abrupt changes in the stress level.

The primary reaction experienced by the hull girder due to global loads is taken by the midship section, the longitudinal members of which contribute to the section modulus. To ensure proper stress distribution over the length, all longitudinal members must be continuous for at least 40% length of the ship. If the longitudinal have to be ended towards the ends, the changes must be gradual. The primary longitudinal members such as intermediate decks, girders, stringers etc. should not be ended abruptly, but with gradual changes of height and cross section. Figure 10.15a shows the ending of a partial deck on a bulkhead with a bracket connection on the other side to ensure continuity. Figure 10.15b shows how the continuity can be maintained in case the main deck girder is not on a straight line. Proper alignment has to be ensured during production or repair stage. Load on a deck can be transferred to the double bottom through a properly aligned pillar connection as shown in Figure 10.15c. If there are more decks in between, the same alignment must be maintained. Figure 10.16a through c show the connections of transverse frames across the deck to maintain continuity. Beam knees are a necessary item of connectivity to maintain continuity in the transverse direction as shown in Figure 10.17a through d. When stiffeners

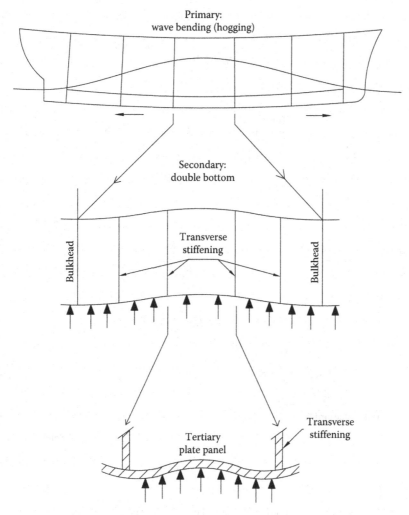

FIGURE 10.14
Primary, secondary and tertiary reactions on ship hull structure.

cross the primary members such as girders, floors or deep beams, connectivity must be done with care so continuity is maintained. Figure 10.18 shows some typical connections. The International Association of Classification Societies (1996–2013) has given standards of connections of joints in ship structures to be used in shipbuilding and ship repairs. If there is a change in the frame spacing or the framing system itself along the length of the structure, it must be ensured that the primary members are changed gradually so that proper continuity can be maintained. Superstructure or a deck house structure presents an unavoidable case of discontinuity. To reduce the effects of discontinuity in the form of stress concentration, one has to do careful layout design. The longitudinal and transverse end bulkheads above the deck should be in line with longitudinal and transverse bulkheads below the deck or the shell side as far as possible. If it is not possible, the deck below must be supported by deep beams and girders in line with the top transverse and longitudinal bulkheads, respectively.

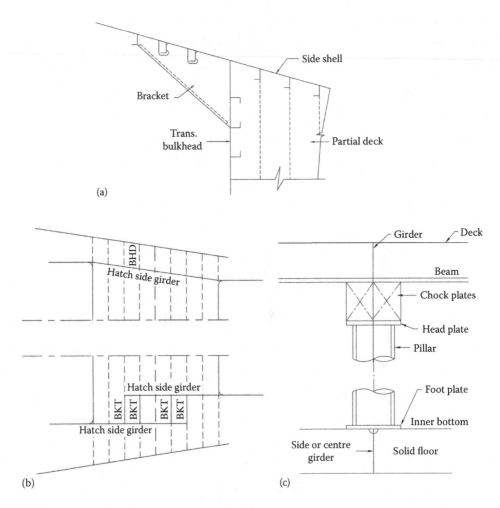

FIGURE 10.15
Examples of structural continuity. (a) Ending of a partial deck with bracket connection, (b) Generation of continuity between deck girders not in line, and (c) Pillar connection showing continuity.

10.3.6 Stress Concentration

Holes must be cut in a marine structure to provide access, cargo handling and passage of engineering systems such as pipes, ducts, etc. Holes are potential sources of structural failure. Openings in structures increase stress in two different ways. Reduction of material to support the load increases stress which can be considered negligible unless the material removed is very high. The other is stress concentration due to abrupt removal of material and hence continuity. In an infinitely long and broad plate, the stress level at the edges of a circular opening is three times the nominal value elsewhere. Care should be taken to see if high stress concentration can be avoided in the first place by ensuring that opening is not in a highly stressed area and it should not have square corners. The degree of stress concentration is primarily a function of discontinuity and even low nominal stress areas can cause high local stress levels leading to plastic deformation and failure. Increase in radius of the corner of a cut results in a sharp reduction in stress concentration. For large openings, 610 mm can be a good approximation for corner radius.

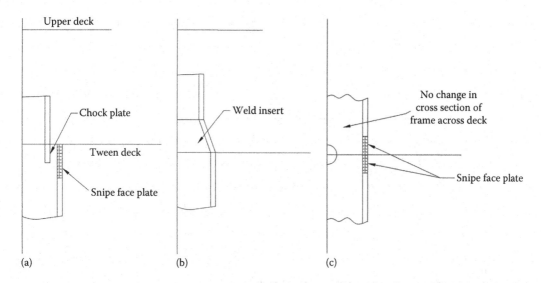

FIGURE 10.16
Transverse frame continuity across deck. (a) Different frame cross-sections across deck, (b) Cranked frame cross-sections across deck, and (c) Same frame cross-sections across deck.

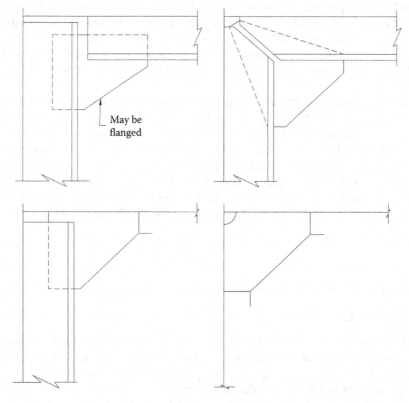

FIGURE 10.17
Beam knee connections.

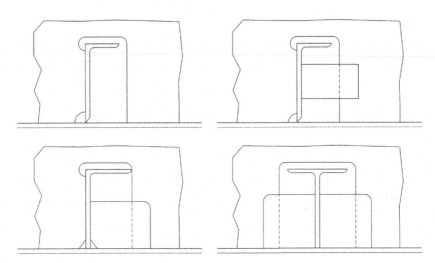

FIGURE 10.18
Stiffener connections with primary members for continuity.

Openings should be so oriented that stress concentration can be reduced. As a first approximation, an opening with a length-to-breadth ratio of 2:1 placed such that the long dimension is parallel to main stress direction can cause reduction of stress concentration by 50%–60% if the opening was oriented otherwise. Accumulation of different welds at the opening should be avoided. Compensation can be provided at the opening so that the structure can withstand the high stress levels. A doubler plate or an insert plate can be provided at the corners in the plane of the plate. A flat bar ring can be provided along the periphery of the opening normal to the plane of the plate. A hatch coaming is a good example of this.

Stress concentration can occur due to any abrupt discontinuity caused either due to design defects or due to faulty production process. Maintenance of continuity as discussed in the previous section is very important. It should be noted that this also means that abrupt change in plate thickness by about 50% or so should be avoided. Any change in plate thickness should be gradual. Misalignment due to design or workmanship should be avoided. Figure 10.19a shows three ways of strengthening the hatch corners to withstand high stress, Figure 10.19b shows design error leading to misalignment and Figure 10.19c shows a misalignment due to production error leading to stress concentration.

10.4 Structural Design

Structural design of a marine vehicle can be done by any of the three following methods: (1) rule-based calculations, (2) direct stress calculations and (3) reliability-based design. Each of the three techniques is very complex and involved. It is beyond the scope of the present work for detailed discussion on all these methods. However, the design philosophy and principles involved in these methods are discussed in the following sections.

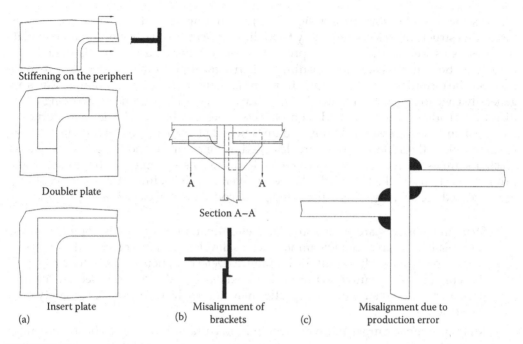

Stiffening on the peripheri

Doubler plate

Insert plate

(a)

A   A

Section A–A

Misalignment of
brackets

(b)

Misalignment due to
production error

(c)

FIGURE 10.19
Examples of stress concentration. (a) Hatch corner strengthening. (b) Misalignment due to design error.
(c) Misalignment due to production error.

10.4.1 Rule-Based Design

Marine structures are subject to global as well as local loads, and the hull girders as well
as the primary, secondary and tertiary members are subject to various stresses due to
tension, compression, buckling, bending, shear and torsion. Further these are also subject
to material degradation due to corrosion and fatigue. Estimation of the loads coming on
each element as well as the combined stresses is rather complex and involved and cannot
be handled regularly in design offices. Therefore, all classification societies have developed rules and codes for structural design of ships and other marine structures. These
are based on experience and understanding of structural reactions to various loads at sea.
Though direct calculation of static still-water bending moment is a part of the rules, most
other loads and their reactions are not specifically calculated in the rules. Further, the
design principles are based on the principle of a single safety factor which is defined as

$$\text{Allowable or working stress} = \frac{\text{Ultimate strength or yield stress}}{\text{Safety factor}}$$

Depending on the design requirement, the ultimate strength (plastic failure) or yield stress
(elastic deformation) is chosen to determine the working stress or allowable stress which
should not be exceeded. If R represents resistance (or material strength of the structure)
and L, the load on the structure and there are a number of loads ($i = 1–m$) acting on the
structure, then the design should aim for

$$\frac{R}{\text{Safety factor}} \geq \sum_{i=1}^{m} L_i$$

This is known as allowable stress design (ASD) or working stress design (WSD). Generally, for marine structural design, the safety factor lies between 2 and 6. Determination of plate thicknesses as well as the sectional properties for stiffeners, girders, etc. is done based on formulations using estimated loading and structural performance as well as experience. So, the formulations vary slightly from one classification society to another. Rules are generally easy to use in a standard design office and save a lot of time by avoiding doing direct calculations. The basis of design of these rules can be stated briefly as design of local structures using standard beam and unstiffened plate theory and estimating tertiary reactions. Next, the stiffened plates are designed using assumed end fixity and secondary reactions are estimated. Scantlings calculated on this basis should be able to withstand primary reactions on the hull girder based on overall ship loading, and if not, scantlings are increased accordingly. This is the principle of code books of classification societies.

- Structural failures are numerous and determination of causes is complex and interrelated. Failures could be minor such as development of a crack and its propagation or major such as hull girder failure. Determination of the exact cause is difficult and may be unknown. So rule-based design calculations based on deterministic loads and structural properties may not be always adequate for safety of the structure.

- Rules are rather simple and work well within certain limits as has been demonstrated by a number of ships being built with these rules which have performed satisfactorily. However, since material resistance as well as loads are uncertain and do not have a fixed value always, the limits of these formulations may be exceeded occasionally causing unsafe performance.

- Due to the use of a standard safety factor without taking into consideration of uncertainties in load and resistance, it is possible that many structural components may be over-designed, leading to increase in overall steel weight, or under-designed, increasing the risk of failure.

- These rules which are based on experience with a large number of products of the same type may not be suitable for new marine structures or structures with non-standard dimensions.

10.4.2 Direct Calculation-Based Design

It is possible to calculate various loads on the ship exactly, for example still-water bending moment and maximum hogging/sagging wave bending moment by superimposing a wave crest/trough at midship. Similarly, slam pressure can be estimated by understanding heaving and pitching motion of the ship. The stresses developed due to various loads can similarly be estimated. As the structure becomes more and more complicated, it becomes difficult to estimate stress by simple beam theory or similar methods. Then it becomes necessary to use numerical methods to estimate stress. The most widely accepted numerical technique used to estimate structural reaction is the finite element method (FEM). With the advent of high-speed computing, it has been possible to estimate stress and deflection of the structure subject to a certain normal or in-plane loading. As has been discussed earlier, it is necessary to estimate the tertiary reactions of an unstiffened plate structure, secondary reaction of a stiffened plate panel and primary reaction of the ship girder. Whereas tertiary and secondary reactions can be estimated relatively easily using FEM since these structures are geometrically less complicated and can be modelled easily in the computer, it is rather difficult to do the same for the entire ship structure. A ship and its structural components, which are large in number, are generally three-dimensional and cannot be defined mathematically. These properties make it suitable for use of FEM where modelling of the entire ship can be done using various different meshing systems. Use of high-speed computers has made it possible to calculate bending stresses due to various types of forces and moments acting on the ship. The resultant stress on the ship girder or a panel or a plate/stiffener can be calculated exactly using FEM.

10.4.3 Reliability-Based Design

There is a certain degree of randomness or uncertainty in our ability to predict the loads (*L*) imposed on the marine structure and its resistance (*R*) which is the ability of the structure to withstand these loads. Some of these uncertainties can be measured and quantified but cannot be adequately controlled at the design stage. These are known as objective uncertainties and those which cannot be quantified are known as subjective uncertainties. Due to these uncertainties, there can be no design solution which can guarantee absolute safety. Thus, marine structural design must not aim at providing absolute safety but adequate safety and proper functioning of the structural components towards giving required performance. Ang and Tang (1990) gives a good description of using probabilistic approach in structural design.

The resistance *R* of the material to withstand load can be the yield stress, ultimate strength, modulus of elasticity (if stiffness is considered) or any other physical property or a combination of a few of these. The uncertainties involved in this could be due to uncertainties involved in the manufacturing process, in the process of handling and plate/stiffener preparation, due to welding and production uncertainties leading to misalignment and stress concentration. The strength of material also reduces due to corrosion and fatigue during service life of the product. The various loads (*L*) coming on the marine structure have been discussed in Section 10.2 and the uncertainties involved in this load estimation could be due to uncertainties in overall loading (draughts), cargo and ballast water loading, cargo operation, sea conditions, motion of the ship and such other causes. These uncertainties or probabilities on *L* or *R* can be modelled using statistical data. But there is an uncertainty or error involved in any statistical modelling. Also analytical models themselves may involve certain uncertainties. Figure 10.20 shows a typical probability

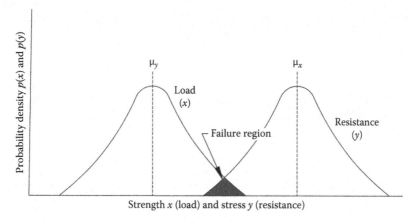

FIGURE 10.20
Probability distribution of *L* and *R* showing the failure region.

density function of working stress (L) and strength (R) and it can be observed that the mean of L is less than the mean value of R, and in ASD method, this can be considered a safe design. However, one can observe that in the shaded area of the figure, R is less than L, leading to a probability of failure.

Reliability of a (structural) system can be defined as its ability to provide the intended performance for a specific time which is normally estimated using probabilities. The complementary event is failure of performance, and therefore,

$$\text{Reliability} = 1 - \text{failure probability}$$

So, to estimate reliability, it is necessary to define the performance criteria and their limit states with significant failure modes. In case of a structural element, performance (g) can be defined as ability (R) to withstand the load (L). Thus, the structure is reliable when $R > L$, or

$$g = R - L$$

and failure can be defined as

$$g < 0.0 \quad \text{or} \quad R < L \text{ (see Figure 10.20)}$$

Reliability can then be defined as

$$g > 0.0 \quad \text{or} \quad R > L$$

The performance criterion g, known as the limit state function, is a random variable which is a function of two random variables L and R. Figure 10.21 shows the probability density of g and the area under the portion of the curve to the left of origin is the total probability

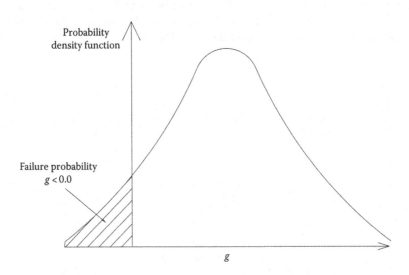

FIGURE 10.21
Probability density function of $g = R - L$.

of failure where $g < 0.0$. The limit state function g can be a linear function of a number of random variables as

$$g(X) = g(X_1, X_2, \ldots, X_n)$$

where X is a vector of basic random variables $(X_1, X_2, \ldots, X_n)$ of L and R. For simplicity let us take a two-dimensional case where the limit state function g is a linear function of two random variables L and R, where both L and R are normally distributed. Then g is also normally distributed. By normalising the random variables into standardised normally distributed random variables, we can draw the limit state function $g = 0.0$ in the space defined by variables L and R (for the time being, consider $R = R'$ and $L = L'$) shown in Figure 10.22. Then the reliability index β can be defined as the shortest distance from the origin to the boundary between safe and unsafe boundaries defined by $g = 0.0$ since this is the most probable failure point P.

Limit state for a marine structure can be based on the following:

- Service or serviceability of the structure which may mean that the structure is unable to provide intended service such as large deflections altering the geometry of the body in question
- Collapse or failure by crossing the ultimate strength limit due to loading
- Reaching fatigue limit state due to low cycle fatigue or high cycle fatigue due to vibration
- Accidental reasons such as collision and grounding

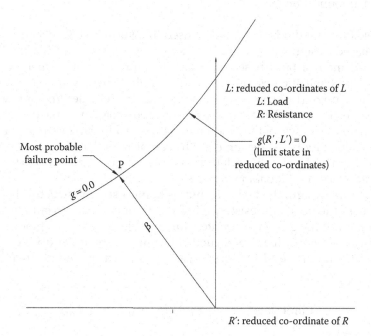

FIGURE 10.22
Limit state failure at $g = 0.0$ and reliability index β.

Failure modes due to crossing limit state could be as follows:

- Tensile/compressive yield of material – elastic deformation by crossing the yield stress limit or plastic elongation and failure by crossing the ultimate strength
- Compressive instability due to buckling where the strength may be considerably less than tensile yield strength
- Failure due to fatigue and brittle fracture

In recent years marine structural design is moving from deterministic design methods towards more rational probability-based design techniques which are normally referred to as limit state design. Information required to do such designs includes a large amount of data for modelling uncertainties in the various random variables defining loads, load combinations and material properties. Based on the data, partial safety factors (PSFs) can be obtained for various random variables. There is a load amplification factor (PSF) for each type of load and load combination which is greater than 1, being higher for higher uncertainties in load estimation. Similarly, there is a resistance reduction factor or PSF associated with each variable of resistance which is generally less than 1. Then,

$$\phi \cdot R = \sum_{i=1}^{n} \gamma_i \cdot L_i$$

where
  ϕ is the resistance reduction factor
  γ is the load amplification factor

First-order reliability methods (FORMs) are used to determine PSFs ϕ and γ.

Reliability-based design can be done at three levels. Level 1 reliability uses PSFs for variables of load and resistance based on reliability approach. But the methods themselves do not require probabilistic description of the variables. The commonly used load and resistance factor design (LRFD) in marine practice is a level 1 design approach (Pike and Frieze 2001, Ayyub et al. 2002). Level 2 reliability-based design uses the probability characteristics (mean and variance) of the random variables and linearization of non-linear limit states. Level 3 reliability-based design uses the complete probability characteristics of the random variables. This level is difficult to attain due to lack of enough data to get the probability characteristics of all random variables.

In ship structure design, the LRFD format is being used for structural design where the designer uses the load and resistance PSFs to account for uncertainties in each variable which may not have been accounted for in ASD practice. In practice, the LRFD format is being used for hull girder structural components under different load conditions. The limit state in the LRFD format can be defined by one of the following two limit states:

Limit state 1: $\phi \cdot R_n \geq \gamma_{SW} \cdot L_{SW} + k_{WD} \cdot \gamma_{WD} \cdot L_{WD}$

Limit state 2: $\phi \cdot R_n \geq \gamma_{SW} \cdot L_{SW} + k_W (\gamma_W \cdot L_W + k_D \cdot \gamma_D \cdot L_D)$

or

$$\text{Limit state 1:} \quad R_n \geq \frac{(\gamma_{SW} \cdot L_{SW} + k_{WD} \cdot \gamma_{WD} \cdot L_{WD})}{\phi}$$

$$\text{Limit state 2:} \quad R_n \geq \frac{(\gamma_{SW} \cdot L_{SW} + k_W \cdot (\gamma_W \cdot L_W + k_D \cdot \gamma_D \cdot L_D))}{\phi}$$

where

ϕ is the strength factor

R_n is the nominal strength such as ultimate tensile strength

L is the nominal load

γ is the load factor

suffices *SW, W, WD* and *D* indicate still-water load effect, wave bending moment load effect, combination of wave bending moment and dynamic effects and dynamic effects, respectively

k_{WD}, k_W, k_D are the load combination factors

Normally, the resistance and load values taken are the nominal values which can be taken equal to design-estimated values. Knowing the load amplification factors and resistance reduction factors (PSF), one can calculate the stress values which should be compared with limit state 1 or 2.

It is possible to calculate the reliability index β using the FORM and compare with target reliability β_0. Classification societies, particularly the IACS, are busy collecting data to establish PSF values for various load and resistance values for different ship types and other structures. So far, success has been limited. However, the IACS has now taken the initiative to establish common structural rules (CSRs) for bulk carriers and tankers and it includes the various PSF values for hull girder strength and structural design method using LRFD technique. Hopefully, reliability-based codes for different and new ship types and structures will be ready soon to be used by designers. Needless to state that primary, secondary and tertiary reactions on hull structure and other components due to loads have to be estimated accurately using direct calculation methods discussed in the previous section.

10.4.4 Corrosion Allowance

Till now we have discussed structural design for a new ship. But as the ship goes into service, the hull gets fouled and corroded. Fouling should be removed regularly and the ship should be newly painted. This does not affect the ship structure. However, corrosion degrades the material. Different types of corrosion are uniform corrosion causing a uniform reduction of plate thickness, galvanic corrosion due to combination of dissimilar metals, erosion corrosion causing thickness reduction, pitting corrosion, crevice corrosion, groove corrosion and bacterial corrosion. Reduction of material can cause reduction of resistance of the material and lead to failure. Also if corrosion takes place in a highly stressed area and the area is subject to fatigue loading, there can be stress corrosion cracking.

The scantlings calculated using any of the methods discussed previously are known as net scantlings which do not take into account material degradation in service life.

It is necessary to add an extra thickness to take into account material degradation due to corrosion. Corrosion is not uniform everywhere. So the corrosion margin to be added to the net scantling should also vary. Classification rules give what corrosion allowance should be provided based on the location of plates or stiffeners. Any extra thickness desired by the ship owner has to be over and above the corrosion allowance. The final scantlings are known as gross scantlings.

10.4.5 Fatigue in Marine Structure

Fatigue constitutes a major source of local damage in marine structures. The local damage can be initiation of a crack which can spread and subsequently form a brittle fracture causing major structural failure. The effects of fatigue are especially severe in regions of high stress concentration. High stress concentration can be due to design inadequacy or due to faulty production process or degradation of material due to corrosion. Fatigue damage can be avoided by taking proper care at the design stage and regular inspection of vulnerable regions for detection and repair of fatigue cracks.

Fatigue can be of two types. In high-cycle low-stress fatigue, the endurance limit of material is reached after the material is subject to several million cycles of loading causing low stress levels. High-cycle fatigue occurs in a portion of a structure subject to fast repeated loads which can be areas close to propellers, machinery, foundations of rotating machinery and areas subject to severe vibration. As an example, a repeated loading at a frequency of 100 per minute can reach 144 million cycles in 10 days. Generally, high-cycle fatigue is of local nature and unlikely to cause major structural failure. The other type of fatigue is low-cycle high-stress fatigue where stress levels are high which cause structural failure due to repeated loading, the total number of cycles before failure being of the order of 10^5. Marine structures like ships are subject to alternating loads leading to generation of high stresses due to passing of waves, alternating between hogging and sagging, at a relatively low frequency. In such a case, the areas of high stress concentration, if subject to these fluctuating stresses, reach their endurance limit at less than a million cycles. Resistance of material to fatigue depends on many factors such as the material itself (physical properties), surface finish (crack initiation is delayed on smooth surfaces), presence of stress concentration or areas having residual stresses and corrosion leading to stress corrosion cracking. Ship hull structure is exposed to randomly varying loads with time, the parts are fabricated by welding and are not precision based, the surface finish is relatively rough and the structure is exposed to harsh and corrosive environment. Therefore, fatigue endurance in ship structures is rather uncertain and cannot be predicted with accuracy.

Initiation of a fatigue crack is generally on the surface and a smooth surface delays crack initiation. A fatigue crack is transgranular; that is, it propagates within the grains rather than along the grain boundaries. If the area around the crack is subjected to further periodic loading, the crack propagates generally in a direction perpendicular to the direction of tensile stress. If further periodic load is applied, stress on the residual cross section increases, leading to further crack propagation ultimately leading to brittle or ductile fracture.

Ideally stress cycles can be basically of three types – (1) pulsating tension cycle where periodic loading is between zero and tension, (2) alternating tension cycle where periodic loading is between negative and positive tension and (3) half tensile cycle where both the maximum and minimum stresses are tensile. The three types of cycles are shown in Figure 10.23.

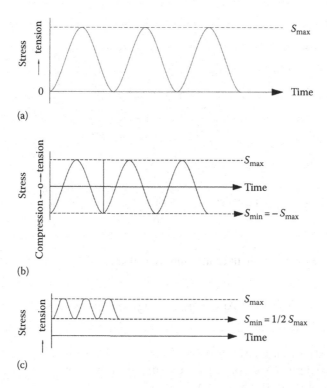

FIGURE 10.23
Different types of load cycles (a) pulsating tension cycle, (b) alternating tension cycle, and (c) half tensile cycle.

The fatigue limit of a material is normally determined by laboratory tests by subjecting a beam to cyclic loading of one of the above three types till failure. Same materials with different surface finish and different types of loading show different stress – number of cycles (S–N) relationships. In a typical S–N diagram, the number of cycles for failure is low for high stress and follows a straight line with a nearly constant slope. However, as cycles increase, the failure stress (in log scale) is nearly constant and low. The sloping portion of the S–N curve can be fitted to an equation given as

$$S_2 = S_1 \cdot \left(\frac{N_1}{N_2} \right)^k$$

where
 S_1 and N_1 are a known failure stress and corresponding number of cycles, respectively
 S_2 is the failure stress of the material at N_2 cycles

Ships use different types of steel rolled plates and sections and so the nature of S–N curve for purposes of marine applications vary. The International Association of Classification Societies (IACS) gives the following equation for fatigue failure in its CSR formulations for tankers:

$$S^m \cdot N = k_2$$

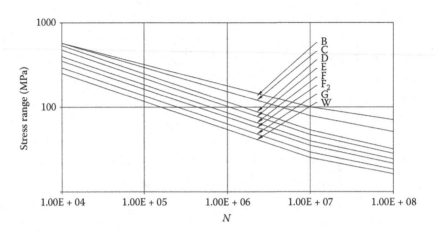

FIGURE 10.24
Basic *S–N* diagrams for steel in air for different loading conditions.

$$\text{And } \log_{10}k_2 = \log_{10}k_1 - 2\delta$$

where
  S is the failure stress range in N/mm^2
  m and k_2 are constants based on the type of material, type of cyclic loading, etc.
  m is the inverse slope of the *S–N* curve
  k_1 is the mean of the *S–N* curve
  δ the standard deviation

Figure 10.24 taken from the CSR for tankers shows the variation of the *S–N* curve with loading shown as different curves *B, C, D,* ... etc.

In a ship considered to be a girder in longitudinal bending, there are hot spots which are subject to failure due to cyclic loading. Such hot spots are generally the joints of primary members with longitudinal stiffeners and other areas of stress concentration such as corners of hatch openings. At such locations, hot spot stresses are assessed. Such stresses are amplified for notch toughness, for corrosive environment, type of local loading such as type of dry bulk cargo, liquids such as fuel oil, water ballast or cargo oil and material based on thickness of plating used. This amplified stress is the equivalent fatigue stress for which corresponding cycles for failure can be determined. Taking the life of the ship into consideration, a damage parameter is estimated for a particular loading condition. This process is repeated for different loading conditions at the end of which a cumulative damage parameter is established which should be within a predefined limit for the life of the product considered.

11

Layout Design

A marine vehicle or structure has limited enclosed and exposed space that has to house all equipment required for the functioning of the product at sea and also provide space for accommodating various items including crew and passengers on board. Since the product is at sea, the space distribution becomes an important part of the design and is formally known as general arrangement (GA), and the drawings depicting this arrangement are called GA plans or GA drawings. Layout design includes defining spaces for various requirements, establishing boundaries for such spaces and also partitioning such spaces, locating equipment and facilities in each space and providing access to areas within each space and egress during emergencies.

About 90% of international trade is carried out by sea. Ships carry this cargo from location to location around the globe, so maximum space must be allocated for carriage of cargo in commercial ships. But in the case of naval vessels and other specialised vessels such as port craft, research vessels or pollution control vessels, space must be allocated for various activities these vessels must perform and specialised equipment they must carry for this purpose. Similarly, offshore platforms, fixed or floating, must have space allocated for equipment required for execution and control of drilling operation including pipe handling, storage of oil or gas and offloading of the same cargo. An underwater vehicle like an AUV, submersible or submarine is space and weight sensitive, and all equipment must be housed inside the body including the propulsion machinery and control mechanism. Thus, layout design must ensure the allocation of spaces for the main activity of the vehicle or structure and spaces for supporting these activities. Generally, spaces must be provided for the following categories:

Cargo spaces

Liquid or tank spaces

Accommodation and associated spaces for crew and passengers

Machinery and working spaces

Any other miscellaneous spaces

Different spaces are separated from each other by steel watertight longitudinal or transverse partitions or bulkheads and horizontal partitions or decks. The interconnectivity is provided by watertight doors on such partitions where required or by top openings such as hatch openings, manholes, access hatches or companion ways. If access openings on transverse watertight bulkheads are below the load waterline (LWL), arrangements must be made to control the openings from above the freeboard deck. The spaces and allocation of equipment and facilities determine the weight distribution, centre of gravity (CG) position and mass moment of inertia. The layout, therefore, must conform to the performance standards of the vessel with regard to stability, both intact and damaged, strength and vibration. It should also conform to ease of transportation and cargo handling, overall safety and ergonomics in accommodation. Structural design and layout must also integrate with the GA so that machinery and equipment can be placed properly on adequate foundation and provide working space around each equipment or facility.

Figure 11.1 shows the internal partitions and enclosures of a 120 m bulk carrier, and in Figure 14.5, similar arrangements are shown for three different ship types: a feeder container vessel, a multipurpose cargo vessel and a product tanker, all of the same length. The sections through the cargo holds in these diagrams have been shown later in Figure 11.2e, h, d and g, respectively.

The uppermost continuous horizontal platform of the ship is known as the upper deck. The ship being a watertight envelop, the topmost continuous watertight deck, is known as the freeboard deck where all openings on deck through which water entry can take place must be closed watertight while at sea. The freeboard deck need not be the upper deck and can be below the upper deck as in passenger ships. The freeboard deck is the reference deck for freeboard measurement. There can be other continuous or intermittent horizontal platforms or decks below the upper deck, and the spaces between any two of these decks are often referred as tween decks. If such decks are above the upper deck, they can be separately designated. If there is a deck above the upper deck at the forward end of the ship, it is known as the forecastle deck and a similar deck at the aft end is known as the poop deck. The vertical height of any deck from the deck below should have a height more than 2.3 m. The required bow height as per International Convention on Load Line Regulations must be maintained at the fore-end with or without the forecastle deck. Longitudinally, the forecastle deck should extend from forward end till the fore peak bulkhead or to the bulkhead aft of the forepeak bulkhead. The reason for having a forecastle deck can be the achievement of minimum bow height, provision of forecastle deck area for anchoring and mooring equipment, adequate volume underneath for storage of anchoring and mooring gear, deck stores and provision of additional cargo space.

Traditionally, ships have been transversely framed where the ship's hull plating is stiffened by transverse frames located longitudinally at short distances from each other. The location of all such frames is defined by appropriately assigning frames from aft end of the ship to the forward end. Today, even though ships or other marine structures may be transversely and/or longitudinally framed, longitudinal location is referred to in terms of frame number which, in turn, is based on frame spacing.

The foremost watertight bulkhead in a ship is known as the collision bulkhead and the narrow space between the fore-end of the ship and the collision bulkhead is known as fore peak which houses the fore peak tank suitable for carriage of ballast water for trimming purposes. Similarly, the aft peak bulkhead is the aftermost watertight bulkhead aft which has the aft peak tank. Classification societies have provided guidelines for fixing frame spacing in mm. The normal frame spacing between aft peak and $0.2L$ form the forward perpendicular (FP) may be taken as

$450 + 2L$ for transverse framing

$550 + 2L$ for longitudinal framing, L being in m

However, it generally does not exceed 1000 mm. Frame spacing is generally not to exceed 600 mm or above, whichever is less in peaks or the foremost compartment and the stern. It is not to exceed 700 mm or above, whichever is less in the region between the collision bulkhead and $0.2L$ from FP.

A ship has a forward collision bulkhead at a distance X_C from the FP such that in case of a forward collision the portion of the ship forward of the collision bulkhead is damaged,

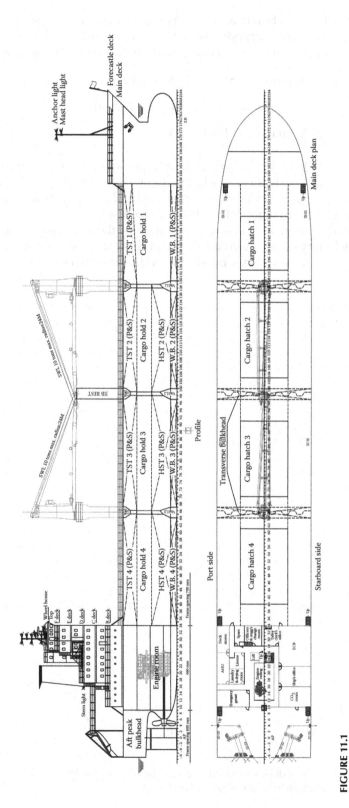

FIGURE 11.1
Location of various spaces in a 120 m bulk carrier.

leaving the space aft of this bulkhead intact. For ships other than passenger ships, the
distance X_C from the forward perpendicular to the collision bulkhead is to be between the
following limits as per the International Convention for the Safety of Life at Sea (SOLAS):

$$X_c, \min = 0.05L - X_R \text{ (m)} \quad \text{for } L < 200 \text{ (m)}$$

$$= 10 - X_R \text{ (m)} \quad \text{for } L \ge 200 \text{ (m)}$$

$$X_c, \max = 0.08L - X_R \text{ (m)}$$

For ships with ordinary bow shape $X_R = 0$ and for ships having any part of the underwater
body extending forward of the FP, e.g. a bulbous bow $X_R =$ the least of $G/2$, 0.015 L and
3.0 (m) where G is the distance from FP to the forward end of the protruded part, and L is
the length on water line at draught of 0.85 depth of the vessel in m.

For most cargo ships today, the compartment containing the main propulsion machinery
and auxiliary machinery or, the machinery space, is normally located forward of the aft
peak bulkhead, between the aft peak bulkhead and the engine room forward bulkhead.

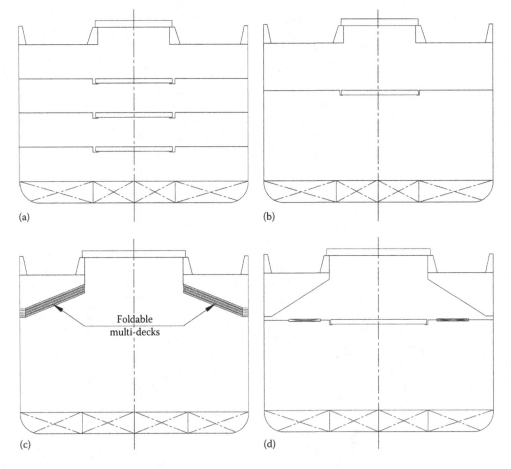

FIGURE 11.2
Cross section of different types of ships. (a) multi-deck cargo ship, (b) tween-deck cargo ship, (c) multi-purpose
cargo ship with foldable decks, (d) multi-purpose cargo ship.   *(Continued)*

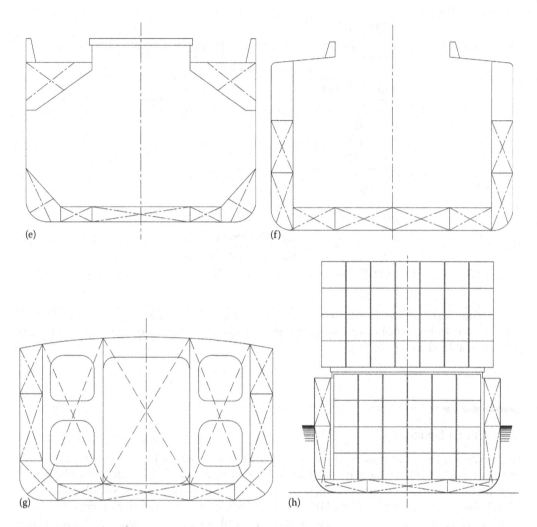

(e)

(f)

(g)

(h)

FIGURE 11.2 (*Continued*)
Cross section of different types of ships. (e) bulk carrier, (f) double-hull bulk carrier, (g) double-hull tanker, and (h) container-feeder vessel.

Cargo space is located between engine room forward bulkhead and forward peak bulkhead. Very rarely, the engine room is moved forward by one cargo hold space, and when this occurs, there could be a cargo hold aft of the engine room. In naval vessels, the engine room may be placed in semi-aft or midship position in which case the propulsion shaft aft of the engine room aft bulkhead is enclosed in casings known as shaft tunnel and the remaining spaces in the aft region may be utilised for other purposes. In merchant vessels, the overall cargo space is divided into a number of cargo spaces or holds by dividing with transverse watertight bulkheads. There could also be longitudinal bulkheads separating the cargo space into transverse watertight compartments. This is done mainly to ensure limited water entry into the ship and limited oil discharge from an oil tank in case of damage, reduce free surface effects in case of partial loading particularly during loading and unloading operations and also provide strength to the ship structure. One way to ensure adequacy of watertight subdivision is to satisfy the probabilistic damage stability

TABLE 11.1

Number of Transverse Watertight Bulkheads

Ship Length (m)	Engine Room Location	
	Aft[a]	Elsewhere
$L \leq 65$	3	3
$65 < L \leq 85$	4	4
$85 < L \leq 105$	5	5
$105 < L \leq 115$	5	6
$115 < L \leq 125$	6	6
$125 < L \leq 145$	6	7
$145 < L \leq 165$	7	8
$165 < L \leq 190$	8	9
$L > 190$	Specially considered	

[a] With aft peak bulkhead forming after boundary of machinery space.

requirements discussed in Section 6.8. Classification societies also provide guidelines with regard to the number of transverse bulkheads in a ship based on length which is given in Table 11.1.

11.1 Cargo Spaces

Unpackaged cargo, but non-containerised cargo (NCC), such as loose cement, grains and ores, is termed bulk cargo, whereas cargo shipped as a unit (bags, bales, barrels, boxes, cartons, drums, pallets, sacks, vehicles, etc.) is termed break bulk. Neo-bulk cargo is a subcategory of general cargo similar to break bulk cargo and container cargo. This cargo is pre-packaged, counted as they are loaded and unloaded, not stored in containers and transferred as units at ports.

About 90% of world trade is carried out by sea. Therefore, transportation of cargo in ships is one of the main activities conducted at sea. Cargoes can be of different types such as general or break bulk cargo, solid bulk cargo, liquid bulk cargo and unitised cargo such as containers, barges or wheeled cargoes on roll-on/roll-off (RORO) ships.

If a cargo compartment is filled with grain cargo, then the space that could be occupied by grain is called the grain capacity which is the moulded volume of the compartment reduced by the volume occupied by structurals in the compartment. It is difficult to estimate the volume of structurals exactly and normally 2%–5% of moulded volume is assumed for this purpose. If the hold was used to carry bales, the capacity that could be utilised is less than the grain capacity since the space between frames and longitudinals with wooden battens and packings on the sides and below deck cannot be utilised for storage of bales. This capacity which can be used to store bales is known as bale capacity and is taken as 5% less than grain capacity. Stowage factor is the ratio of weight to stowage space required under normal conditions usually expressed as cubic metres per tonne (m³/tonne) or cubic feet per ton (ft³/long ton). Broken stowage is the space between packages which remain unfilled. This percentage varies with the type of cargo and with the shape of the hold. It is large when large cases have to be stowed in an end hold. On an average, 10% of space can be taken as broken stowage.

11.1.1 General Cargo

All types of break bulk and neo-bulk cargo come under the category of general cargo. Break bulk cargo includes all packed goods in crates, bags (grain, rice, etc.), barrels, boxes, pallets, drums, lumber, steels, papers, cars, wagons, trucks and refrigerated cargoes such as fruits and meat and canned food. Neo-bulk cargoes are goods that are pre-packaged, are counted as they are loaded and unloaded on ships and include heavy machinery, bundled steel, scrap iron, wastepapers, cars and other pre-packaged items. General cargo can be very light like cars, cotton or hemp in bales and can also be very heavy like steel plates and rolled sections. The stowage factor of general cargo varies not only based on the density or specific gravity of cargo but also based on the packaging method such as bundles, bales, pallets, chests or crates. Table 11.2 gives stowage factor of some general cargo items normally carried in ships. If general cargo ships provide liner service between two or more initially identified ports, the type of general cargo may be known a priori. But if a general cargo ship works as a tramp ship, cargoes and ports are not known a priori and it may be required to carry general cargoes, standard containers, some amount of bulk cargoes and other varieties of cargoes. Such a ship is known as multipurpose cargo ship. Such vessels are generally limited in size based on cargo availability and restricted in dimension based on port requirement. Generally, such ships are limited by deadweight up to 35,000 tonnes.

TABLE 11.2

Stowage Factor of Some Cargoes Commonly Carried by Sea

Item	Stowage Factor (m^3/tonne)
Iron ore	0.4–0.53
Bauxite	0.85–0.93
Copper ore	0.4–0.63
Fertiliser	0.74
Cement	0.65
Sugar	1.3
Salt	0.99–1.13
Coal	1.22–1.56
Grain (light)	>1.60
Grain	1.25–1.70
Crude oil (light)	1.07
Crude oil (heavy)	0.95
Woodchips	2.5
Cars	4.0
Loaded containers	1.6–3.0
Cotton (compressed) bales	2.4
Hemp (compressed) bales	3.4
Tea chest	1.83
Pig iron	0.28–0.36
Steel scrap in bundles	0.56–1.11
Steel plates and sections in bundles	0.4–0.6

Multipurpose cargo ships carry non-standard cargo sizes such as reefer cargoes or cars or non-standard cargoes and therefore require large floor space or cargo deck space. The bottom floor area is the double bottom of the ship providing a continuous floor space. To provide more deck area, a number of intermediate decks are generally provided in such ships. The height of the hold space and each tween deck space should be adequate for the maximum height of cargo in that deck space. Versatility of carrying different types of cargo is one of the main requirements of such ships. Therefore, decks are so arranged that such ships can carry general cargoes, bulk cargoes and also containers. For this purpose, many times the intermediate decks are made foldable so that volumetric space can be made available for carriage of bulk cargoes. Figure 11.2a to d shows the cross sections of hold space of general cargo ships with different hold arrangements.

Generally, vertical loading and unloading is adopted in such ships using single or twin cranes. Providing cranes on board ships ensures that the ship does not depend on port facilities for loading and unloading. It also means that large weight is placed at a higher level on deck. Further adequate strengthening has to be done to place heavy cranes on deck. The crane boom should be of such length that it must have an outreach of 4–5 m beyond the ship side for picking up load from or unloading cargo on the quay side. For accessing the hold or tween deck spaces for loading or unloading cargo, hatch openings are provided on all decks keeping the clear openings vertically in line so as to access the hold space and all tween deck spaces easily. Hatch openings are generally rectangular in shape with width around half of the beam of the ship such that there is no structural weakness introduced in the ship girder. The length of a hatch opening can be slightly less than the length of the hold such that enough space is available on deck, on either side of the hatch opening, to stack the hatch cover panels. If the hold is served by one crane hook, the crane hook at one end should reach the other end of the hatch opening. If the hold is served by two hooks, one from either end, each hook should be able to reach at least the middle of the hatch opening. The upper (freeboard) deck hatch opening must be closed by watertight hatch covers sitting on hatch coamings of adequate height satisfying ILLC recommendations. The hatch covers on other decks need not be watertight and should form a continuous surface when closed for ease of cargo storage. The hatch covers are generally in panels operated (raised and lowered) remotely by mechanical means by wire ropes or by hydraulic or electro-hydraulic automatic mechanism. These could be of end-folding or side-rolling type. Buxton et al. (1978) gives a good description of various types of hatch covers and their requirements for installation.

Sometimes, if horizontal loading facility is to be incorporated, the ship must have side doors above the LWL, which must open in such a manner that access from quay side is available to the tween deck level so that loading vehicles such a as forklift trucks can move smoothly in and out of the ship. Such side doors must be arranged such that these can be held watertight when closed and do not open accidentally due to water pressure from outside while at sea. The forklift trucks could also move inside the ship arranging the cargo properly. The ship could also have cargo lifts to lower and raise cargo from one level to another.

The spacing of bulkheads can be nearly equidistant giving equal length holds. Alternately, bulkheads could be so placed that holds at the ends are of longer length than those at the middle, giving nearly equal volume holds. One can also arrange a longer hold in the midship region with other holds of smaller length so that long units of cargo such as steel plates and rolled sections can be accommodated in the long hold.

11.1.2 Solid Bulk Cargo

Solid cargo that is transported unpackaged in large quantities in granular or particulate form such as grain, ore, coal or gravel is called solid bulk cargo. Bulk cargo occupies the shape of the compartment in which it is loaded and the volume it occupies includes the intergranular air space. Therefore, the stowage factor of bulk cargo is close to the inverse of density. Bulk cargo can be very light like loose grain or very heavy like iron ore. Table 11.2 gives the stowage factor of some bulk cargoes carried in ships across the seas. Since freight for carriage of bulk cargo is charged based on weight carried, the cargo space in a bulk carrier should be arranged to carry maximum amount of cargo by weight. The maximum amount of cargo or deadweight that can be carried in a ship is equal to the displacement of the ship up to the freeboard draught reduced by the lightweight of the ship. The volume of cargo space required for this deadweight is equal to the cargo weight (= deadweight – weight of consumables, crew and their effects) multiplied by the stowage factor. Thus, while carrying grain, volume may be maximised, whereas for carrying ore, volume is not important. Bulk carriers are normally designed as deadweight carriers where the design draught is also the freeboard draught.

The total cargo volume available in a ship is the volume enclosed between the double bottom and the upper deck and between the forward collision bulkhead and the engine room forward bulkhead plus the volume of the space enclosed within the hatch coamings. Since bulk cargo requires volume, these ships are normally single deck ships where the upper deck is the freeboard deck. Sometimes, the volume between the forecastle deck and the upper deck is also used for carriage of cargo increasing the volume of cargo space. The size of the holds can be based on equal length, equal volume, or specific (typically alternate) holds long and short. The cargo should be so distributed that the strength of the vessel is not impaired and no undue stresses are developed due to cargo distribution. The CG of the cargo should be at such a height that the stability of the ship is not impaired and the rolling period is within acceptable range. If the bulk carrier is a dedicated ore carrier, the double-bottom height can be raised substantially above the minimum height required from other considerations, thus increasing the cargo height and the cargo CG. On the other hand, if it is a dedicated grain carrier, double-bottom height can be kept at a minimum. In such a case, to increase the volume of cargo space, the ship may be provided with forward and aft shear. But generally, a bulk carrier is required to carry different types of bulk in different legs of the voyage, i.e. in forward voyage and return voyage. In such a case, the bulk carrier is designed for light cargo with provision that heavier cargo can be loaded in specific or alternate holds. Special care has to be taken to ensure that the strength of the vessel for such loading is adequate.

When bulk cargo is loaded in a compartment, small or big, it forms a cone shape on top and the angle between the horizontal plane and the cone slope is known as the angle of repose. The angle of repose of different types of cargo varies, lighter cargoes having lesser angle of repose than heavier cargoes as shown in Table 11.3. Granular cargo, when loaded in a compartment, settles as the ship moves in the sea way and reduces in volume by about 2% due to reduction in intergranular air space. Further, granular cargo carried in a ship is likely to shift from one side to another due to heeling of the ship increasing heal of the ship causing loss of stability. The International Maritime Organisation (IMO) has recommended requirements for carriage of granular cargo to avoid loss of stability and thus safety of ship. As per IMO recommendation, granular cargo having an angle of repose less than 35° is liable to shift. Bulk carriers are, therefore, provided with top side sloping tanks of about 30° inclination so that empty space is reduced when cargo is loaded in the ship, and therefore shifting of grain from one extreme side (port or starboard) to the other

TABLE 11.3

Angle of Repose of Different Bulk Cargoes

Cargo Type	Angle of Repose (°)
Grain (wheat)	27
Granular urea	27
Granite	35–45
Dry sand	34
Iron ore	36

is avoided reducing grain shifting moment. In addition to top side sloping tanks, bulk carriers are often provided with bottom side sloping tanks, called hopper tanks, the angle of inclination being between 40° and 45°. During the unloading process, cargo automatically slides down towards the centre of the hold facilitating unloading using grabs of cranes. Figure 11.2e shows a bulk carrier cross section with top and bottom side sloping tanks.

A traditional bulk carrier is a structure limited by the sloping tanks, top and bottom, and the double bottom. This leaves a portion of the side shell as single skin. So this portion is open to damage due to collision and, occasionally, grounding. Seawater enters the hold and causes loss of stability due to damage. This entry of water damages the cargo and exerts excessive pressure on end bulkheads which could rupture one after the other, leading to capsize within a short time leading to loss of ship and life. Study of bulk carrier fatalities between 1978 and 1998 showed that 45.8% had occurred due to side shell failure. Therefore, it is being recommended now that bulk carriers should be of complete double-hull construction by having a double skin on the single side shell portion. The distance between the outer and inner side skins should be such that access is possible for a surveyor to enter this space for regular inspection, a recommended value being a clear space of 0.6 m at the narrowest portion. A double-hull bulk carrier cross section is shown in Figure 11.2f. It can be observed that the need for top and bottom sloping tanks is eliminated.

11.1.3 Liquid Bulk Cargo

Different types of liquids have varying physical and chemical properties which can cause pollution due to voluntary discharge such as tank cleaning or involuntary discharge such as damage due to accidents and grounding. The liquid could also be noxious or toxic and, if discharged to sea, can cause severe damage to marine ecosystem. Since accidents and grounding take place near the shore, large-scale discharge can cause severe damage to shoreline and affect human settlement. Carriage of liquids may also cause safety problems due to inflammable nature of many cargoes. There could also be issues of purity, solubility and temperature sensitivity. Gases such as natural or petroleum gas, ammonia or similar material need to be compressed, preferably liquefied, to be carried by sea so that volume required can be made available in a ship. Therefore, liquids can be crude oil and other petroleum products, chemicals, vegetable oils or liquefied gases such as LPG, LNG, ammonia and similar products.

11.1.3.1 Crude and Product Oil

Properties of crude oil vary based on the location from which crude is extracted and properties of petroleum products vary with the type of product. Machineries and systems required for loading and unloading of cargo, for tank cleaning, for control of fire hazard, etc. are discussed in Chapter 13. But all such products have a density nearly equal to that of water. So if a cargo compartment was damaged and oil came out, the empty space would be filled with water which may not cause much sinkage and trim and loss of stability. Therefore, in the past, tankers used to be of single skin construction even without double bottom and were considered to be very safe ships. Till the 1970s, tanker size increased taking advantage of scale and so very large and ultra large crude carriers were built. But subsequently, tanker size was restricted due to reduction in oil transportation by sea. Further, a number of tanker and offshore oil platform disasters caused large oil pollution of the sea.

The oil pollution severity came to focus with *Torrey Canyon* disaster in 1967 where the tanker ran aground due to navigational error and nearly 31 million gallons of oil was spilled on the coastal waters of England and France. In 1978, the tank vessel *Amoco Cadiz* carrying 227,000 tonnes of oil ran aground off the coast of Brittany, broke into two and polluted 360 km of French coastline. In 1989, Tanker *Exxon Valdez* ran aground spilling 11 million gallons of oil into the biologically rich waters of Prince William Sound, Alaska, United States. *M.V. Braer* carrying 84,700 tonnes of oil was grounded in 1993 off Shetland Island, Scotland, in 1993 spilling its contents. *M.V. Erika*, an oil tanker, broke into two in heavy weather in 1999 off the coast of Bretagne, France, carrying 30,000 tonnes of furnace oil. The oil tanker *Prestige* carrying 77,000 tonnes of oil suffered a 50 m gash in November 2002 and broke into two off the coast of Spain. The largest oil platform disaster is that of *Deepwater Horizon* in the gulf of Mexico which spilled 780,000 m³ of crude oil before capping.

Tanker disasters have become a major concern of the international community and the IMO has been discussing this issue of pollution avoidance at sea since then, and the convention on this is known as MARPOL, recommendations of which have been updated from time to time. It is not the intention here to discuss these in detail with regard to the design of oil-carrying tankers. Suffice it to say that the major effect has been the development of double-hull tanker. The tanker today is no more of single skin construction but is constructed with double skin. The other major change has been development of segregated ballast tanks, or ballast water can be carried only in designated tanks separate from designated cargo oil tanks. Though tankers are assigned Type A freeboard (Section 5.3) enabling these vessels to carry large amount of cargo by weight, due to segregated ballast tanks, cargo space is reduced. Therefore, it is not possible to load a tanker up to freeboard draught anymore.

IMO recommends that wing tanks or shell side tanks should extend for the full depth of the vessel or from top of double bottom to the upper deck and that cargo tanks should be arranged inside of these side tanks.

For vessels of 5000 tonnes deadweight (dwt) and above, the width of the side tank w is

$$w = 0.5 + \frac{\text{dwt}}{20,000} \text{ (m)}$$

or 2.0 m, whichever is the lesser, with a minimum value of $w = 1.0$ m.

For vessels of less than 5000 tonnes deadweight (dwt)

$$w = 0.4 + \frac{2.4 \times \text{dwt}}{20,000} \text{ (m)}$$

with a minimum value of $w = 0.76$ m.

Similarly, for double-bottom space at any cross section, the depth of each double-bottom tank or space shall be such that the distance h between the bottom of the cargo tanks and the bottom shell plating is not less than specified as follows:

Oil tankers of 5000 tonnes deadweight and above $h = B/15$ (m) or 2.0 m, whichever is the lesser, with a minimum value of $h = 1.0$ m

Oil tankers of less than 5000 tonnes deadweight, with a value of $h = B/15$ (m), but in no case less than 0.76 m

Oil cargo tanks generate residues which settle at the bottom and corners between plates and sections on the sides. These require to be cleaned occasionally to maintain cleanliness and carrying cargoes of slightly different compositions. The cleaning is normally done by crude oil washing system under strictly laid guidelines for fire prevention. Even though there are enough designated ballast tanks giving required forward and aft draughts, sometimes during heavy weather at sea in a ballast voyage, it may be necessary to load seawater to a ship's hold to increase the draught for ship's safety. This generates oily water mixture that cannot be discharged to sea. This oily water mixture is known as slop and this must be loaded in a separately designated tank as slop tank. This slop is then discharged to a port facility in the next port of call or is passed through an oily water separator such that oil can be removed from the slop and the separated ballast water is clean (containing less than 15 ppm) and can be discharged to sea. The volume of the slop tank need not be less than 3% of the total cargo volume.

Thus, cargo space and slop tank are located between the forward collision bulkhead and engine room forward bulkhead in the tanks located inside the inner hull. This space is divided into a number of holds by placing transverse bulkheads as discussed earlier. Further, these compartments could be divided transversely into side and centre tanks by placing longitudinal bulkheads as required. Since cargo tanks are generally pressed full (up to 98% of available volume) when loaded or are empty when in ballast voyage, there is no negative effect of free surface on stability. However, division of tanks should be such that cargo loading and unloading can be controlled and done uniformly without any adverse free surface effect in partial loading condition.

In an offshore oil exploration or production facility, there may be requirement of storage of oil on the same or another floating or fixed platform. Such storage facility may be located in a drill ship, a floating storage and offloading facility or a floating production, storage and offloading (FPSO) facility. The space allocated for storage follows the same principles of cargo space in a tanker. This oil is required to be transported across the sea to a land-based facility which can be done by being offloaded either to a ship or to a pipeline on the seabed if the distance of transportation is not very long.

11.1.3.2 Chemical Cargo

Chemical cargoes cover an extensive range from noxious or hazardous heavy chemicals such as sulphuric acid, phosphoric acid, nitric acid, chlorhydric acid, caustic soda and ammonia to other chemicals such as coal tar products including benzene, phenol and napthalene; food products such as molasses and alcohols; and edible oils and fats such as palm oil, soybean oil, sunflower oil, palm kernel oil, peanut oil, olive oil, cottonseed oil and coconut oil.

IMO has recommended a four-category categorisation system for noxious and liquid substances as follows: Category X, noxious liquid substances that, if discharged into the sea from tank cleaning or deballasting operations, are deemed to present a major hazard to either marine resources or human health and, therefore, justify the prohibition of discharge into the marine environment; Category Y, noxious liquid substances that, if discharged into the sea from tank cleaning or deballasting operations, are deemed to present a hazard to either marine resources or human health or cause harm to amenities or other legitimate uses of the sea and therefore justify a limitation of the quality and quantity of the discharge into the marine environment; Category Z, noxious liquid substances that, if discharged into the sea from tank cleaning or deballasting operations, are deemed to present a minor hazard to either marine resources or human health and therefore justify less stringent restrictions on the quality and quantity of the discharge into the marine environment; and other substances (OS), substances that have been evaluated and found to fall outside Category X, Y or Z because they are considered to present no harm to marine resources, human health, amenities or other legitimate uses of the sea when discharged into the sea from tank cleaning or deballasting operations. The discharge of bilge or ballast water or other residues or mixtures containing these substances are not considered to cause any marine pollution.

IMO has further recommended three types of ships for carriage of chemicals by the sea as follows:

Ship Type 1 is a chemical tanker intended for the transportation of products considered to present the greatest overall hazard. The quantity of cargo required to be carried in such a ship should not exceed 1250 m^3 in any one tank. In such a ship, the chemical is carried in spaces away from the shell sides or bottom, or the ship is a double-hull chemical tanker in the cargo spaces. The width of the side tank on such a ship is not to be less than $B/5$ or 11.5 m whichever is less. Similarly, the double-bottom height is not to be less than $B/15$ or 6.0 m whichever is less, and in no case, it is to be less than 760 mm. This requirement is not applicable to slop tank carrying the slop due to tank washing in such a ship.

Ship Type 2 is intended to transport products with appreciably severe environmental and safety hazards which require significant prevention measures to preclude escape of such cargo. The quantity of cargo required to be carried in such a ship should not exceed 3000 m^3 in any one tank. Such a ship need not have double hull, but double bottom. Double-bottom height is not to be less than $B/15$ or 6.0 m whichever is less, and in no case, it is to be less than 760 mm. This requirement is not applicable to slop tank carrying the slop due to tank washing in such a ship.

Ship Type 3 is a chemical tanker intended to transport products with sufficiently severe environmental and safety hazards. These products require a moderate degree of containment to increase survival capability in a damaged condition. There are no filling restrictions for chemicals assigned to be carried in such a ship.

Table 11.4 shows some sample categories of various chemicals and the ship types in which these are to be carried in ships. Based on the type of ship, total cargo space is identified. This space is then subdivided into number of cargo holds such that, apart from meeting IMO-recommended requirements, the damage stability is adequate for survival in the event of damage. Figure 11.2g shows a typical cross sections of a double hull tanker carrying oil as well as chemicals (side tank width and longitudinal partitions may be adjusted as per requirement).

Another way of classifying the type of chemical tankers is based on the carriage of parcels or bulk chemicals. Parcel tankers are chemical tankers designed to carry small quantities (as little as a few hundred tonnes) of a large number of products (as many as 50 or 60). A parcel tanker gives to those who wish to transport small quantities (*parcels*) of a product the opportunity to use a large ocean-going ship, which is safer and more economic. These tankers follow the liner trade pattern and are known as liquid liners. Some of the parcels in

TABLE 11.4

Category of Chemicals and Ship Types in Which These Are to Be Carried – Samples Only

Product	Polution Category	Ship Type
Acetic acid	Z	3
Acrylic Acid	Y	@
Sulphuric Acid	Y	3
Phosphoric Acid	Z	3
Nitric Acid	Y	2
Hydrochloric Acid	Y	2
Methyl Alcohol	Y	3
Phenol	Y	2
Sulphur (molten)	Z	3
1,2,3 Trichlorobenzene (molten)	X	1
Turpentine	X	2
Urea Solution	Z	3
Vinyl Acetate	Y	3
Vinyl Ethyl Ether	Z	2
Sodium Carbonate Solution	Z	3
Vegetable Protein Solutions (hydrolysed)	Z	3
Citric acid (70% or less)	Z	3
Coal Tar	X	2
Coal Tar Pitch (molten)	X	2
[a]Cocoa Butter	Y	2
[a,b]Coconut/Corn/Cotton Seed/Sunflower seed/Olive/Rapeseed/Palm/ Ground Nut/Fish Oil	Y	2

[a] These items are subject to further regulations.
[b] These items have a free fatty acid limitation.

such ships may be hazardous and some may not be so. Consequently, parcel tankers have a large number of cargo tanks and each tank generally has its own separate piping, pumping and venting arrangements. This ensures that the different products remain in their different tanks and do not get mixed while flowing through pipes and being pumped, and even the vapours of the different products do not mix.

Bulk chemical tankers are designed to carry large quantities (as much as 10,000 tonnes) of a small number of products (1–6). A chemical tanker that carries only a particular type of product, e.g. vegetable oils, may be described as a specialised chemical tanker, and one that is designed to carry a specific product may be described as a dedicated chemical tanker. Bulk chemical tankers normally carry what are described as 'easy chemicals', products that are not very dangerous, do not require many stringent precautions to be taken and do not result in extremely severe consequences in the event of an accident. Chemical tankers that can pass through the Panama Canal have deadweights of about 50,000 tonnes. Sometimes, even larger chemical tankers having 60,000–70,000 tonnes dwt are required. Occasionally, chemical tankers are designed to work as oil product tankers if required.

11.1.3.3 Liquefied Gas

If gases are required to be moved by sea, these must be liquefied to reduce the volume so that tonnage moved can be profitable. The most significant cargoes moved are LNG

TABLE 11.5

Physical Properties of Liquefied Gases Carried by Sea

Material	Boiling Point(°C) at 1atm	Vapour Pressure at 45°C (bars abs.)	Practical Carriage Condition	Flash Point (°C)	Flammable range by % vol in air.	Auto-ignition Temp (°C)
n-butane	−0.5	4.3		−60	1.8–8.5	365
i-butane	−12	5.2	Fully pressurised	−76	1.8–8.5	500
Butadiene	−5	5.1	Fully pressurised	−60	2–12.6	418
Vinyl chloride	−14	6.8	Semi-pressurised or FR	−78	4–33	472
Ammonia	−33	17.8	Semi-pressurised or FR	−57	16–25	615
Propane	−43	15.5	Semi-pressurised or FR	−105	2.1–9.5	468
Propylene	−48	18.4	Semi-pressurised or FR	−180	2–11.1	453
Ethane	−89	Above critical temp	Semi-pressurised or FR	−125	3.1–12.5	510
Ethylene	−104	Critical	FR	−150	3–32	453
Methane/LNG	−161	Temp.	FR	−175	5.3–14	595

(mostly methane), LPG (butane, propane and mixtures of these) and ammonia. Other cargoes moved by sea are butadiene, butylene, ethylene, propylene and vinyl chloride. Apart from ethylene and methane/LNG, all these gases can exist as liquids at normal ambient temperatures in pressurised form. These can therefore be transported in pressurised cargo containment systems at any temperature up to the highest expected ambient temperature. There is a critical temperature above which the gas cannot be transformed into a liquid at any pressure and must therefore be refrigerated for shipboard carriage. The critical temperatures for ethylene and methane/LNG are below normal ambient temperatures. Thus, carriage of ethylene, ethane and methane/LNG requires semi-pressurised or fully refrigerated (FR) cargo containment. LNG is normally carried in FR containment system. Table 11.5 gives the physical properties of various liquefied gas cargoes carried by sea.

As can be seen from Table 11.5, the boiling point of various gases varies over a wide range. Those gases having a higher boiling point can be carried in liquefied form in pressurised vessel up to a pressure of 17.5 bars. But as the boiling temperature at normal pressure falls, it becomes difficult to carry liquefied gas at normal temperature and refrigeration becomes necessary. So gases can be carried in semi-pressurised or semi-refrigerated condition, i.e. partially pressurised at refrigerated temperature. But as the boiling point falls further, like methane or LNG, it is not possible to pressurise anymore and gases are carried in liquefied form by full refrigeration. The liquefied gas tankers of today have cargo containment systems that may have five parts – primary barrier, secondary barrier, inter-barrier space, thermal insulation and hold space. A container for a liquid must have a primary barrier to keep the liquid inside and prevent its flowing out. The primary barrier is the boundary between the inside and the outside of the container. The function of the secondary barrier is to contain the liquefied gas and prevent it flowing out in case of a failure of the primary barrier. The space between the primary and secondary barriers is the inter-barrier space. If the gas has been liquefied by cooling with or without an increase in pressure, the temperature of the liquefied gas is lower than that of the surroundings. Heat flows from the higher temperature outside to the cold liquefied gas inside and causes it to evaporate or *boil off*. This is minimised by providing thermal insulation. These components of the cargo containment system are usually placed in the holds of the ship, and the region in each hold around these components is called the hold space and is regarded as a part of the cargo containment system. The primary barrier is an essential part of a cargo containment system; the other four parts may or may not be present. Based on these systems, the lique-fied gas tanks to be used in gas carriers can be of different types as discussed here:

Independent tanks: These tanks are completely self-supporting and do not form a part of the ship's hull nor do these contribute to the strength of the ship. Depending on pressure, there are three types of such tanks:

Type A tanks: These tanks are prismatic tanks and are internally divided by a cen-tre line liquid-tight bulkhead to reduce free surface effect and maintain stabil-ity. These tanks can be designed and built for carriage of LPG and also LNG and material of construction can be chosen accordingly.

Type B tanks: These tanks are normally spherical and welded to a vertical cylindri-cal skirt, which is the only connection to the ship's hull.

Type C tanks: Theses tanks are cylindrical pressure tanks mounted horizontally on two or more cradle-shaped foundations. The tanks may be fitted on, below or partly below the deck and both longitudinally and transversely located. To improve the poor utilisation of the hull volume, lobe-type tanks are commonly used at the forward end of the ship. This containment system is used for LPG and LEG.

Membrane tanks: Membrane tanks are not self-supporting tanks. They consist of a thin layer (membrane), normally not exceeding 1 mm thick, supported through insulation by the adjacent hull structure. The membrane is designed in such a way that thermal and other expansion or contraction is compensated for, and there is no undue stress developed in the membrane. Membranes could be pri-mary barriers as well as secondary barriers. These tanks are used for the car-riage of LNG.

Semi-membrane tanks: Semi-membrane tanks are not self-supporting. These consist of a layer which is supported through insulation by the adjacent hull structure. The rounded parts of the layer are designed to accommodate thermal expansion and contraction without developing undue stresses. The semi-membrane design has been developed for carriage of LNG.

Integral tanks: Integral tanks form a structural part of the ship's hull and are influ-enced by the same loads and stresses which the adjacent hull structure is sub-jected to. This form of cargo containment is not normally allowed if the cargo temperature is below −10°C. This containment system is partly used on some LPG ships dedicated to the carriage of butane.

Internal insulation tanks: When thermal insulation is provided to minimise the flow of heat from outside to the cold liquefied gas inside the tank, the insulation is usually outside the primary barrier. There is a type of containment system used in liquefied gas tankers in which the insulation is inside the tank and acts as the primary barrier, i.e. the insulation keeps the liquefied gas contained within it. Such tanks are known as internal insulation tanks, and these are of two types: Type 1 and Type 2. In an internal insulation tank Type 1, the thermal insulation acts as the primary barrier, while the hull structure (decks and bulkheads) acts as the secondary barrier. In a Type 2 internal insulation tank, the insulation is divided into two parts, one acting as the primary barrier and the other as the secondary barrier.

There are six types of liquefied gas carriers that move in the international waters. These are briefly discussed here:

1. *Fully pressurised ships*: These ships are the simplest of all gas carriers in terms of containment systems and cargo handling equipment and carry their cargoes at ambient temperature. Type C tanks, pressure vessels, fabricated in carbon steel with a typical design pressure of 17.5 bars or slightly higher up to 20 bars, corresponding to the vapour pressure of propane at 45°C, are used in such ships. Because of their design pressure, the tanks are extremely heavy. As a result, fully pressurised ships tend to be small with maximum cargo capacities of about 4000 m³, and they are used to carry primarily LPG and ammonia. Because these ships utilise Type C containment systems, no secondary barrier is required.

2. *Semi-refrigerated/semi-pressurised ships*: These ships are similar to fully pressurised ships in that they incorporate Type C tanks designed as pressure vessels designed typically for a maximum working pressure of 5–7 bars. Compared to fully pressurised ships, a reduction in tank thickness is possible due to the reduced pressure, but at the cost of the addition of refrigeration plant and tank insulation. These ships range in size up to 7500 m³ and are primarily used to carry LPG. Tanks on these ships are constructed of steel capable of withstanding temperatures as low as −10°C and can be cylindrical, conical or spherical in shape.

3. *Semi-pressurised/FR ships*: These SP/FR gas tankers use Type C pressure vessel tanks and therefore do not require a secondary barrier. The tanks are made either from low-temperature steels to provide for carriage temperatures of −48°C which is suitable for most LPG and chemical gas cargoes or from special alloy steels or aluminium to allow the carriage of ethylene at −104°C. The SP/FR ship's flexible cargo handling system is designed to be able to load from or discharge to both pressurised and refrigerated storage facilities. This type of gas carrier has evolved as the optimum means of transporting the wide variety of gases, from LPG and vinyl chloride monomer or VCM to propylene and butadiene. The ship size can also vary from 1,500 to 30,000 m³.

4. *FR LPG ships*: FR ships carry their cargoes at approximately atmospheric pressure and are generally designed to transport large quantities of LPG and ammonia. Four different cargo containment systems have been used in FR ships: independent tanks with double hull, independent tanks with single side shell but double-bottom and hopper tanks, integral tanks and semi-membrane tanks, both these latter having a double hull. The most widely used arrangement is the independent tank with single side shell with the tank itself a Type A prismatic free-standing unit capable of withstanding a maximum working pressure of 0.7 bars. The tanks are constructed of low-temperature steels to permit carriage temperatures as low as −48°C. FR ships range in size from 10,000 to 100,000 m³. A typical FR LPG carrier would have up to six cargo tanks. The tanks are usually supported on wooden chocks and are keyed to the hull to allow expansion and contraction as well as prevent tank movement under static and dynamic loads. The FR gas carrier is limited with respect to operational flexibility. Where Type A tanks are fitted, a complete secondary barrier is required.

5. *Ethylene ships*: Ethylene ships are built for specific trades and have capacities ranging from 1,000 to 30,000 m³. This gas is normally carried FR at its atmospheric pressure having boiling point of −104°C. If Type C pressure vessel tanks are used, no secondary barrier is required. Type B tanks require a partial secondary barrier

and Type A tanks require a full secondary barrier, and because of the cargo carriage temperature of −104°C, the hull cannot be used as a secondary barrier, so in this case a separate secondary barrier must be fitted. Thermal insulation and a high capacity reliquefaction plant are fitted on this type of vessel. As mentioned, many ethylene carriers can also carry LPG cargoes, thus increasing their versatility. A complete double hull is required for all cargoes carried at temperatures below −55°C whether the tanks are of Type A, B or C.

6. *LNG ships*: LNG carriers are specialised vessels built to transport large volumes of LNG at atmospheric pressure with boiling point of −163°C. These ships are now typically of between 120,000 and 130,000 m^3 capacity. These ships are normally built with membrane tanks or independent Type B tanks. All LNG ships have double hulls throughout their cargo length, which provides adequate space for ballast. The membranes have a full secondary barrier with a drip-pan type protection. Another characteristic common to all LNG carrying ships is that they burn the cargo boil off as fuel. To date, reliquefaction plants have been fitted to a few very large LNG vessels with diesel engine propulsion system. The necessary equipment for reliquefaction being costly it has been more economic to burn the boil-off gas in steam turbine propulsion plants. However, due to the rising cost of oil fuel and increasing value accredited to LNG, future designs of LNG ships tend towards the provision of greater tank insulation (to reduce boil off), a reliquefaction plant and diesel engine main propulsion. Figure 11.3 shows the cross section of gas carriers having different types of tanks discussed earlier.

Density of liquefied gas reduces sharply as temperature increases because of high coefficient of expansion. All gases carried at sea in liquefied form have a density much less than 1. Propane, the main constituent of LPG, has a density of 0.59 at −50°C, and methane, the main constituent of LNG, has a density of 0.42 at −161°C. As has been already observed, there is much loss of cargo earning space due to specific tank geometry and insulation. Therefore, a liquefied gas carrier carries much less load compared to its capacity. The design draught is much less than the assigned freeboard draught. This is evident from the high D/T ratio of liquefied gas carriers shown in Section 5.3.

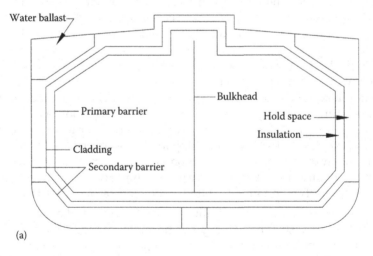

(a)

FIGURE 11.3
Cross section of liquefied gas carriers. (a) free standing self-supporting prismatic tank.   (*Continued*)

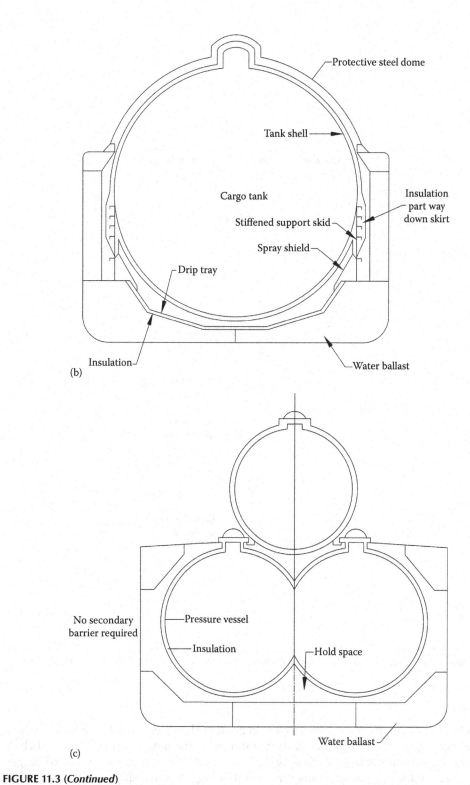

FIGURE 11.3 (*Continued*)
Cross section of liquefied gas carriers. (b) free standing self-supporting spherical tank (c) semi-pressurised/
semi (or fully) refrigerated tank. (*Continued*)

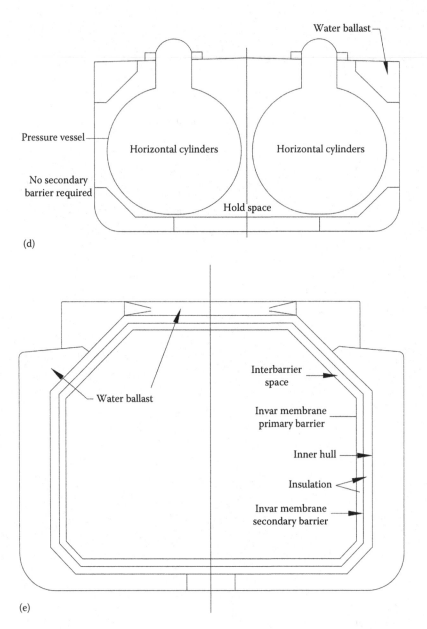

FIGURE 11.3 (*Continued*)
Cross section of liquefied gas carriers. (d) Fully-pressurised tank (e) Membrane tank in LNG carrying ships.

11.1.4 Unitised Cargo

It would be convenient if all cargoes carried on board a ship were standardised to a single unit size. There have been a number of such standard units moved across the seas such as containers and standard barges such as BACAT and SEABEE barges. The standard barges are used directly for intermodal transportation moving on board ships across the oceans and being towed (or, self-propelled) in river waters to go to the hinterland. Containers, on the other hand, are standard boxes that can be carried on ships, wagons moving on rail roads, trucks on roads and highways and in boats on rivers. Container movement across the oceans has become a major component of international trade and a separate class of ships, container ships, has evolved and grown.

11.1.4.1 Containers

A container is a (ISO) standard box, the most common being the 20 ft unit having an extreme length of 6.058 m or 20′, a width of 2.438 m or 8′ and a height of 2.591 m or 8′6″. There are other ISO standard containers of 12.192 m or 40′ and 13.716 m or 45′ length. Normally, there are three variations with regard to height: half height, standard (2.591 m or 8′6″) and high cube (2.896 m or 9′6″). There are more variations of size prescribed in the ISO standards. The most common container carried by sea is the 20 ft unit and the container carrying capacity of a ship is given in numbers of 20 ft equivalent unit (TEU). Containers are made of mild steel frame work, steel-corrugated sheet side walls, back panel and roof with double doors in the front made of the same material and wooden floor. These boxes are generally watertight. There are variations to container make-up to facilitate carriage of cargo that cannot be put in a closed box which include open top containers, ventilated containers, refrigerated containers with refrigerant as liquid nitrogen or electrical plug-in system of refrigeration, container flats with folding sides and tank containers. All containers have corner fittings suitable for interlocking of containers or lashing to floor and sides of the compartment where these are carried. Containers are manufactured to ISO standards and generally, dimensions are maintained within 10 mm accuracy. The tare weight of an empty container varies based on the manufacturer and the variations in container type and use. Figure 11.4 shows the diagram of a TEU and Table 11.6, taken from Interfreight, shows the dimensions 4 standard sea containers of which 20′ and 40′ units are the most common. The dimensions and tare weight as well as gross weight are only approximate varying from manufacturer to manufacturer. Since containers can be closed watertight,

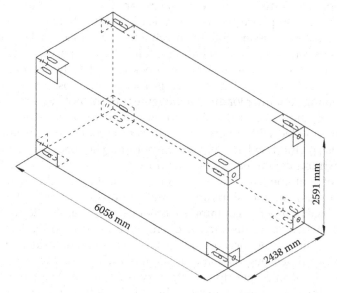

FIGURE 11.4
Twenty feet equivalent unit container dimensions.

TABLE 11.6

Dimensions and Weights of Standard ISO Containers

Freight	Length		Breadth		Height		Tare wt.	G. wt.	
Container	(ft)	(m)	(ft)	(m)	(ft)	(m)	(kg·wt)	(kg·wt)	Equiv. TEU
20' unit	20	6.058	8	2.438	8'6"	2.591	2200–2940	30,400	1
40' unit	40	12.192	8	2.438	8'6"	2.591	3800–4800	30,400	2
40' high cube	40	12.192	8	2.438	9'6"	2.896	3900–4800	30,400	2
45' high cube	45	13.716	8	2.438	9'6"	2.896	4800	30,400	2.25

almost any kind of cargo can be carried in it such as general cargo, machinery, bulk cargo and perishable food items (in refrigerated containers). This also ensures safety of cargo against pilferage. Since containers can be moved by sea, rail, road or river, handling is standardised and minimised. Thus, door-to-door transportation is possible and for this very reason, transportation by containers is very popular. Carrying large number of containers makes the container ship very big with large draught and so such ships can call only at limited ports. Such ports that handle large number of containers and can berth large container ships (up to 10,000 TEU or more) are called hub ports. Smaller container ships of up to 1000 TEU carrying capacity take smaller number of containers from these ports to smaller, limited draught container ports. Such vessels are called feeder container vessels.

A loaded container can have a weight up to 20 tonnes, and therefore the lifting device must have at least 20 tonnes lifting capacity. Normally, gantry cranes handle containers on board ship and at ports. A containership can have its own crane or it can be gearless and use the crane facility available at the calling port. If a ship has its own handling facility, the gantry cranes are usually rail mounted with rails running on extreme port and starboard sides of the ship. This device is very heavy and located above the upper deck, thus raising the CG of the ship. Containers being large in size have to be vertically loaded or unloaded through hatch openings only. Further containers can be stacked on top of the one below and therefore container ships are generally single deck ships. To maximise container stowage under the deck, it is necessary that the hatch opening is made as large as possible and in multiple units of container size (including intercontainer spaces). Therefore, container ships have large hatch openings, making the deck almost fully open and having hatch width up to 80% breadth of ship. Figure 11.2h is a schematic cross-section of a feeder container vessel showing container loading arrangement in the hold and above deck. Torsion boxes are designed and fitted at the sides to withstand torsional stresses. Occasionally, twin hatches are provided with pillared structure at the centre to withstand torsion.

Container ships are fitted with normal double bottom. The space between shell sides and the sides of the extreme containers cannot be utilised for carriage of cargo. And to facilitate proper stowage of containers and provide enough ballast space, the shell sides are made double sided and thus container ships are normally double-hull ships. These ships are fine form ships and large number of containers cannot be stored under deck. A large number of containers are carried above deck in the open, sometimes up to six tiers, the tiers being limited by the visibility of the forward side from the navigation bridge or the blind zone being limited to two times length of ship or 500 m whichever is less. Since containers are carried on decks, CG of ship goes up and stability is impaired. Therefore, loading of containers should be done in a manner such that heavily loaded containers are at a lower level than lightly loaded or empty containers. Further ballast water adjustment should be done in such a manner that CG is below the maximum allowable CG.

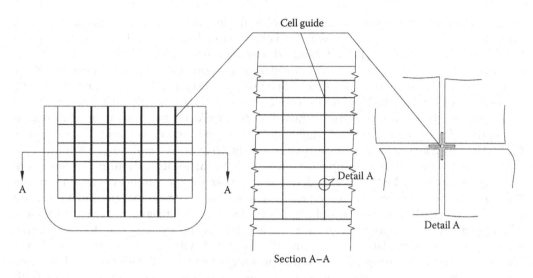

Cell guide

Detail A

Detail A

Section A–A

FIGURE 11.5
Cell guides in cellular container ship.

Holding of containers to the ship in a sea way is an important part of container ship design so that containers do not escape to sea due to ship motion. Container fittings play an important part in this respect. The containers at the lowest level are fixed to the double bottom of the ship by twist locks fitted at the bottom container corners and at double-bottom level. The containers at the top tiers are fixed to the bottom tier by similar twist locks. As tiers go up to two, three or more, it is necessary to lash the containers to ship's bottom and sides by wire ropes and tightening locks. But if the tiers go up above three or four inside ship's hold, lashing is not enough and some rigid restraint must be introduced. One such rigid restraint is cell guide. Cell guides are pillar like structures with recesses at four sides of the cell guide to fit container corners and are erected as permanent structure on ship's hold, welded to ship's double bottom (Figure 11.5). Such a ship cannot carry any other cargo since cell guides remain even when the ship is not carrying any container, and therefore it is called a cellular container ship. The containers on top of the deck are held by lashings and, if necessary, a rigid restraint system by fitting horizontal guides at the ship sides.

Container ships do not make full ballast voyages except going to dry dock for maintenance and repair. In light condition they carry empty containers. In such a case and since containers are watertight and almost the entire hold space is filled with containers, there is no need of watertight hatch covers on freeboard deck. If such a ship is designed with fully open deck without hatch covers, containers can be loaded on top of each other from double-bottom level till top. Such a ship is known as hatchless container ship. In such a ship due to ship motion in heavy seas and during rain, there is likelihood of water ingress into the ship's holds. Such ships must be extensively tank tested for their behaviour in a sea way and must be fitted with pumps to remove entrained water from the holds.

11.1.4.2 Roll-On/Roll-Off Cargo

RORO cargo, as the name indicates, is on wheels which roll onto a ship and roll off the ship through roadways or ramps which are lowered from the ship onto the quay side of the dock. There are different kinds of wheeled vehicles or trailers which are carried on board ships over short sea or oceanic distances. Cars of different sizes and makes are one of the most common wheeled vehicles carried on ships, the length varying between 4.0 and 5.5 m, breadth 2.2–2.5 m and height 2.0 m with load on each axle being about 1.0 tonne. Trucks and trucks with trailers carrying loaded standard containers (generally TEU, known as high and heavy [H&H] cargo carried on RORO vessels) have a length of 10, 12 or 15 m, breadth 2.8–3.1 m and depth 4.7 m with axle load 12 tonnes on single wheel axles and 15 tonnes on double wheel axles. CG of the vehicle with cargo can be taken at 3.0 m above the base. There could be non container cargo (NCC) on trucks/trailers such as pallets with approximate dimensions as length 1.25 m, breadth 1.0 m and height 1.8–2.0 m or specialised containers to carry paper (SECU) having a length 13.8 m, breadth 3.6 m, height 3.6 m and loaded weight 90 tonnes.

There are different types of ships to carry wheeled vehicles for various purposes. There are pure car carriers which are multi-deck vessels to carry large number of cars only across the seas. The deck heights are limited to car height with sufficient margin. Then there are pure cars and truck carriers where the upper decks of limited height are used for carriage of cars and lower decks for heavier cargo such as trucks and trailers. The ships that carry passengers over short distances are normally equipped with facility for carriage of passenger cars. These are commonly known as passenger ferries or ROPAX vessels. The lower decks have the capacity for carriage of passenger cars that the passengers may bring with them to be transported to the desired destination where these can be driven off by the disembarking passengers. The passengers are located on the upper decks. Then there are hybrid RORO vessels. A ship designed to carry containers and wheeled vehicles is known as a ConRO vessel where containers are carried on the upper deck and wheeled vehicles are carried under deck. These vessels normally do not carry cranes for loading and unloading containers. These are handled by shore facilities. The vessels that carry general cargo as well as wheeled vehicles are known as GenRO vessel where general cargo space and RORO cargo space are separately marked. Whereas RORO cargo is loaded/unloaded through normal horizontal transfer, general cargo is handled by cranes located on board the ship for this purpose. These vessels are also known as roll-on/lift-off vessels.

The most common access to a RORO ship is by providing a stern ramp or side ramps for moving cargo into or out of the ship. The ramp when raised serves as a watertight enclosure and when lowered serves as a roadway connecting the ship and the quay side. If access is to be provided from the front, a bow door with a bow ramp or the bow itself serving as a forward ramp can be provided. These ramps lead the vehicle to the internal lanes inside the ship on a particular deck. For moving cargoes to other decks, there must be internal ramps with/without hoistable decks and lifts. The external ramps should not have an inclination more than 12° and the width to be between 7 and 12 m. The ramps should be designed to take concentrated moving wheel load and the surface should be anti-skid. The inclination of internal ramps should be between 7° and 8°, i.e. 1:7 and 1:8. These ramps as well as the vehicle decks should be made rough so as to provide anti-skid properties. The deck space on which vehicles are parked should have a width as multiple of single lane width which is taken as 2.9 m normally. The total lane length in each deck is equal to single lane length on that deck multiplied by the number of lanes. The number of vehicles that can be parked on a deck is given as

$$\left(\frac{\text{Length of each lane}}{\text{Length of the vehicle with margin}} \right) \times \text{Number of lanes on the deck}$$

The total number of vehicles that can be loaded equals the sum of the number of vehicles on all decks. Vehicles are loaded on the vessel on first-in-last-out principle which means that proper planning of vehicle loading must be done based on the sequence of ports of call.

RORO vessels carry cargo that has large volume compared to its weight. Also, there is a lot of empty space inside the ship to allow vehicle movement and housing ramps and their handling facilities. Thus, these are capacity-oriented vessels having low draught. The freeboard deck is fitted at the lowest possible level satisfying the minimum freeboard (Type B) requirement. The under deck space is required to be subdivided with the required number of transverse bulkheads as per Table 11.1. If vehicles are loaded below this deck and doors are provided on the transverse bulkheads, these must have watertight doors which can be operated by remote control from above deck by providing sliding and locking arrangement. If there is a bow ramp, there must be an access door on the forward collision bulkhead for vehicle access, and this door must be closed watertight after the vehicle loading/unloading is completed at any port. On the decks above, it is difficult to provide transverse bulkheads since these would restrict vehicle movement to a large extent. As has been mentioned before, most of the vehicles are loaded on multiple decks above the freeboard deck. So in the deck space being very large, any water entry causes loss of stability due to excessive free surface effect. To reduce this effect, the deck(s) above freeboard deck may be subdivided by providing a centre line deck to deck longitudinal bulkhead. To have adequate intact stability, it is necessary to load heavier cargo at lower levels and lighter cargo on top decks. There must be enough ballasting arrangement to lower the CG adequately. Before the bulkhead positions and internal ramps and lift positions are finalised, it must be ensured that the vessel is a two-compartment standard vessel and probabilistic damage stability requirement is satisfied with adequate margin. The loss of RORO ferries such as the Herald of Free Enterprise, Estonia and others in recent years has led the IMO to look at RORO vessel safety very critically. Apart from the stability requirements stated earlier, there are a number of operational requirements such as alarms, securing of loads properly, and damage control which also must be satisfied.

11.2 Liquid Non-Cargo Spaces

Fuel oil is carried in ships and other floating and fixed platforms. These may be heavy fuel oil as the main propulsion fuel in ships, oil refined to different degrees used in machinery of various kinds such as diesel engines, steam and gas turbine machinery for motive power or electric power generation. The capacity required to be stored in storage tanks must cater for the required number of days before refuelling can be done. Normally, on a marine platform, oil-consuming machinery works non-stop continuously. Then the capacity can be given as

$$\text{Capacity} = \text{Fuel consumtion per day} * \text{number of days} + \text{margin}$$

where margin can be between 10% and 15%. In case of a ship, if the range of travel is R nautical miles and V is the service speed in knots, then

$$\text{Number of days} = \frac{R}{(24*V)}$$

and fuel consumption per day $= \sum \text{SFOC} * P * 24 \text{ in g}$

SFOC being specific fuel consumption in g/kWh and P in the continuous service rating of all diesel engines of the main propulsion plant. Thus, in a ship, for any fuel type, capacity can be estimated as

$$\text{Capacity} = \left[\frac{R}{(24*V)}\right] * \left[\sum (\text{SFOC} * P * 24)\right] * \frac{(1.1-1.15)}{10^6 \text{tonnes}}$$

If an engine works partially or works in ports, the capacity can be suitably estimated.

The capacity of lubricating oil is estimated based on the recommendation of the engine/machinery manufacturer. Fresh water capacity is calculated based on total requirement of fresh water and the fresh water generating capacity on board.

Tank space is to be provided for all the liquids stated earlier for their estimated capacity. All these tanks contain consumable liquids and as time progresses after fully loading the liquids in the tanks, liquids get consumed and in the platform of a floating structure, free surface effect reduces stability of the platform.

Provision of ballast water is provided in all floating platforms for loading the platform when required. The use of ballast water can be one or more of the following:

1. Increased displacement of the platform increasing draught of the platform.
2. Controlling trim of the platform by selectively loading the ballast water tanks, typically in a ship; frequently the aft draught is more than the forward draught.
3. In a partially loaded or no-load voyage when the forward draught is very low, the platform or ship may be subject to excessive slamming load in heavy seas causing impact loading and vibration. So ship must be loaded with ballast water so that the vessel can have a minimum forward draught to avoid excessive impact loading and reduce frequency of slam.
4. Similarly, adequate draught must be available in the aft end to ensure full propeller immersion which is achieved by ballast water loading.
5. The CG of the vessel can also be controlled by selectively loading ballast water which is a critical requirement in passenger vessels, RORO ships and container vessels.

Based on these requirements, the ballast water capacity is estimated. In bulk carriers and tankers where the vessel operates half of its life in ballast voyages, ballast capacity is approximately 30%–35% of the assigned deadweight.

Over the years, environmental issues have become important and various governmental and international bodies have discussed issues related to discharge of oil to sea. Recommendations and regulations have evolved regarding carriage of these liquids at sea. The first requirement of carriage of these liquids at sea is that all tanks for each liquid are to be segregated from other liquid tanks. In other words, a tank carrying a particular liquid cannot carry any other liquid. Since pipelines and their fittings are likely to leak during operation, pipelines carrying liquid of one type should not be taken inside a tank carrying any other liquid. For the same reason of accidental pollution, oil tanks should not have a boundary next to the sea or should not be next to the shell side. Similarly, to maintain purity, fresh water tank should not have a boundary with a tank carrying any other liquid or shell side below LWL. Normally, fresh water tanks are provided above waterline. A fresh water tank and any other tank must be separated by a cofferdam of at least one frame space width.

Liquid takes the shape of the container in which it is carried and handling is done by pipelines and pumps, and so no separate openings for handling is required. Therefore, the usual place for ballast water tanks is the double-bottom space, the shell side tanks, top and bottom sloping tanks in bulk carriers, the forward peak tank and the aft peak tank. The fuel oil tanks can be provided as deep tanks inside the engine room or just outside it. Tanks should be located such that piping length is minimal and fluid flow is with minimum bends and valves. Tanks should be divided to cater for aforementioned needs. Since ballast water tanks are either fully pressed or empty, they should be divided longitudinally so as to maintain trim as required for vessel operation. Due to the requirement of ballast water exchange at sea, nowadays, ballast water tanks can be considered generating free surface during ballast water exchange. Therefore, tanks have to be divided transversely such that stability is not impaired during ballast water exchange at sea. Tanks carrying consumable liquids generate free surface and careful arrangement of such tanks should be made by transversely dividing the tanks that the loss of metacentric height due to the combined effect of all tanks is limited and does not hamper stability of the vessel.

11.3 Working Spaces

A marine structure or vehicle has to stay at sea, at port or under water in a particular disposition to perform the task it is intended to do. For this purpose, the platform or ship must have enough space to house the various machinery, equipment and facilities and also provide convenient working area keeping in view ergonomics as a means of efficient and effective working. Machinery requires spares and stores items to be used for maintenance and repair and so stores of various kinds are grouped here under working spaces.

11.3.1 Machinery Spaces

In a ship or submarine, it is the space where the main propulsion engines including the electric power generation system and all auxiliaries required for running the main engine are located. The pump room in an oil tanker and the electric motor in a bow thruster room are also rooms containing machinery. Similarly, in an oil rig platform, the spaces required to house the equipment for drilling and oil production are also machinery spaces. The design steps for such spaces are discussed in the following.

The location of machinery space should be away from accommodation or cargo space, and if located next to either, these should be enclosed with watertight bulkheads on the sides and steel horizontal platforms such as decks and double bottom on top and bottom. In a ship, the propulsion machinery is located in the engine room located forward of the aft peak bulkhead and aft of all cargo spaces. The location should be such that people can have easy, covered and protected access from sleeping accommodation area to come for work.

It is necessary to identify all machinery and equipment to be housed in each such space and estimate the space required for each with respect to its geometry, its weight and its functionality. Mechanical power in the form of rotary motion is generated by one or more low-speed or medium-speed diesel engines, steam turbines, gas turbines or electric motors. For ship propulsion, this rotating power is transmitted to the screw propeller(s) or to the impeller in a water jet propulsion system or other hydrodynamic devices to change rotating power to forward thrust by hydrodynamic action. In a low-speed engine, one could have a direct drive system, whereas in a medium- or high-speed engine, the power must be transmitted through geared system. The shafting runs from the engine and gear end to the propeller outside the ship. Further, the shafting should be at least in two pieces or more: the one connecting the propeller is known as the propeller shaft and the others, intermediate shaft. The shaft pieces are joined by couplings at either end and supported on ship structure through bearings. The bearing at the engine end is a heavy bearing known as thrust bearing which transmits the propeller thrust to the ship's hull through strong foundations so that the ship can overcome resistance to forward motion and move at a steady speed forward. The thrust bearing also isolates the engine from being subject to a forward force due to a propeller thrust. The aft-most bearing is the stern gland through which the shafting comes out of the ship and water is not allowed to get in. The stern gland is an oil-lubricated bearing. All the auxiliary systems which have been discussed in Chapter 9 including their pumps, heat exchangers, piping and valves are also housed in the machinery space or other spaces housing the associated machinery. The pumps, piping and associated equipment for ship's ballast water system, fire and general service system and seawater cooling system are also provided in the engine room. Normally, pumps and valves required for liquid cargo handling are provided in a separate room called the pump room separate from engine room though that is also a space housing certain machinery. The dirty oil generated in the engine room is in the form of oily water mixture and sludge. Both of these systems have been discussed in Chapter 9 and later in Chapter 13. The tanks, pumps and all associated equipment are located in the machinery space.

All the equipment discussed earlier work with electric power. So electric power generation, either from diesel generator sets or from shaft generator systems, forms a major activity of the ship's machinery space. This electric power is also necessary during port operations, for cargo refrigeration, cargo handling (cranes), repair works on board the ship and, most importantly, domestic use. All equipment required for this purpose must also be housed in the machinery space.

Machinery spaces are generally confined spaces, i.e. these spaces are large enough for workers to enter, but small for continuous occupation. These spaces have limited entry and exit facilities, are poorly and nonuniformly ventilated and may have sloping walls. Since these spaces contain fossil fuels, it is possible for fossil fuel gas and air mixture to be generated forming an explosive mixture. Therefore, there must be provision for firefighting adequately. In such a case, it is necessary to close all openings to cut off oxygen supply and produce an inert gas atmosphere to douse the fire. This is done by flooding the engine room with carbon dioxide. Thus, there must be a CO_2 bottle room outside the engine room having entry from outside for operation when necessary. Different confined spaces may generate other gases such as carbon monoxide due to incomplete combustion and toxic gases due to decomposition of organic matter such as hydrogen sulphide and methane. Care has to be taken to provide sufficient ventilation at such spaces.

Machinery in a maritime product or vehicle has to work continuously over a sufficiently long period of time without major breakdown. Therefore, they have to be under continuous (preventive) maintenance at sea. For this purpose, it is necessary that they carry sufficient spares. Also for the same purpose, sufficient redundancy in the needed equipment and also equipment for emergency purposes is provided in machinery space or near it based on the requirement. Many of the machinery spaces have become automatic and unmanned with only a control room to monitor the work of different equipment. This room must also form a part of the machinery space if located inside it.

All equipment require floor space where a platform can be erected to house the equipment. If the space is voluminous with sloping walls like the engine room of a ship, it is not possible to house all equipment on the engine room double bottom. If sufficient height is available, one or more platform decks in machinery space may be erected providing larger floor space as one goes up in height. The exhaust from all diesel engines and boilers must be laid out through a funnel of sufficient height such that the exhaust is discharged to atmosphere without affecting the work or people on main deck. These could also be used for aesthetic beauty. The platform decks should have sufficient size of opening in line with or more than the funnel opening on the main deck. The funnel should be so located that the exhaust pipes do not have large length or many bends. The location of other equipment should be so arranged that piping length is minimised. There should be sufficient space around and height above each equipment for working around it or lifting a part to be worked upon or replaced. For this purpose, machinery space must be provided with some lifting arrangement and access to outside.

11.3.2 Working Spaces on the Open Deck

Anchoring and mooring is one of the most important activities of a marine platform at the port or in shallow waters. The crew of the ship normally do the needful in this respect using ship's mooring ropes, anchor chains, associated machinery such as anchor windlasses, mooring winches, capstans and mooring fittings on deck such as bitts or bollards and fairleads and chocks of various kinds. Normally, on a ship these deck machinery and fittings are provided in the front of the ship on the forecastle deck and in the aft on the poop deck or, in its absence, on the upper deck itself. The space in these areas marked for these operations should be enough, not only to house the machinery and equipment, but also provide enough space for ropes and chain to pass conveniently for proper working and enough working space. For an anchor chain, it is necessary to layout the path of the chain going from the windlass to the hawse pipe to the anchor pocket to drop down to ocean bottom without hitting the ship hull. Similarly, on the other side when the anchor is lifted the anchor chain must drop straight down to the chain locker to be stored. Accordingly the chain locker location and size must be allocated below the forecastle deck. The mooring ropes are stored on the deck or in spaces below the forecastle deck designated for the same purpose. Similar arrangements are made in the aft end. Figure 11.6 shows a typical space layout on the front and aft ends of a 120 m long ocean-going vessel as an example of such layout design. An oil tanker or a shuttle tanker or any other vessel may be required to be moored to a single point mooring (SPM) system. A shuttle tanker may be required to be moored to a platform or to an FPSO directly. Offshore support vessels may be required to be moored to a fixed or floating oil rig for supply and discharge. Therefore, mooring fittings and associated space must be provided on any floating platform based on its activities.

In a merchant vessel, if cargo handling gear is provided on the upper deck, space must be provided for housing and working of the cargo machinery such as cranes and derricks and also space for hatch cover panels when open. Arrangements should be such that the hatches can be opened and closed conveniently as per the recommendations of the supplier so that cargo movement (loading and unloading) can be done efficiently. Access of crew members from accommodation to work space on deck must be available in all weather conditions.

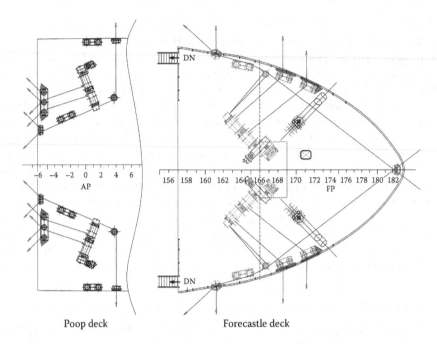

Poop deck   Forecastle deck

FIGURE 11.6
Arrangements on open deck fore and aft.

In a trawler or fishing vessel, large deck space towards the aft must be available for operating fishing gear, to unload the trawl net, to handle the fish and, depending on the type of vessel, to process the fish if required. Similarly, in a tug or offshore supply vessel, large deck space is required for handling the tow rope and storage of items to be supplied. In a research vessel, large deck space is required for handling scientific equipment in the open for lowering to the seabed and sample collection. Of course there has to be large covered or sheltered deck space for scientific study of sample collected. Similarly, depending on vessel's purpose, sufficient deck space must be available at convenient locations on the open deck.

11.3.3 Navigation and Control Spaces

Good space must be provided for housing all the navigating equipment for efficient navigation of the ship from one port to another. Equipment required for this purpose include position and speed indicating equipment such as GPS or DGPS or Carrier DGPS for accuracy, heading direction indication using gyro compass, magnetic compass in an open deck, depth indicating echo sounder or hydrophone, electronic chart indicating route and equipment for safe navigation such as radar, light and sound signalling equipment, speed log, communication equipment, etc. In some ships physical charts are still used for routing the vessel. Control is achieved through the wheel of the steering system which transmits orders to the steering gear to control the rudder. The connection from the wheel to the steering gear is normally done by a telemotor system. Steering gear is located in the aft end of the ship below the upper deck such that the rudder stock can match with the steering gear properly. The steering gear area is also the working area where emergency steering arrangement should also be housed. The navigation and control area is located on the top deck of the ship above accommodation which is conventionally known as navigation bridge or the wheel house. The navigation bridge deck houses the wheel house and a chart room behind the wheel house for storage of charts, a battery room or radio communication room if required. Apart from navigating equipment, the wheel house also contains other control equipment such as communication system including radio communication system and satellite navigation including weather forecast from Inmarsat and emergency position indicating radio beacon, Automatic Identification System, internal communication system, navigational lights, navigational flags, sound signals and flares. Safety control system includes fire detection and alarm systems and fire control. If there is no separate room, the wheel house is also used for cargo control and record keeping. Engine control command is also issued from the wheel house from where the massage of control is transmitted to the engine room from where the engine is controlled. For efficient navigation, it is necessary to have all-round view of the outside from the wheel house. There should be free deck space available on all sides for quick movement to the outside for any reason. The higher the wheel house, the less will be the blind zone for the navigator. Figure 11.7 shows a wheel house arrangement on the navigation bridge deck of a conventional 60 m long coastal research vessel where the wheel house is located inside the navigation bridge deck giving access to both side wings weather protected by awnings and also to fore and aft on the same deck to have a good all-round view. Also note the space layout inside the wheel house providing work spaces around various electronic and communication equipment including the battery room.

A floating oil rig platform can have many complex control systems compared to a fixed platform. Since large amount of electric power is consumed in an oil rig platform, the power systems require to be controlled. Control of individual thrusters of any DP system installed for position keeping is also required. Production process requires to be monitored. Safety and backup systems require to be operated when necessary. Also there must be an information system based on networking of all control systems. Large space is required to house all these individual control systems which should be so located that various signals should be received easily and feedback provided wherever necessary. Signals must be received from various sensors from different equipment as well as other monitors such as wind and current sensors, gyro compass, DGPS or CDGPS, readings from vertical reference units indicating heave, pitch and roll, accelerometers indicating surge, sway and heave, rpm and torque from thrusters and many such other sensors.

11.3.4 Space for Stores and Spares

As has been discussed before, it is necessary to carry spares for various machine parts on board the ship or platform. The spares should be properly labelled and stacked in predesignated racks so that these can be found easily when needed. Normally, spares should be carried in space identified for this purpose near the machinery space. Items which should normally be carried on board, but are not used regularly should be stored in properly designated locations. Rope and paint stores and deck stores are the normal store spaces allocated in ships and platforms. These stores can be single or multiple stores depending on amount of material to be used and the location where these are likely to be used. In ships, these stores are provided in the upper deck space below the forecastle deck around the chain locker and deck spaces below the crane housing. In platforms, these are provided at convenient locations.

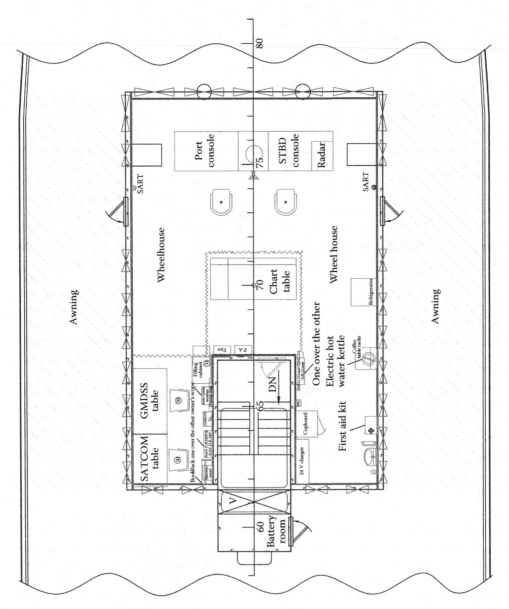

FIGURE 11.7
Navigation bridge deck.

11.4 Accommodation Spaces

Total number of personnel for whom accommodation is to be provided on a fixed or float-ing platform at sea and on ships depends on a number of factors, the minimum standard being as per the recommendations of IMO and International Labour Organisation (ILO) providing for limited working hours per day for the crew on board. The factors include

The size and type of the vessel or platform

Number and type of the main engines and auxiliary engines

Equipment installed and level of automation

Method of maintenance

Type of cargo carried and handled

Frequency of ports of call, length and nature of voyage

Trading areas and nature of waters of operation

Agreement between the flag state government and the concerned seamen's associations

Generally, an ocean-going ship may have about 15–20 crew members on board which may include all or some of the following designations:

Deck side:

Master or captain, 1

Chief mate or chief officer. 1

Deck officer or officer in charge of navigational watch, 1

Deck seafarers, 3

Engine side:

Chief engineer, 1

Second engineer, 1

Engineer officer, 1

Engine seafarers, 2

Others:

Radio communication and radio officer, 1

Electrotechnical officer, 1

Electrotechnical rating, 1

General-purpose rating (includes cooks and catering crew), 2

Provision of accommodation for at least two deck cadets and two engine cadets may also be provided. In a small installation or ship, duties of two or more personnel may be combined to be executed by one person subject to limitations of working hours. The number and skill

TABLE 11.7

Floor Area Requirement for Ship Crew and Officers

Single room for crew	4.5–7.0 m² without toilet space
Floor area per person	2.8–3.6 m² in two, three or four bedded rooms for crew
Single room for junior officers	7.5–14 m² without toilet space
Bedroom for senior officers	10–14 m²
Senior officers (master and CE)	35–40 m² which includes bedroom, day room and toilet

of personnel on board varies based on specific requirements as has been aforementioned. In a passenger ship, a large number of general-purpose ratings and catering officers need to be on board. In a drilling platform, a large number of drill crew need to be on board. On an offshore platform there are three types of extra personnel for whom requisite facilities need to be provided on board. The first type is personnel who come for a specific purpose and return to shore on the same day. No sleeping accommodation need be provided for such personnel though sitting and catering arrangements are to be provided. The second type are personnel who come for a day's work and are likely to stay overnight for whom sleeping accommodation is to be provided. The third category is personnel who came on an assignment to the platform for a few days for whom adequate sleeping accommodation needs to be provided.

Space or floor area required for providing sleeping accommodation for officers and ratings on board depends on space availability, duration of voyage and other considerations such as type of trade, size or GT of the vessel and minimum requirement of floor area that has been specified by 'accommodation for crew convention' of ILO. Generally, large ocean-going ships provide a much higher standard of accommodation than that recommended by ILO. An indication of floor area to be provided is given in Table 11.7 which is indicative only.

Each single room should contain a bed, a clothes locker, a drawer, a writing table or desk with adequate sitting arrangement, a small cabinet for toilet requisites and a book rack. In passenger ships, floor area per passenger is decided based on duration of voyage. In short voyage trips of duration 4–8 h, there is no need to provide sleeping accommodation. Only seating arrangement needs to be done by providing space of about 1.5 m²/person. But for longer voyages lasting more than 8 h requiring one- or two-night stay sleeping accommodation needs to be provided. In such a case the room need not have a large floor area nor be very heavily furnished. In ships where passengers stay for a few days, the floor area is to be provided based on the status of the passenger – dormitory, ordinary class, second class, first class and deluxe class. The floor area may be provided for each class of passenger separately ranging from about 3 m² for dormitory accommodation to 40 m² (bedroom, sitting room with a lobby and attached toilet) for deluxe class passengers.

As far as possible each single sleeping accommodation should be provided with an attached toilet containing one water closet, one wash basin and one shower or bath. Sometimes a toilet can be provided between two single rooms with opening and locking arrangements from either room. In case provision of attached toilet is not possible, a common toilet may be provided where the number of water closets, wash basins and showers/baths is to be on the basis of one set for six persons. Such a toilet should be easily accessible from the sleeping accommodation of people using this facility. In addition there may be a requirement for providing a toilet with all facilities to be used by shore personnel when the ship is at port. One engine crew change room is to be provided near (but outside) the engine room to facilitate changing of dress for engine crew and officers. This room must contain adequate locker space for dirty linen, clothes hangers, wash basin with mirror and a water closet.

Mess rooms with attached pantry must be provided with floor space at the rate of approximately 1.5 m²/person for as many people that are expected to eat at a time. Depending on flag state requirement, there may be two mess rooms, one for officers and one for crew. The mess rooms must be easily accessible from sleeping accommodation as well as the galley with facility for carriage of cooked food.

There must be a galley with adequate platform space and shelves complete with kitchen equipment consisting food preparation facility and cooking facility. ILO Convention No. 92 and convention no. 133 had addressed the issue of food storage and subsequently the Maritime Labour Convention in 2006 has laid down the minimum requirement of refrigeration. Deep freezer rooms should not have a temperature more than −12°C where food can be stored up to 1 month and not more than −18° in case food is required to be preserved for more than 1 month. Chill cabinets carrying vegetable and dairy products should not have a temperature more than 5°C. Galley is to be supported by a daily store, a dry provision store apart from cold stores. Normally, ships and platforms are provided with a cold stores complete with entry from the passage through a lobby to a meat room and a vegetable room. This is done so that the refrigerating system can be optimised and the use of insulation can be maximised. The dry provision stores should be with adequate shelf space and lockers to hold dry provisions for the longest duration intended with adequate margin. Access to both cold stores and dry provision stores must be convenient from the galley.

There must be laundry with facility for housing one or more washing machines, drying machine and ironing facility complete with clothes hangers. Since a lot of linen is carried on board, there must be adequate locker space for clean linen and also for dirty linen conveniently located near the laundry.

Since ship's staff or staff on an offshore platform does not have access to shore based recreational facilities, the same must be provided on board. There must be provision of a smoke room or lounge next to the dining space. There must adequate open deck space and a swimming pool for relaxation. The bigger the ship and duration of voyage, the larger are the recreational facilities.

Office space must be provided near the accommodation area. In ships there must be a deck office and an engine office to cater to the needs of the office work. Sometimes these two offices may be combined, and therefore there is a single ship's office.

If the platform has 15 people and has voyage duration more than 3 days, the ship must have a hospital with at least one bed and a small dispensary with easy access.

Once the extent of accommodation is determined, the layout design can be done keeping in mind certain basic principles. No accommodation should be provided below the LWL. In exceptional cases of a passenger ship, limited crew accommodation may be provided below the LWL. Further, accommodation may be provided in the aft or midship. Forward accommodation is generally discouraged since motions are high in the forward end. Whereas minimum deck height from top of flooring to the top of deck longitudinal is to be maintained at 2300 mm, clear head room in accommodation space must be about 2000 mm.

Accommodation arrangement must be such that it ensures individual privacy and also group privacy. This is ensured by providing single room accommodation as far as possible. Providing ratings, junior officers and senior officers accommodation in separate decks, group privacy is also ensured. The toilet space and the dining and smoke rooms may be provided in the respective decks.

11.5 Ergonomics in Layout Design

Ergonomics is the art and science of designing and arranging things people use so that people and things interact most efficiently and safely. In design of ships and platforms at sea, this can be translated as designing for lighting, comfortable interior environment, vibration, noise and access arrangement.

11.5.1 Lighting and Visual Comfort

General layout design for lighting and visual comfort should be such that proper illumination is provided to facilitate

Proper and adequate visibility for performance of task at hand

Easy movement between working and habitable areas

Operation, maintenance, inspection and survey of spaces including closed spaces

Safety by detection of hazards or potential hazards

Visual comfort and freedom from eye strain

Aesthetically comfortable visual environment

Good design practice should include integration of natural and artificial lighting such that proper and uniform illumination is available for the works at hand without any visual strain such as glares bright spots and shadows. All artificial lighting should be so arranged that the illumination is not obstructed by structural, pipe, trunking or any other obstruction causing shadow effects and dark spots. All artificial lighting arrangement should be easily maintainable.

All living rooms including rooms of common usage such as dining spaces should have at least one boundary as a part of ship's external boundary wall. This wall should be provided with a window or port hole properly located so as to provide sufficient natural light to the room and also provide a view of outside from inside. All other rooms and work spaces that have one boundary as externally exposed surface should also be provided with natural lighting through windows or port holes. As far as possible all major work spaces should have provision of some natural lighting through windows, port holes. Sky lights with glass closures may be provided on the top enclosure of enclosed work places such as engine room.

11.5.2 Interior Environment

Interior environment should be maintained at a comfortable level with adequate thermal protection so that operator can remain vigilant and alert during operation, maintenance and inspection of facilities and equipment in manned spaces. Safety of personnel should be ensured while entering or working in enclosed spaces with removal of poor or dangerous quality of air and removal of heat from spaces where spontaneous combustion hazard may exist. Generally, environment should ensure provision of comfortable and pleasant living environment continuously in commonly occupied spaces.

The interior environment at various locations is maintained by designing and installing heating, ventilation and air-conditioning (HVAC) on board. This includes natural and mechanical ventilation, air-conditioning and heating to control the interior environment.

Ventilation in manned spaces should be provided permanently and include living quarters (accommodation, recreation, offices, dining areas) and work areas (control rooms, bridge, machinery spaces, offices and void spaces). Ventilation in unmanned spaces should be temporary and include tanks, small holds, infrequently occupied enclosed spaces to which entry may be required only occasionally.

The HVAC system should provide adequate heating and/or cooling for onboard personnel, provide uniform temperatures and temperature gradients (the inside and outside could be separated by a lobby), maintain comfortable zones of relative humidity, provide fresh air (air exchange) as part of heated or cooled return air and provide clean filtered air free of fumes, particulate matter or airborne pathogens. In the HVAC system there should be provision for monitoring gas concentration (CO, CO_2 and O_2). The air flow should be easily adjustable by onboard personnel, ensure minimum contribution of ventilation noise to living and work spaces, provide sufficient velocity to maintain exchange rates while not being noisy or annoying, provide means to use natural ventilation, provide/assess safe air quality while working in enclosed spaces. The comfort level of internal environment depends on the ambient condition outside and so varies from location to location, typically from tropics to temperate to cold climate. Table 11.8 gives the range of comfortable internal environmental conditions.

Basic Layout design from HVAC point of view should take the following aspects into consideration at the initial stage:

- Natural ventilation design should be provided to spaces within the accommodation by consideration of compartment layouts and specifications. Natural ventilation should be provided to cargo spaces if required. Ventilation fans for cargo spaces should have feeders separate from those for accommodations and machinery spaces. Typical natural ventilation devices include mushroom ventilators, gooseneck ventilators and ventilators with weather proof covers on open deck.

- Forced or mechanical ventilation should be provided typically in main engine room, in working spaces which are not air-conditioned. Spaces such as machinery rooms or working rooms should have mechanical supply and natural exhaust. Spaces which may generate air of poor quality, such as galley, toilets and laundry should be provided with mechanical exhaust even if conditioned air is provided to such spaces.

TABLE 11.8

Recommended Ranges of Interior Environment Conditions

Item	Recommended Criterion
Air temperature	18°C–27°C (68°F–77°F).
Relative humidity	The HVAC system should be capable of providing and maintaining a relative humidity within a range from 30% minimum to 70% maximum.
Vertical gradient	The acceptable range is 0°C–3°C (0°F–6°F).
Air velocity	Not exceeding 30 m/min or 100 ft/min.
Horizontal gradient (berthing areas)	The horizontal temperature gradient in berthing areas should be <10°C (18°F).
Air exchange rate	The rate of air change for enclosed spaces should be at least six (6) complete changes per hour.

- All the air-conditioned spaces should be grouped together. As far as possible, non-air-conditioned rooms and heat generating rooms containing machinery items should be outside the air-conditioned space. Even if the accommodation area is located above the engine room the funnels carrying hot exhaust gases should be kept aft of the accommodation space. These requirements, if fulfilled, can generate an optimised HVAC load.

- For HVAC load calculation, the external ambient condition considered should be the worst or extreme condition. In tropical climate this can be 45°C dry bulb or DB temperature at 100% humidity where as in temperate climate it can be 15°C at 0% humidity. In cold climate the extreme temperature can be as low as −15°C. It is necessary to find the extreme condition in which the ship or platform has to work.

- Ductwork (particularly elbows and vents) should not contribute to excessive noise to a work or living space. Though high velocity air ducting reduces duct diameter and is good for layout in confined and space-limited locations, care should be taken to see that it does not generate excessive noise. Ducts should not discharge directly on people in their living or working locations such as on sleeping berths and sitting areas. Ductwork should not interfere with the use of means of access such as stairs, ladders, walkways or platforms. Fire dampers should be provided wherever necessary to contain the spread of fire.

- At least two manholes and other accesses should be provided in each tank or confined space for accessibility and ventilation to points within.

- Air Intakes for ventilation systems should be located forward of the main engine exhaust to minimise the introduction of contaminated air from sources such as exhaust pipes and incinerators.

11.5.3 Vibration

Vibration of ship structures adversely affects shipboard operations which may even lead to injury or risk of health of personnel involved in operation, maintenance and inspection of manned spaces. Continued vibration may lead to higher structural stresses leading to local structural cracking and failure. This may also lead to fatigue failure, local or global, leading to minor or major structural failure. The vibratory effects felt in sleeping accommodation and work spaces cause acute discomfort and mental stress. Classification Societies have their guidance notes which deal with ship board vibration and remedy in detail. It is not intended to deal with these in this section but to understand the vibration phenomenon in ships and structures at sea and develop design guidelines to reduce vibration.

Vibration on a floating or fixed structure or a ship at sea can be global where the ship can be considered a free beam or girder. Vibration and its effects can be limited to a region of a particular substructure such as double-bottom panel between two bulkheads in engine room, the fore body structure and the aft superstructure. There could also be local vibration where a steel girder or a steel panel could vibrate due to local excitation.

Vibration could be vertical vibration where a structural girder or beam oscillates around its longitudinal axis in the vertical plane, or could be horizontal where the girder oscillates around its longitudinal axis in the horizontal plane. Similarly, longitudinal vibration of the girder means oscillation in the longitudinal plane and torsional vibration is periodic twisting of the girder around its longitudinal axis. Vibration could be in single mode or multi-mode. The number of nodes of vibration is generally one more than the number

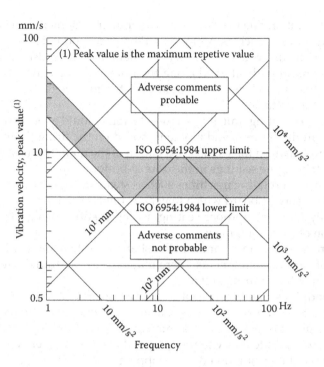

FIGURE 11.8
Limits of tolerance of vibration.

of modes. A node is a location on the girder where there is no displacement of the body during vibration. When there is no continued external exciting force, a body can vibrate at its natural frequency, f_0. This is called free vibration. When the body is subject to an external periodic exciting force, the body vibrates at the exciting frequency f and this vibration is known as forced vibration. It can now be observed that if f_0 and f (the natural frequency and frequency of excitation) coincide, resonance occurs inflating the effects of vibration.

Figure 11.8 shows in a convenient graphical diagram, relationship between acceleration, velocity, displacement and frequency of vibration. Based on ISO standard ergonomic practice, the uncomfortable range of vibration is marked in this diagram so that vibration limits for human acceptance can be maintained. As per ISO 6954 (2000), adopted by IMO (report 132, December 2013), the whole body vibration exposure is considered comfortable if vibration acceleration is limited to 180 mm/s² (or, velocity 5 mm/s) in accommodation area and 215 mm/s² (or, velocity 6 mm/s) in working area within the frequency band of 1–80 Hz. The vibration is the RMS value of vibration measured in all three x, y and z axes using the weighing scale w as given in the ISO document.

Vibration excitation can happen due to various reasons, a few of which are mentioned below. Ship has a highly curved surface and flow around this body varies in direction as well as magnitude all around the body. An example of this is the flow around the propeller disc discussed in Section 7.2.3 and shown in figure 7.14. Propeller excited vibration at the stern of single screw ship has been discussed in Section 7.2.3. To avoid its harmful effects, designer has to do propeller design (Figure 7.13) after selecting the number of blades as per the hull resonance diagram as shown in Figure 7.11 and stern aperture design (Figure 7.12).

In a twin screw ship, the variation of velocity field is not so high since wake values are low. In a vessel with a tunnel at its bottom, where the propeller or thruster is located, there may be cause for vibration due to improper structural arrangement in a varied wake field. There may be appendages such as A-brackets, P-brackets and nozzles around propellers. The tunnels of bow thrusters and any other thrusters units may also be a source of vibration unless these are properly aligned along the flow of water. If flow is not streamlined properly around the appendage, there could be vibration excitation on the hull surface where appendages are attached.

The main power generating unit, if it is a diesel engine, could generate unbalanced vertical, lateral or longitudinal forces and moments based on the firing timing and number of cylinders. These unbalanced forces cause excitation of the main frame of the engine which is transmitted to the engine seatings in the double-bottom frame.

There are a number of rotating machines such as steam and gas turbines, electric motors, pumps and compressors on board which may have unbalanced forces and moments internally or externally (due to improper seating arrangement) causing vibration at the frequency of rotation of such equipment.

Based on accuracy of alignment of propeller shafting system and mass moment of inertia distribution of various components, there may be vertical, transverse, longitudinal or torsional vibration of the shafting system.

Adverse sea conditions could be a major source of harmful vibration. The forward end of the ship may experience adverse motions causing bow emergence, impact loading due to slamming and shipping of green seas. Slamming causes whipping vibration of the lower portion of the ship. The extent of vibration depends on the frequency and extent of bow emergence. There could be vibration due to slapping which is an impact force on the fore-end of the ship without bow emergence. Sometimes there is springing vibration of the whole ship caused due to resonance of natural frequency of hull girder vibration and the wave encounter frequency.

Ship operating conditions such as loading conditions, shallow draught conditions and hard over manoeuvres, can also aggravate the vibration conditions.

Vibration on an existing ship is a nuisance since removal of vibration effects is rather difficult and very expensive. It is difficult to implement vibration removing measures in an existing ship. So care must be taken at the basic and detailed design stage to avoid harmful effects of vibration. The following are some basic guidelines to be implemented at the design stage:

- Design of the ship stern (discussed in Section 8.1.6.8) to generate a uniform wake field in the propeller disc supported by flow calculations and model tests.

- Design of appendages similarly supported by model test and flow simulation studies.

- To select propeller RPM (in case of geared drive) and number of propeller blades based on hull resonance diagram (Figure 7.11) to ensure that propeller excited vibration frequency ($=n * z$) does not resonate with any mode of natural frequency of hull girder vibration.

- Once the structural design is completed a numerical simulation of hull girder vibration, regional vibration and vibration of local panels should be carried out to determine natural frequency and amplitude of vibration. If necessary, redistribution of material may be done to change the natural frequency. Particularly, this may be useful for open deck vessels such as container ships against torsional vibration. The funnel which is a long, weak and cantilevered structure, must be analysed for vibration, and the structure should be modified if required.

- Selection of propulsion engines and other auxiliary machines and rotating machines and their proper installation such that no unbalanced forces and moments are transferred to the main structure.

- Use of vibration dampers or resilient mountings on seatings of rotating machines to isolate vibrations due to rotating parts. But such mountings have to be chosen carefully so that the frequency required to be isolated matches with properties of the dampers.

- Structural panels can be mounted on flexible attachments so that the structural vibrations need not be transferred to the panels intended to be isolated.

- Design of fore-end of ship to reduce slam impact and frequency and also to reduce probability of impact of lower side of ship flare. This has been discussed in Chapter 10.

- Proper loading of the ship in all operating conditions such that there is enough forward draught to minimise impact and frequency due to slamming, frequency of deck wetness and slapping.

11.5.4 Noise

Noise on board at different locations can affect human beings in many ways such as interfere with speech communication, mask audio signals, interfere with thought processes, distract sleep and affect habitability by causing discomfort. Noise in places of work can affect productive task performance, increase fatigue, contribute to hearing loss and permanent or temporary impairment of hearing.

Noise is defined as sound made of various non-harmonic vibrations. Sound is a vibration phenomenon which, due to an elastic shock to an elastic medium such as air or water, spreads out within this medium in the form of periodic variation of pressure. This vibration spreads out from the source to the human receiving organ in the form of a longitudinal wave causing pressure variation over and above existing pressures such as atmospheric pressure in air or hydrostatic pressure in water. The simplest form of sound wave is sinusoidal and is represented by a frequency and amplitude. Actual sound is never a single wave, but a multiplicity of waves of different frequencies with respective amplitudes. The spectral composition of this sound is known as tone. The velocity of sound depends on the medium in which it travels and some common useful values are given in Table 11.9.

TABLE 11.9

Velocity of Sound in Different Media

Medium	Speed of Sound (m/s)
Steel, iron	4700–5200
Glass	5000–6000
Wood	1000–5000
Cork	450–540
Rubber	40–150
Air	330 at 0°C
Air	340 at 20°C
Fresh water	1460
Seawater	1540 (see Section 2.2.6)

Sound intensity is expressed in decibels and is defined as

$$L(\text{dB}) = 10\log\left(\frac{I_1}{I_0}\right)$$

where

I_1 is the intensity of considered sound
I_0 is the referenced intensity which is 10^{-12} W/m^2

Sound intensity is proportional to the square of the acoustic pressure and can be defined as

$$L(\text{dB}) = 10\log\left(\frac{P_1^2}{P_0^2}\right) = 20\log\left(\frac{P_1}{P_0}\right)$$

where

P_0 is the reference acoustic pressure $2 * 10^{-5}$ m^2
P_1 is the acoustic pressure of considered sound

Audible sound for human beings is between 20 and 20,000 Hz frequency range. The intensity levels of sound that human beings can experience range from 0 to 120 dB. Typically, a quiet forest has a sound intensity of 15 dB and an airplane taking off has a sound intensity of 125 dB at a distance of 100 m.

Sound measuring instruments (sonometers) normally have an electric filter which generates a weighted average of sound pressures across frequencies. Ponderation A, which is the most commonly used filter, has a lower weight for lower frequencies from 0 till 500 Hz and also above 4000 Hz. Ponderations B and C, which are also used in industrial practice have a higher weightage in lower and higher frequency ranges compared with Ponderation A. Steady noise is a noise which varies within 5 dB(A) in a 1 min duration of measurement. Impulse noise is a noise which is an isolated event or as one of a series of events of less than 1 s duration or occurring less than 15 times/s. Noise that rises and falls or fluctuates more than that in steady noise and is not an impulse noise, is known as fluctuating noise. Equivalent continuous sound level, L_{eq}, is a notional average sound level over a time period T which is equal to the A-weighted sound energy in that time and is given as

$$L_{eq} = 10\log\left[\frac{1}{T} \cdot \int_0^T \left\{\frac{P_1(t)^2}{P_0^2}\right\} dt\right]$$

where

T is the time period
$P_1(t)$ is the instantaneous A-weighted sound level

The tolerance of noise level varies from location to location and duration of exposure. In maritime sector, IMO-recommended sound levels, as per Resolution A 468(XII) adopted in November 1981, are maintained. These levels are given in Table 11.10. Human endurance of noise is also dependent on time of exposure and use of ear protection such as ear muffs or ear plugs. Based on these quantities, IMO has suggested different noise zones and their tolerance levels (Figure 11.9). The effect of noise on human endurance is also related to the frequency levels of noise. In order to make possible a better comparison of noises without

TABLE 11.10

Limits of Noise Levels at Various Locations

Space/Location	Maximum Noise Level (dB)
Work spaces	
Machinery space (continuously manned)[a]	90
Machinery space (not continuously manned)[a]	110
Machinery control rooms	75
Workshops	85
Non-specified work spaces[a]	90
Navigation spaces	
Navigation bridge and chartrooms	65
Listening post, including navigating bridge wings and windows	70
Radio rooms (with radio equipment operating but not producing audio signal)	60
Radar rooms	65
Accommodation spaces	
Cabins and hospitals	60
Mess rooms	65
Recreation rooms	65
Open recreation areas	75
Offices	65
Service spaces	
Galleys, without food processing equipment operating	75
Serveries and pantries	75
Normally unoccupied spaces	
Spaces not specified	90

[a] Ear protectors should be worn when the noise level is above 85.

losing the advantage of global evaluation, ISO (ISO/R 1996) has given a system of Noise Rating (NR) which is shown in Figure 11.10. Sometimes even if the limit of global (A) noise levels are exceeded, NR may be within acceptable limits.

Rotating machinery is the major source of noise in a marine platform. The following is a list a few of these:

- Low-, medium- or high-speed diesel engines
- Turbo charger of the diesel engine
- Diesel generator sets
- Exhaust pipes of diesel engines – depending on the open ends and bends of the pipe
- Steam or gas turbines
- Air compressors
  - Refrigerating machinery
  - Ventilation and air-conditioning fans and ducts
  - Pumps – cargo oil and hydraulic and other pumps
  - Propeller

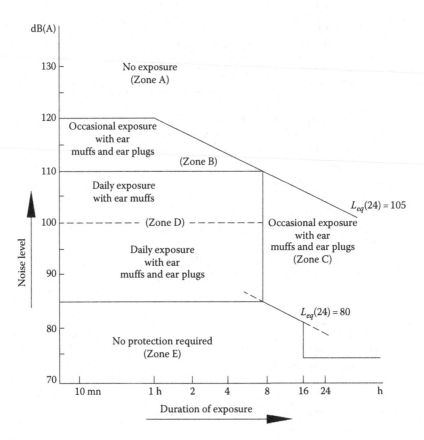

FIGURE 11.9
Allowable daily and occasional noise exposure zones. (From International Maritime Organisation (IMO), Reduced/ No B.W. Ship Designs, GloBallast Monograph 20, IMO, London, U.K., 2011.)

Noise is transmitted from its source through air or structure. Airborne noise is transmitted from the noise source by air. Structure-borne noise is transmitted from the noise source through structural vibration of the seatings and foundations of the noise source. This becomes noise when this vibration is transmitted to air as sound waves. This is known as primary structure-borne noise. When airborne noise encounters a structure, the structure vibrates and gets transmitted as structure-borne noise to various spaces. This is known as secondary structure-borne noise.

Design principles which should be adopted for reducing noise can include the following:

- The location where sound is to be reduced should be placed as far away from the noise source as possible. Airborne noise intensity reduces with distance.
- The distance between the sound source and the location should be divided by walls. Partition walls generally have sound absorption properties which may be increased by suitably choosing the wall material.
- The sound source itself can be hooded or covered so that the airborne noise is reduced at the immediate vicinity of the source itself.
- Partition walls between accommodation spaces or sound generating rooms can have sound insulation so that sound transmission from the room or into the room

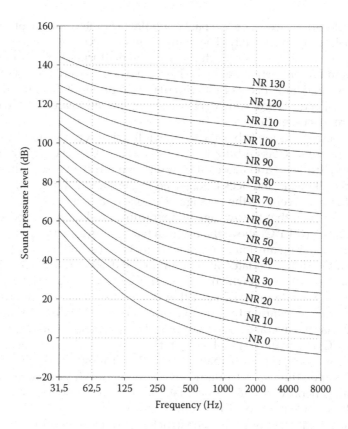

FIGURE 11.10
Noise rating based on sound pressure and frequency.

is reduced. Normally, partition panels have heat insulating material which should also work for sound insulation.

- Selection of suitable machinery such as low draught fans for low sound generation at the source.
- Proper design of machinery foundation so that structure-borne noise is reduced and quickly damped. If necessary, machinery should be mounted on resilient mounting as has been mentioned in the previous section.
- Proper structural design avoiding stress concentration and discontinuity.
- Design of ducting for air flow and piping for liquid flow should be done with care by laying minimum ducting and piping in the accommodation area and reducing bends in the layout.
- Propeller design and stern design should be done with care so that the fluctuating force transmission to the aft body structure is minimised.

11.5.5 Access and Egress

Access and egress means general access between rooms and spaces within a particular area such as accommodation space, working space or service space, between rooms in different spaces, from different spaces to outside and to escape decks such as boat decks.

Access and access structure in crew spaces should facilitate movement of personnel in working and habitable areas safely. The components of access and egress arrangements include passage ways, ladders, ramps, stairs, working platforms, hatches and doors including hand rails and fall protection devices. These have been covered adequately by various regulatory bodies in detail such as MSC Circular 1120 under SOLAS regulations of IMO. Only the layout aspects of access and egress in accommodation area for the main staircase and passages are discussed here.

All bulkheads and decks are to be of one of the following types of bulkheads: A-60, A-30, A-15, A-0, B-15, B-0 or C. Class *A* bulkheads or decks are of steel or equivalent metal construction, suitably stiffened and made intact with the main structure of the vessel, such as shell, structural bulkheads and decks. Subjected to the standard fire test, these bulkheads would be capable of preventing the passage of smoke and flame for 1 h. In addition, they are to be insulated such that the average temperatures on the unexposed side would not rise more than 250°F (121.1°C) above the original temperature, nor would the temperature at any one point, including any joint, rise more than 325°F (162.8°C) above the original temperature, within the time stipulated as 60 min for A-60, 30 min for A-30, 15 min for A-15 and no time for A-0 bulkheads which means without any insulation. Class B bulkheads are made of incombustible material such that if subjected to the standard fire test, they would be capable of preventing the passage of flame for 1/2 h. In addition, they should be so insulated that the average temperature of the unexposed side would not rise more than 250°F (121.1°C) above the original temperature, nor would the temperature at any one point, including any joint, rise more than 405°F (207.2°C) above the original temperature in 15 min for B-15 and no time for B-0 bulkheads. All other bulkheads and decks are made of C class material, generally incombustible material but having no restrictions with regard to passage of flame and smoke or rise in temperature.

Superstructures and deckhouses are constructed of steel and should be in continuation with the main hull structure, main bulkheads and causing no major structural discontinuity. Generally, bulkheads and decks are so laid out and chosen that in the event of a fire in one locality, its spread is limited to a limited area without hampering movement and escape of people from the location of fire. Class A bulkheads and decks generally prevent spread of fire from one side to the other or from one level to the next or lower level. If there is an opening in a class A bulkhead, it must be closed by a fire proof door and should be self-closing if it forms a part of the egress to escape. Long accommodation areas are split into vertical fire zones, each fire zone being limited to maximum 40 m length enclosed by class A bulkheads. Each fire zone has its own passages and staircase and other means of escape. Stairways are enclosed by class A bulkheads having self-closing fire doors at each landing and at the beginning and end of the staircase. This is necessary to prevent spread of fire to other decks and allow escape from the deck under fire. Boundaries between control stations, elevators and stairways, galley, store rooms, workshops, machinery spaces, etc. must be of class A material. Boundaries between washrooms, toilets, corridors and passages, state rooms and public spaces such as dining room, smoke room and dancing rooms are generally made of class B material. Division between two similar spaces, such as partition between sleeping accommodation or partition between two store rooms, can be of C class material.

Passages and staircases should be so located and designed that these give easy access to people in working spaces, in service spaces, common areas and sleeping areas easy means of escape to boat embarkation deck from within the deck house/superstructure or from outside. Ladders and staircases should be provided for movement of people from

TABLE 11.11

Guidelines for Main Stairways in Ships

Main stairways inclination to horizontal plane	38°–45°
Tread depth	≤280 mm including tread nosing of 25 mm
Tread width	≥700 mm one way and ≥900 mm two ways
Main and intermediate landing	Width = tread width, length ≥ 915 mm
Intermediate landing for stretcher	Width = tread width, length ≥ 1525 mm
Head room width of passages	2130 mm
Single person	≥710 mm
Two persons	≥915 mm
Emergency escape	≥1120 mm

one open deck space to another open deck space or from one deck to the other inside the accommodation area or between open deck space and inside easily. There are a number of architectural standards available for the design of ladders, staircases, passages and other access facilities. Table 11.11 gives certain guidelines for stairways and passages for marine platforms.

12

Design for Safety

All human activities involve risk. In marine environment a vehicle or structure has to accept a certain amount of risk to life and property, which includes the vehicle or structure as well as its cargo. The risks can be due to varied and multifarious reasons. An accidental hazard may take place due to one or more risk factors occurring simultaneously causing loss of life and property. A design with absolute safety or with zero risk may be possible but could be prohibitively costly and unaffordable. Therefore, it is necessary to define an acceptable safety level such that increased safety may not be worth the cost. Further, this safety level must be acceptable to all the stakeholders such as the builder, the owner, statutory authorities, classification societies, operating personnel and society at large. With fast-changing technologies as well as trade demands and patterns leading to larger and sophisticated ships, platforms for oil and gas production moving into deeper and dangerous waters and the demand for high-performance vehicles for various applications and novel naval application vehicles, satisfaction of conventional safety requirements may not be adequate and new safety features may have to be incorporated in the design of the product. Though details related to safety issues can be highlighted at the later stage of design, the safety-related design features must be established at the concept design stage itself. The feasibility of various safety features, their cost implications and their integration with all ship systems must be done as a part of concept design. Generally, the ship designers consider safety as a constraint on the design objective to be satisfied, an example being that the selected main particulars must satisfy stability requirements. However, in the present context of safety, it is no more a constraint and becomes one of the objectives of design that the marine platform must be safe up to a predefined level with acceptable cost implications. In this chapter, first, the various safety issues related to operation of marine vehicles and platforms are discussed and their design implications are highlighted. Following this, a short discussion on rule-based design is done highlighting the need for risk-based design (RBD), and finally, the formal safety assessment (FSA) methodology leading to RBD is discussed.

12.1 Safety at Sea and Design Application

The safety of lives on board as well the property such as the ship and cargo involves many issues such as codes for ensuring safety at the design stage, escape in the event of a hazard such as fire or loss of ship and stability. Some of the issues have been discussed in the previous chapters. In this section, some of the major safety issues are highlighted.

12.1.1 Personal Safety on Board

Crew members or passengers can have accidents while on a marine platform, a cruise ship or a general cargo ship due to an accidental fall from a height or slip on a platform while walking leading to grievous injuries. Sometimes accidental slips on open deck may lead to fall from the side onto the sea a few metres below. The sea, in such a condition, could be warm like in the tropics or cold and freezing in temperate and cold waters in which survival may be difficult even for a few hours. Design features that must be incorporated to reduce such risks must include the following, mostly covered under regulations conforming to International Load Line Convention, 1966, and subsequent amendments:

- Bulwark or railings of minimum 1.0 m height on the entire open deck sides of the ship or platform and, in case of railings, stanchions placed not more than 1.5 m apart.
- All ladders and staircases must have handrails with solid grip of between 0.9 and 0.965 m height.
- All openings on freeboard deck such as doors and companion ways must have a minimum sill height, coamings of hatch openings and manholes must have a minimum height and ventilator trunks and air pipes must have a minimum height above the freeboard deck with appropriate weathertight closing devices.
- Enough facilities such as scuppers and freeing ports for discharge of water on deck, particularly in the event of green sea coming on board, must be provided.
- Any opening below the freeboard deck and above the load water line (LWL), such as scuttles, seawater discharge openings from inside the hull and side ports for cargo operation, must have proper closing devices and fitted with non-return valves where necessary.
- There is a possibility of spilling oil and foamy substance on the upper deck of tankers. To avoid skidding of personnel on such vessels, a separate walkway from the fore to the aft of the ship at the level of the first deck above the upper deck should be provided having a minimum width of 1.0 m complete with railings on both sides.
- In passenger ships, all open deck spaces should have anti-skid flooring of rubber or synthetic material non-degradable due to water immersion.
- A suitable mechanism must be provided to rescue a person gone overboard taking care that the rescuer should be properly protected.
- First aid and dispensary facility must be available in the event of an accident. Depending on the number of people on board and the duration of stay in the marine environment, a hospital with medical assistance may also be provided.

12.1.2 Stability and Safety

The stability of an intact ship has been discussed in detail in Chapter 6. But it must be understood that this stability assessment is based on a statically poised ship. It has been shown in the same chapter that the ship's stability characteristics change drastically if the ship is poised on a wave. In a seaway, with the ship rolling and pitching, the stability characteristics change momentarily which could lead to taking in of water and eventual loss of ship. The loss of passenger ferry *MV Estonia* in 1994 brought to the fore the case of RORO ships with no transverse bulkheads resulting in the loss of stability in the event of damage. Assessment of stability in damaged conditions has been elaborated in Section 6.7 and probabilistic damage has been discussed in Section 6.8. The other issues related to stability include free surface effects on board a floating platform such as a swimming pool, a passive roll stabilisation tank, a partially filled liquid tank due to consumption of liquid during voyage and a partially filled cargo tank. Grain shifting in a light bulk cargo carrier or carriage of grain in a multipurpose cargo carrier may be a cause for concern. At the design stage, the following care must be taken:

- Sufficient reserve buoyancy must be provided by ensuring the adequacy of freeboard.
- Preliminary and detailed assessment of intact and damaged stability must be done.
- Damaged equilibrium and stability assessment must be done for different transverse and longitudinal bulkhead positions and selection of the one which ensures the lowest probability of immersion of any deck opening or immersion of the fore end of the ship.
- Connection may be considered between port and starboard side tanks such that unsymmetrical flooding can be avoided to reduce heel in the event of damage. This is important for tanks situated at the ends where damage effects may cause deck immersion due to the combined effects of heel and trim.
- The watertight closing devices on the hull below LWL must be continuously monitored to ensure no accidental failure and subsequent opening leading to water entry.

12.1.3 Motions and Safety

Ships and floating platforms exhibit six degrees of freedom in the seaway. Three of these, surge, sway and yaw, do not have restoring force or moment to bring the ship back after the force applied is removed. The body subject to such forces changes its position and orientation. This change in body location can cause harm to the operation of the platform such as moving away from the oil production site causing stress on the drill pipe. It can also cause undue stress on the mooring ropes on the single point mooring systems or anchor chains of a multi-point anchoring system or on the tethers of a tension leg platform (TLP). It may be advisable to keep the body in position by providing position stabilising forces with additional equipment such as dynamic positioning (DP) systems. Heave, roll and pitch motions are oscillatory in nature. Roll and pitch being rotational motions can cause high linear accelerations at points far away from the rotational axis. Parametric rolling, particularly in container ships, can cause similar high accelerations. Typically, in a container ship, the linear acceleration and therefore the force on the top container are the highest compared to a container inside the hold. Stresses on container fastening devices are pretty high at these locations which may break sometimes leading to container loss at sea. In general cargo ships, if cargo is not fastened properly or adequately, cargo can shift causing damage to the hull structure. In one instance, the steel plates carried inside the hold of a ship shifted piercing the side shell causing damage and water entry into the hold. The following design considerations may be studied:

- Preliminary and detailed study of motion characteristics and estimation of extreme motions.
- Consideration of provision of DP systems consisting of two, four or more thrusters.
- Consideration of active or passive roll stabilising devices.
- Evaluation of stresses in probable extreme sea conditions and design of locking devices of containers accordingly.

12.1.4 Controllability and Safety

During manoeuvres in a crowded seaway, channel or canal, the vessel should have good controllability to avoid collisions as discussed in Section 7.4. The design of the control surface (rudder) plays a major role in avoidance of collision. Additional manoeuvring devices such as bow and stern thrusters may be considered for this purpose, which is effective even at low speeds, particularly for quick manoeuvres.

Ship handling in ports and at shallow waters through anchoring and mooring is also an important aspect of ship control. Ships very often impact against the quayside. Small vessels like tugs and supply vessels frequently impact against ships and platforms at sea. The vessels must be protected against such impact to prevent structural damage. Fenders are the devices required to absorb the impact through compression. There are different kinds of commercially available fenders, wood, natural rubber or synthetic rubber, which may be selected and placed at required locations. Similarly, tension in mooring ropes and anchor chains may be estimated and anchors, chains and mooring ropes may be selected accordingly such that there is no failure during operation. The following design aspects may be considered:

- Manoeuvring characteristics of ship and control surface together
- Type and location of fenders
- Mooring and anchoring loads and anchors, anchor chains and mooring ropes including their stowage arrangements

If there is an equipment failure in the seaway, there could be total loss of control and major safety issues. Therefore, there have to be emergency facilities and redundancy in equipment.

12.1.5 Fire

Fire on board a ship or any marine platform can be a potentially dangerous hazard quickly leading to loss of life and ship. In the accommodation area, the furnishings, fittings, ceilings and wall partitions and linings are all fire prone, and when the accommodation area catches fire, it can spread quickly throughout the accommodation area and then outside this area. The main machinery room and other auxiliary machinery rooms contain a lot of fossil fuels, such as fuel oil and/or natural or petroleum gas, which are inflammable. If any such room catches fire, it becomes inoperative hampering operation and loss of control of the ship or platform. Depending on the intended operation, particularly in ships, an inflammable material may be carried in cargo holds. Solid cargo such as coal carried in bulk can self-ignite and cause fire in a cargo hold if the heat generated is not dissipated by adequate ventilation. Crude oil and petroleum products catch fire if ignited in the presence of air or oxygen. Even if a cargo hold is empty, the oily vapour which fills the space is combustible. Similarly, escaped natural gas or petroleum gas is combustible. Some other gases, liquids and chemicals carried in ships can catch fire. If a cargo hold catches fire, it could explode causing a catastrophic event leading to ship damage and capsize.

One of the ways of preventing fire is to reduce the presence of combustible material. Generally, the designer should ensure that combustible materials such as wood and carpets are avoided in the accommodation area. The ceilings and panels should be of fire-retardant synthetic material. Paint used in the interior should not be easily combustible.

In working spaces and machinery spaces or cargo spaces, it is not possible to eliminate the presence of oil, oily vapour or gas. However, it should be ensured that the heat generated due to working machinery is quickly dissipated through proper ventilation and does not raise the temperature to such a level that self-ignition can take place.

In the event of a fire, it should be contained within a limited zone. Different types of bulkheads and fire zone extents in the accommodation area have been discussed in Section 11.5.5. All cargo and machinery spaces must be separated from each other by steel bulkheads which are fire-resistant A class bulkheads. Any passage through these bulkheads must be complete with a self-closing fire-tight bulkhead. If this opening is below LWL, the door should be watertight as well to prevent flooding of adjacent compartments.

Once a fire starts, it must be fought immediately. The first step in this matter is fire and smoke detection in fire-prone spaces and notification of such event to the officer on watch on the bridge deck. Therefore, the fire detection system must raise an alarm, both visual and audio, in the bridge or wheel house for immediate action. Firefighting facilities and equipment have to be provided adequately so as to fight small fire, electrical fire, fire in accommodation, fire in machinery room or fire in cargo space.

Water is the most commonly used fire extinguishing medium which douses fire by cooling it or removing the heat when the fire catches on plastics; plastic-based substances such as paints, flooring and lining materials; and organic materials such as wood, linen, paper and materials made of these items. Thus, water is the best suited material for fighting accommodation fire. To provide a continuous flow of water on the weather deck and the decks above, the fire remains connected to a fire pump(s) in the engine room run on port and starboard sides of these decks with hydrants provided on either side at regular intervals. Provision is made to accommodate shore side standard connections for supply of water in dry docks, ports, etc., in case of a fire. For decks that are located above a fire pump in the engine room or accommodation decks, fire mains are provided on either side of the spaces. The fire mains or fire hydrants have suitable openings (with closing devices) to connect water hose reels fitted with nozzles such that water can be directed to any portion of the accommodation in the event of a fire. This could also be fitted with a water sprinkler system, which could be automatic with an electric motor or manual, so that water can be sprayed over a large area simultaneously.

Often, foam is used to douse oil fire, particularly fire on the upper deck of an oil tanker due to oil spill on the upper deck covering a large area. If foam is sprayed on such a fire over a large area, the foam separates the oil from oxygen or air and the fire is extinguished. An oil tanker could carry a foam making machine in its engine room where foam lines could run on the port and starboard sides of the upper deck complete with provision of hoses and nozzles.

Carbon dioxide or any other inert gas containing less than 5% oxygen is the best fire extinguisher for oil fire engulfing a large volume such as the main machinery room or an empty cargo space filled with fuel–air mixture. For this purpose, either an inert gas plant is installed or a carbon dioxide bottle bank is provided in the ship such that it is accessible for operation and maintenance without entering into the engine room. The inert gas pipelines run into the engine room and fill the space such that the fire is starved of oxygen and is extinguished. Needless to mention that care must be taken to ensure escape of all personnel before gas penetration to avoid any suffocation casualties. In case of ships carrying inflammable chemicals and gases in liquid form, it may be necessary to give an inert gas cover on top of the liquid surface to prevent air coming in contact with the cargo by injecting IG into the tank on top of cargo. In case of oil tankers, in a ballast voyage or in port, it may be necessary to clean the tank for which the tank with crude oil itself using the crude oil washing system or, even gas free the tank before cleaning. In such cases, inert gas must be injected into the tank to prevent fire. All tankers carry an inert gas generating plant, IG pipeline system with requisite valves and other controls.

Static electricity generated during loading or unloading of a flammable chemical can be a potential source of fire hazard. Some chemicals known as static accumulators have very low electrical conductivity, and static electricity accumulates in such liquids forming a potentially hazardous atmosphere inside the tank. Providing an inert atmosphere inside the tank reduces the risk of hazard. Such an arrangement must be provided in these chemical tankers.

All platforms at sea carry a number of portable fire extinguishers fitted at specific locations in the platform which can be used immediately in the event of a fire. These could be

Water or water spray extinguishers for fire on plastics, wood, linen, paper, etc.

Water mist extinguishers which can also be used for electrical fires and fire due to fats

Powder extinguishers for fire on fats, grease, oil, paint, petrol, etc.

Dry special powder to be used on fires on lithium, magnesium, sodium and aluminium powder

Foam extinguishers to smother fire due to oil, paint, etc.

Carbon dioxide extinguisher used in computer rooms and electrical equipment

Wet chemical extinguishers for fire due to cooking oil and fats

Other firefighting equipments include fire buckets located at pre-designated locations to be used for dispensation of water or sand as required, fireman's axe to break open a closed space required for entry or rescue and fireman's suit to enable a certified fireman to enter into the zone of fire for rescue of personnel.

12.1.6 Hazardous Cargo: Liquefied Gas and Chemical Tankers

Liquefied gas and chemical tankers carry potentially hazardous cargo since they carry flammable and toxic material. Cargo containment areas and construction of these ships requiring special attention have been discussed in Sections 11.1.3.2 and 11.1.3.3.

12.1.6.1 Gas Carriers

The International Convention for the Construction and Equipment of Ships Carrying Liquefied Gas in Bulk, the IGC Code, describes the cargo containment and precautions to be taken to avoid hazardous situations at sea while carrying liquefied gas.

The special areas in gas tankers can be defined as follows. The full length of the cargo holds and the deck on top over this length from side to side is called the cargo area. The cargo compressor room and other such compartments in which cargo liquid or vapour may be normally present should be within the cargo area. In liquefied gas tankers, separating spaces such as cofferdams if provided at the ends of the line of cargo tanks are not regarded as a part of the cargo area. A space in this context is an enclosed space while a zone is an open space or semi-enclosed space. A gas-dangerous space is a space that is not arranged and equipped in an approved manner to ensure that its atmosphere is maintained at all times in a gas-safe condition, i.e. there is a possibility of a dangerous vapour being present in the space. A gas-dangerous zone is similarly defined.

Gas-dangerous spaces include the following:

Cargo containment systems including hold spaces

Cargo pump rooms and cargo compressor rooms

Spaces separated by only a single gas-tight steel boundary from a hold space containing a cargo containment system requiring a secondary barrier

Enclosed spaces outside the cargo area through which any piping containing cargo liquid or vapour passes or in which it terminates

Gas-dangerous zones include the following:

An open space or a semi-enclosed space on the deck within 3 m of any cargo tank outlet, gas or vapour outlet, cargo pipe flange or valve or entrances and openings to gas-dangerous zones

The open deck over the cargo area and 3 m forward and 3 m aft of the cargo area up to a height of 2.4 m

Zones within 2.4 m of the outer surface of a cargo containment system above the deck

Enclosed or semi-enclosed spaces in which pipes containing cargo are located. Spaces that have approved gas detection equipment or in which boil-off gas is utilised as fuel are, however, excluded

Compartments for cargo hoses

An enclosed or semi-enclosed space having a direct opening into any gas-dangerous space or zone

Gas-safe spaces are enclosed spaces that are not gas dangerous, i.e. they are so equipped and arranged that there is no possibility of a dangerous vapour being present in them.

The IGC Code has specified the requirements of proper cargo containment. The hold space must be separated from other spaces by a cofferdam, fuel oil tank or A-60 partition if there is only a primary barrier for cargo or A-0 partition if the cargo is contained in tanks having secondary barriers also. Accommodation, machinery rooms, boiler rooms and control rooms must be outside the cargo area. The safest way to achieve this is of course to have the engine room and accommodation fully aft of the entire cargo space. If the cargo carried has a temperature less than –10°C, the ship must have a double bottom, and if the temperature is below –55°C, then the ship must be of double-hull construction. Access from gas-dangerous space to gas-safe space is not permitted. Access from the gas-dangerous zone to the gas-free space must be via an airlock. The entire cargo pipeline and cargo handling system must be segregated from any other system and must be contained within the cargo space. However, the ballast water pumps could be located in the ship's machinery room. In chemical tankers where there is a possibility of leakage of cargo, the ballast pumps also must be located in the cargo space. Gas-dangerous spaces and cargo operation spaces, which are gas-dangerous zones, should be mechanically ventilated to avoid accumulation of dangerous vapour by facilitating more exhaust than intake. The spaces which are entered only rarely must be ventilated properly prior to entry. In general, the IGC Code has suggested three categories of cargo for proper containment from a hazard point of view, which is given in Table 12.1.

TABLE 12.1

Protection of Cargo – Liquefied Gas and Chemicals

Ship Type	Measures to Prevent	
	Level of Safety Hazard	Escape of Cargo
1G	Extremely severe	Maximum
2G/2PG	Appreciably severe	Significant
3G	Sufficiently severe	Moderate

12.1.6.2 Chemical Tankers

To avoid potential hazards at sea, IMO has set standards through the International Code for the Construction and Equipment of Ships Carrying Dangerous Chemicals in Bulk (the IBC Code) for proper cargo containment and precautions to be taken at sea while carrying chemicals in bulk.

Products transported in chemical tankers may have the hazards listed as follows: safety hazards—flammability, toxicity, corrosivity, reactivity and hazards due to environmental pollution or pollution of the sea due to discharge of noxious liquid substances (NLS).

Chemical tankers have some special features that allow them to carry hazardous cargoes safely. That part of the ship occupied by the cargo tanks, slop tanks, cargo pump rooms and the compartments separating the machinery space and the forepeak from the cargo tanks and the above deck spaces over the whole length and breadth of these spaces is known as the cargo area. However, if the aftermost or the forwardmost cargo tanks are 'independent tanks', the separating space at that end is excluded from the cargo area. The cargo area in a chemical tanker is that part of the ship in which the hazards of the cargo in the ship are present and against which appropriate precautions must be taken. The ship is designed to ensure that as far as possible the hazards are confined to the cargo area and do not go beyond its limits. Hazards are based on the type of cargo carried.

Cargoes and cargo residues are segregated (kept apart) from accommodation spaces, drinking water spaces, service spaces, stores for human consumption and machinery spaces. Cargoes and cargo residues are segregated from these spaces by providing buffer space in between which may be cofferdams, void spaces, cargo pump rooms, empty tanks, any similar spaces and fuel oil tanks which cannot have toxic cargoes next to them. No pipe containing cargo liquid or vapour may pass outside cargo area.

Chemicals take part in chemical reactions, but all such reactions are not hazardous and it is necessary to define a hazardous chemical reaction. A hazardous chemical reaction produces one or more of the following: (1) a large amount of flammable vapour such as hydrogen, (2) a large amount of a toxic vapour such as chlorine, (3) a large amount of a corrosive vapour such as sulphur trioxide and (4) a large amount of heat (strongly exothermic reaction). Such hazardous reactions may take place between a cargo in the ship and air or water or another cargo in the ship, and then there are chemicals that react dangerously with themselves. If a chemical tanker is to carry two chemicals that can result in a dangerous reaction if they mix, e.g. nitric acid and acetic anhydride, steps must be taken to ensure that the two chemicals do not come into contact, i.e. a cargo, cargo residue or cargo mixture that reacts dangerously with another cargo; the cargo residue or cargo mixture must be segregated from each other by one of the aforementioned spaces. If a ship is carrying two cargoes, A and B, that react dangerously with each other, then between a tank containing A and a tank containing B, there must be a double barrier, i.e. two watertight bulkheads. This can be done most conveniently by providing a tank containing a third cargo C that does not react with either A or B. A cruciform joint is accepted as a double barrier by most classification societies. It is also necessary to provide separate piping and pumping systems for the tanks containing A and B and separate tank venting systems. If a pipe from a tank containing A passes through a tank containing B, or vice versa, that pipe must be enclosed in a steel tunnel to ensure a double barrier between A and B. In this context, a cargo mixture is a mixture of cargo and water, i.e. the wash water from the tank that contained A must not be put into the same slop tank as the wash water from the tank that contained B.

Direct access from the open deck must be provided to compartments in the cargo area, e.g. cargo tanks, ballast tanks and cofferdams. It should not be necessary to go through a cargo tank to gain access to any other compartment. The access opening should be large enough to allow a person wearing protective gear and self-contained breathing apparatus to enter and exit and also large enough to allow an injured person to be evacuated. Access to double-bottom spaces may be through a cargo pump room, deep cofferdam, pipe tunnel or similar spaces provided that all these spaces are properly ventilated. There can be no opening in the accommodation facing the cargo area. If there are windows in the accommodation facing the cargo area, they must be of the non-opening type.

Each cargo tank in a chemical tanker must have venting arrangements appropriate for the cargo, independent of the venting systems of all other compartments and designed to prevent cargo vapour accumulating on the deck or entering spaces such as the accommodation and, in the case of flammable vapours, entering spaces that have a source of ignition. Each cargo tank may have an independent vent pipe or the individual vent pipes may be combined into a common header if the tanks contain the same product. Compartments in the cargo area, other than cargo spaces, may have the hazards associated with the cargoes being carried in the ship, and the ventilation of such compartments must be provided adequately and with proper precautions such as not locating the electrical fans in the same area. The ventilation system should provide at least 30 air changes/h. For toxic cargoes, a minimum of 45 air changes/h should be provided. For spaces that are not normally entered, such as cofferdams, void spaces and double-bottom spaces, 8 air changes/h should be provided if the ventilation system is permanently fitted and 16 air changes/h if a temporary portable ventilation system is used.

12.1.7 Life-Saving Appliances

In case it is required to abandon the ship, life-saving appliances (LSA) are to be used to save lives. In the event of evacuation, it is necessary to sound general alarm and use the public address system to inform all lives on board to assemble at specified locations to abandon ship. The following are the main LSAs used on board ships:

- Lifeboats – Lifeboats carry people in distress till such time that they can be rescued by outside help such as other ships, coastal services and helicopters. Therefore, lifeboats must be provided with adequate ration at the rate of 10,000 kJ·cal/person and water for survival for more than 48 h, must have a means of communication (one way) such as emergency position indicating radio beacons (EPIRBs), search and rescue transponder (SART), visual signals such as flares and smoke signals. These must have the capability to withstand heavy motion and move forward. Generally, lifeboats are self-propelled with a speed of about 6 knots and the engine is capable of working in heavily listed condition also. Lifeboats must have inherent stability to survive at sea including capability of unsinkability with adequate buoyant material. The total capacity of the lifeboats in a merchant ship should be enough to cater to all lives on board when launched from either side. In a cruise ship, it is not possible to provide a large number of lifeboats, and so lifeboats on each side cater to 50% of lives complemented with life rafts on either side for the remaining 50% of lives. There are three main types of lifeboats in use today:

- Fully enclosed lifeboats generally required for escape from toxic atmosphere, fire or extreme environmental condition – normally recommended for tankers, bulk carriers and ocean-going merchant ships. These lifeboats are generally heavier than other types and made of aluminium or fibre reinforced plastic (FRP). These lifeboats are provided with a self-contained air support system to provide air support when all openings are closed, particularly when passing through a toxic atmosphere. Such lifeboats in tankers are also made fireproof with a water spray system outside the boat surface so as to reduce transmission of heat inside during passage through fire.
- Partially enclosed lifeboats are those with some openings on top, i.e. about 20% of length from either end is fully covered and the remaining portion could be covered with rigid and foldable canopy. This lifeboat is required for protection against weather.
- Open lifeboats are conventional lifeboats used in short sea trips and coastal voyages.

Lifeboat launching arrangement can be by davits and winches where each lifeboat hangs on the port/starboard side of the boat deck on the ship. After all the persons embark on the boat, the davit is swung around a fulcrum by means of winches and the boat is lowered into the water. Such arrangements are normally used for partially enclosed and open lifeboats. Fully enclosed lifeboats can be held on sloping rails on board the ship and can be fully loaded, can be closed and, by means of a release mechanism, can be released to slide down the rails and fall in the water. Such a system is called the free fall system. This can be conveniently located on the ship centre line at the stern and the boat can be released to fall towards the aft of the stern. This avoids duplication of boats on both sides of the ship.

- Life rafts – Life rafts are platforms which float on the sea surface carrying the loads of number of lives on it. These could be rigid or inflatable.
- Rescue boats – These boats are normally carried on board ships to be used for rescuing people fallen into the sea. These boats could be rigid or inflatable or a mix of both. Normally, the rescue boat capacity is to accommodate a maximum of five persons and a space for carrying a person on a stretcher.
- Lifebuoys – Each lifebuoy should hold its occupant afloat and should contain self-igniting lights or self-activated smoke signals known as man over board (MOB) markers.
- Life jackets – Each life on board should have access to a life jacket in the event of an emergency. Life jackets should also be available for babies and infants. Life jackets could also be inflatable providing adequate buoyancy.
- Immersion suit/anti-exposure suit – The suits protect the human body from hypothermia (loss of body heat) while in cold seawaters, usually after abandoning the ship or during times of rescue at sea.

- Signals – There are different types of signals used by the ship and lifeboats to attract the attention of the rescuing parties. Every ship is provided with one or two EPIRBs with a hydrostatic automatic release mechanism, to send a unique signal identifying the ship and few other particulars over to satellites for distant coast stations that carry out search and rescue. Others are visual signals which include rocket/parachute flares, SART, hand flares and buoyant smoke signals and radar reflector.
- Miscellaneous equipments include portable very high frequency (VHF) radio communication devices on marine channel 16, line-throwing appliances, general alarm, public address system and short training facility for use of the aforementioned appliances.

In the event of an emergency, normally passengers and crew become panicky and there could be a scramble of people towards the embarkation deck leading to accidents even before embarkation which may also hamper smooth abandonment of the ship. The design must cater to the smooth movement of a large number of people from anywhere in the ship to the boat embarkation deck. The access and egress arrangements discussed earlier play an important part in this. Figure 12.1 shows a typical 3D view of access and egress in a small passenger vessel between two fire-resistant bulkheads in one fire zone. Both the views shown are for the same one with an L-portion removed to show the forward stairwell in one view and, in the other, a U-portion removed to show the aft stairwell. Note the egress to boat deck and also to other decks.

12.1.8 Machinery Failure

Marine platforms are generally equipment-intensive facilities. A marine platform, being away from land, has to have all facilities for the operation intended as well as for facilitating living on board and for safety. A mechanical, electrical or electronic equipment can develop defects or stop working altogether based on manufacturing defects, use and handling. This may lead to malfunctioning hampering safe working of the activity intended from the equipment or the system of which the equipment is a part. Also, in the event of damage and flooding of a compartment, the equipment in the compartment may not be available to give the intended service. If the ship's control becomes defective, this becomes a major safety issue. Equipment and technical systems, the sudden operational failure of which may result in hazardous situations, must be identified at the design stage and provide for specific measures aimed at promoting the reliability of such equipment or systems. The common measure is to provide enough redundancy in the components of the equipment, the equipment themselves and the total system. Three steps are taken at the design stage to take these aspects into account:

1. Ships or platforms must carry spares as recommended by the manufacturer/supplier of the equipment on board which must be used if required either during preventive maintenance or breakdown maintenance. Space for storage of spares and workshop for regular maintenance activities must be provided.
2. It would be ideal if spare equipment itself could be carried on board which would be redundant in normal working conditions, but comes into action during a major breakdown. Statutory requirements and classification society guidelines provide for certain redundancy in equipment in ships.
3. Whereas it is not possible to carry a spare main engine on board due to cost, space and installation issues, a large number of standby equipment, properly installed and connected to its system, can be provided such that these can come into action automatically in the event of a major breakdown.

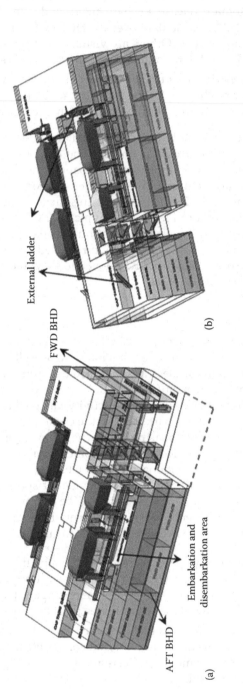

FIGURE 12.1

Access and egress arrangements in a passenger vessel. (a) L-section viewing forward stairwell for a particular fire zone of a passenger vessel. (b) U-section indicating aft stairwell for the same particular fire zone of the passenger vessel.

Emergency activities may be carried out by the redundant or standby equipment to avoid a hazardous situation. Also, such an equipment may come into action in the event of a failure leading to certain operational activities resulting in financial loss. As an example, failure of certain equipment may lead to slow steaming and increase in sailing time. A damaged equipment or system may demand more port time for repair and maintenance leading to loss of income. Thus, a ship need not carry redundant equipment only as per statutory requirements. The amount of redundancy is thus decided by considerations of avoidance of a hazardous situation and operational requirement of stoppage of work and financial loss depending on financial demands for such equipment and company policy. Some examples of redundancy normally provided for in diesel-driven merchant ships can be as follows.

- Electrical power is essential for the ship's operation and survival in emergency situations. Therefore, a standby (redundant) diesel generator set must be provided properly connected to the main switch board so that, in the event of failure of a generator set, this standby set comes into operation automatically and immediately without hampering any operation.
- Emergency compressed air bottles and an emergency air compressor for starting the diesel engines of the diesel generator sets in the event of non-availability of compressed air from the main air bottles.
- An emergency generator having its own drive system must be provided outside the engine room with an emergency switch board outside the engine room/main switch board or control room, so as to provide electrical power to the emergency equipment.
- Emergency battery for uninterrupted functioning of navigation, the ship's internal communication, fire alarm and engine automation and protection alarm system and radio equipment with a 24 V DC system.
- A redundant mechanical/manual steering control system located in the steering gear compartment which can be used in the event of steering control breakdown from the bridge. Provision should also exist to steer the vessel in case of failure of one set of hydraulic ram and pump in case of hydraulic steering gear.
- A redundant ballast pump and a redundant fire pump which may be combined into one since both these operations are not required at the same time.
- Adequate redundancy in alarm and emergency shutdown systems.
- Adequate redundancy in maintaining fuel oil system integrity and cargo (liquid) system integrity.
- Adequate redundancy in components of life-saving equipment, firefighting equipment, communication systems and navigation equipment including their starting and alarm systems.

12.2 Design for Maintenance

It is possible to take steps at the design stage to help maintain the ship's hull and structures properly so that the probability of structural failure can be reduced. Such design methodology is known as 'design for maintenance'. The following are some of the points which may be kept in mind as design for maintenance:

- There could be substantial thickness reduction and material degradation due to corrosion and erosion. Viscous flow around ship hull causes tangential hydrodynamic shear stress on the hull surface. In areas of high wall sheer stress, particularly at the shoulder regions and bilges, there is a probability of higher paint degradation leading to corrosion. Figure 12.2a shows the computational fluid dynamics (CFD) analysis of flow around a tanker hull form showing the wall shear stress distribution, and Figure 12.2b shows a photograph of paint deterioration on the same ship surface

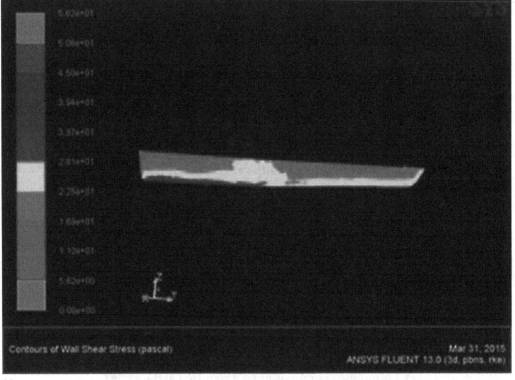

(a)

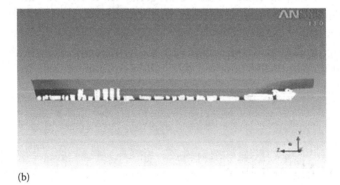

(b)

FIGURE 12.2
Relationship between calculated wall shear stress and paint deterioration in a tanker. (a) Wall shear stress with CFD calculation in full load at a speed of 4 m/s. (b) Photograph of tanker in dry dock showing paint depletion marked in white.

during a dry dock survey. This ship initially had a constant paint thickness application over its entire hull surface. If the painting specification during the design stage could include varied paint thicknesses on the hull surface, areas subject to faster paint degradation could have a thicker layer of paint to protect the plates.

- Care should be taken to see that a water pocket does not accumulate inside the ship, particularly in inaccessible areas where corrosion can start and progress, sometimes rapidly, without being noticed. It is possible to make arrangements so that water, particularly bilge water, could accumulate at the lowest point of the tank so that it can be removed totally at regular intervals.

- As has already been discussed, fatigue could be a major source of material failure. In most of the cases a crack develops over a period of time and also spreads over a period of time. Ships and platforms generally undergo a dry dock survey or an extensive underwater hull survey at regular intervals. There are also afloat inspection surveys in between two regular survey periods. The purpose of these surveys by a competent classification society surveyor is to identify areas of thickness reduction, identify initiation of cracks at welded joints at any internal location and suggest remedial measures such as plate or weld replacement and, sometimes, alter the local structural layout. The areas to be surveyed have to be easily accessible so that these can be easily inspected by the surveyor or even by the ship's operating staff and remedial measures taken. If the area is not easily accessible, there could be error of judgement or even the possibility of non-inspection. Cracks generally occur at the stiffener joints in the double bottom, at the bottom of shell side tanks or at the bracket/beam knee connections above the double bottom inside side tanks or hold spaces. Due to minor collisions, the fore end of the ship may be deformed if not damaged and this may lead to weld failures at the internal joints inside the forward ballast tank. All these places are difficult to access because of lack of height, fully covered and dark locations and too many obstructions (structural components). Therefore, it is necessary to design the ship or any other marine structure for ease of inspection and subsequent maintenance. It is necessary to prepare an access plan showing manhole and ladder positions in the ship for accessing all the tank and hold spaces (including void spaces). For each enclosed space, tank or void space, there must be at least two manholes which must be opened and ventilated before human entry. For very small tanks where provision of two manholes is difficult, one manhole access must be provided. Figure 12.3a shows such an access plan for a small vessel consisting of ladders and staircases from the top deck up to the double-bottom level. Figure 12.3b shows the manhole distribution on the double bottom for all the tank spaces below.

- To do preventive maintenance and plan for long-term maintenance and also dry docking, it may be necessary to install a hull condition monitoring system on board. It is possible to analyse voyage data recorder (VDR) and data of the ship's log book on a continuous basis to identify excess power consumption due to hull and propeller surface deterioration. Similarly, it is possible to examine the health condition of the machinery system to identify possible examination of engine components and replace components well in advance of a major failure. It is possible to install instruments such as motion recording units (MRU) to get accelerations in longitudinal and transverse directions at forward, aft and midship regions from which it may be possible to identify dangerous hydrodynamic loading on the ship. Installation of strain gauges at sensitive locations should give an idea of stresses at various locations prompting preventive maintenance of the hull structure.

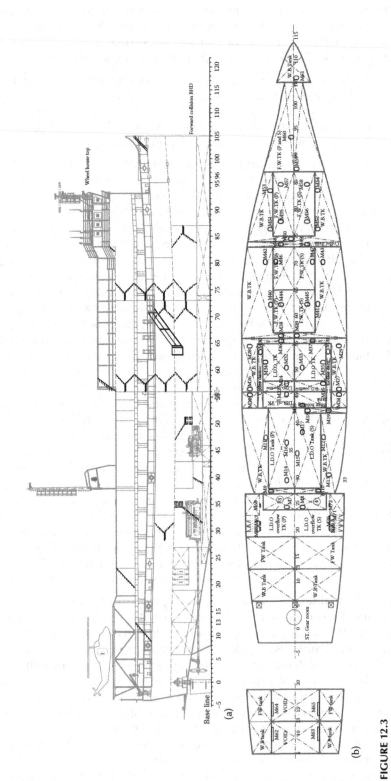

FIGURE 12.3
Access plan to ship's interior. (a) Ladders – access between decks. (b) Manholes at the double-bottom level.

12.3 Rule-Based Design

A ship or platform is registered in a port of registry and then the flag state is the state to which the port of registry belongs. Then it becomes the responsibility of the flag state to ensure the safety of life and property of the marine platform. For this purpose, the flag state generally sets guidelines as rules which vehicles and structures have to follow. Ships moving in international waters go away to other ports and they have to conform to the rules and codes of the port state also. Further, in high seas, ships and platforms have to adhere to some codes to ensure safety during operation. So it has been necessary to set standards and codes for various issues related to safety, pollution, security and similar issues of common interest at an international level. The International Maritime Organization (IMO) under the aegis of the United Nations is the body which discusses various maritime issues at an intergovernmental level and lays down recommendations for implementation at the national level as national law. The IMO has issued recommendations related to safety which have been mandatory on signatory governments. Apart from the IMO, other bodies of the United Nations including the International Labour Organization (ILO) have issued recommendations which are mandatory to be implemented. If a platform does not go to international waters and is bound by operation in territorial waters or in rivers, it does not have to follow all recommendations of the IMO but all acts and regulations of the state. Apart from these regulatory bodies, a platform may be registered with a classification society which has its own codes for structural design and machinery and electrical standards. A platform may be registered for design, construction and operation in which case the codes have to be followed in the design. All the international classification societies have formed a consortium known as International Association of Classification Societies (IACS) to bring uniformity of codes across all societies. The IACS also interacts with and advises the IMO regarding safety and other issues. Thus, the goal-based structural design principles, known as Common Structural Rules (CSR) for bulk carriers and tankers, which have been accepted by the IMO, have evolved. National naval establishments generally have their own rules and practices for naval vessels and platforms, and they are not obliged to follow IMO or class recommendations. But with increasing demand and complexity of naval vessels, the naval establishments are taking guidance from the IMO recommendations and IACS deliberations. The designer of a platform has to satisfy these rules and regulations for the ship or platform that has to operate in marine environment; one can also say that the design gets constrained by these requirements. It is not the intent here to discuss the rules and regulations in detail and how they affect the design. The designer has to go through the detailed rules relevant to his or her platform and implement the same. The following are the main rules and regulations applicable to marine platform safety at sea:

- United Nations Convention on the Law of the Sea (UNCLOS), 1958, on territorial seas and contiguous zone setting rights of nations on territorial waters.
- UNCLOS, 1982, declaring exclusive economic zone (EEZ) for each coastal state and declaring the sea as common heritage of mankind.
- IMO – International Load Line Convention, 1966, setting recommendations for freeboard and closing appliances.

- IMO – International Convention for the Safety of Life at Sea (SOLAS), 1974, and amended subsequently. A list of chapters dealing with various safety issues is given in the following:
  - Chapter II-1 sets forth intact stability criteria, probabilistic damaged stability criteria and subdivision, bilge pumping requirement and essential machinery and electrical installations. It also deals with goal-based standards for tankers and bulk carriers against structural failure leading to loss of ship due to loss of strength, watertight integrity and stability.
  - Chapter II-2 deals with fire protection, fire detection and extinction.
  - Chapter III deals with LSA and arrangements.
  - Chapter IV deals with radio communications.
  - Chapter V deals with safety of navigation which makes it mandatory for ocean-going ships to carry VDR and conform to the Automatic Identification System (AIS).
  - Chapter VI deals with safe carriage of cargo (all cargoes except liquids and gases) to avoid hazards related to the cargo including stowage and securing of cargoes and cargo units (containers) and complying with the International Grain Code for carriage of grain.
  - Chapter VII deals with carriage of dangerous goods. Part A contains carriage of dangerous goods in packaged form following the International Maritime Dangerous Goods (IMDG) Code; Part A-1 deals with carriage of dangerous goods in bulk; Part B concerns carriage of liquid chemicals in bulk following the International Bulk Chemical (IBC) Code; Part C deals with carriage of liquefied gases in bulk and specifies the (IGC) code applied to gas carrier construction; Part D deals with carriage of packaged nuclear material complying with the code for carriage of package irradiated nuclear fuel, plutonium and high-level radioactive waste on board ships (INF Code) which must also comply with relevant portions of the IMDG Code.
  - Chapter VIII deals with nuclear ships and basic requirements against radiation hazards.
  - Chapter IX contains management for the safe operation of ships.
  - Chapter X deals with safety measures for high-speed craft (HSC Code).
  - Chapter XI-1 contains special measures to enhance maritime safety.
  - Chapter XI-2 contains special measures to enhance maritime security.
  - Chapter XII deals with additional safety for bulk carriers, mainly structural requirements.
  - ILO – crew accommodation standards and habitability issues such as noise.

Rule-based design is easy to implement requiring less decision making on the part of the designer. Take, for example the intact stability rules. If a particular design satisfies the requirements, the designer need not look for higher or additional stability features. The safety-related equipment such as LSA, fire fighting appliances (FFA) or redundant equipment can be given as have been prescribed in the rules; there is no need for the designer to critically evaluate the necessity or alterations or additions in these equipments which may incur additional cost. Thus, compliance of rules ensures satisfactory external configuration of the marine platform, general arrangement and space layout, structural design, machinery and equipment selection, electrical and piping layout, outfit configuration, miscellaneous equipment selection satisfying various performance criteria such as stability, weight, trim, speed and power and hydrodynamic performance within reasonable cost with satisfactory operation during its life. Rule-based designs have generally satisfied all the stakeholders such as designers, builders, owners, classification society surveyors and statutory bodies, and reasonable safety in maritime transportation has been achieved.

Technology innovation and advancement has been leading to more sophisticated ships and platforms with varied configurations and performance criteria. This has been coupled with rapid changes in the business and economic environment leading to different demands for transportation. Marine platforms like oil exploration and production platforms are moving from shallow to deep waters and from smooth to not-so-smooth waters. Under such circumstances perhaps it is necessary to visit the principles on which rule-based design is perceived:

- Existing rules are complied with till a major disaster takes place. Only then the community wakes up to look at the issue and new rules are framed.
- The new rules are thus based on existing and up-to-date data set of major accidents which would work well for platforms which conform to such data. For a different platform, these rules may not be effective and the designer may be unnecessarily forced to satisfy these rules.
- The rule-making process itself is a consensual process where a large number of stakeholders have to come to an unjustifiable consensus to ratify and implement the rules where certain important aspects of safety may be compromised.
- Rules provide a minimum requirement of safety and generally the stakeholders are happy if the minimum requirements are met. Hopefully, adequate safety is achieved and there is no need to give higher safety standards with the premise that 'safety does not pay'.
- In rule-based design, generally life cycle issues of safety are not addressed.
- Since rules are the same for all the bidders for a contract, there is little scope for competition at the technology level for a vessel or platform.
- Rules are easy to implement where these work as a constraint on the design process. As an example, the dimensions and weights are finalised subject to a constraint that minimum metacentric height should be greater than 0.15 m.
- Existing rules may not cater for a new and different vessel or platform with different mechanical and electrical systems having different operational requirements.

Is it possible to change the scenario related to safety? As designers, can we say that the design should have maximum safety within the cost constraints? Then safety becomes one of the objective functions of design rather that a constraint. To achieve this, one has to go for a goal-based design for safety where risk is minimised. This is also known as RBD.

12.4 Risk-Based Design

Risk is defined as 'a situation involving exposure to danger, a hazard or a dangerous chance, the possibility that something unpleasant or unwelcome may happen'. Risk does not mean damage or an accident but a chance that occurrence of such events is possible. Ships and maritime structures are always exposed to dangers (hazards) such as fire and flooding. In pure quantitative terms, risk is expressed as

$$\text{Risk} = (\text{Probability of a hazard occurrence})$$

$$* (\text{quantified consequence from the hazard occurrence}).$$

Also, its logarithmic transformations can be used for the sake of simplicity as

$$\text{Log(risk)} = \text{Log(probability of hazard)} + \text{Log(consequence)}.$$

where hazard is an undesirable event, usually dangerous, like an accident that causes damage, injury or loss of life. Hazard may also be defined as a dangerous attribute of an activity, which may result in physical, economic or psychological loss.

Safety describes the completion of a task without being restricted or controlled by hazards, which may include employment of suitable controls to minimise hazard occurrence. It shows how well the activity is positioned and carried out, away from a hazard occurrence. Kuo (2007) discussed about safety management in maritime applications. The employment of management techniques, with an aim of highest safety for crew, systems and equipment that effect maritime operations and maintenance, is referred to as a safety management system (SMS) which is now mandatory in the operation of ships and mobile offshore drilling units (MODU) under SOLAS Chapter IX of the IMO.

Risk analysis and management becomes necessary to be carried out at the design stage when choosing system components and installation of the same for ensuring safety during operation and in the shipping business. Vassalos (2009) dealt with risk-based applications in ship design. In the case of complex systems and during building of large ships and ocean structures, the designer frequently faces the problem of choosing one over the other system based on cost without sacrificing safety standards. The decision-making process in design in such circumstances for each unit of job can be analysed as illustrated in Figure 12.4. The domain of risk and the level of risk, which best suits the system must be determined to make it a viable and competitive model without sacrificing the cost aspects. Discussed in the following are the steps shown in Figure 12.4 leading to a cost-benefit analysis.

12.4.1 Step 1: Hazard Identification

Hazards should to be identified exhaustively for every unit segment of the system. Based on the practical knowledge of the designer, one can go through the following two stages:

1. Identify all the possible hazards like fire, flooding, explosion (physical aspects) or injury, death, fatigue (human aspects) or monetary loss, loss of production, downtime (economic aspects) or damage to environment (pollution, disturbance to birds and other marine or terrestrial species).
2. Identify the component of the system that can be related to these occurrences.

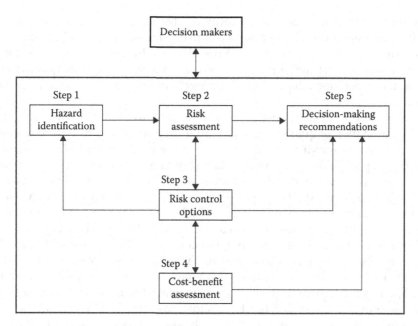

FIGURE 12.4
Decision-making process in design for safety.

For example, consider a cargo line system on a petroleum or chemical tanker ship. The line could contain pressurised flammable fluid which could be a pollutant, toxic chemical, expensive cargo and a hazard that may lead to loss or contamination, damage or loss of human life, damage to expensive pipelines and structures and severe loss of time for repairs (in case of a failure). Moreover, an incident in a port can effect a large population, damage to fisheries (loss of livelihood and damage to ecology), damage to the port facilities or adjacent vehicles, hindrance to traffic, etc. These could be possible hazards identified as per (1) earlier.

The components that can be instrumental in causing such hazards could be pipeline material failure, failure of a joint or expansion joint, instrumentation pocket leaks, wrong operation or sabotage (human aspects), terrorism, failure of monitoring and control systems, failure of pressure relief devices and a potentially dangerous surrounding for the piping through which it is routed.

Hazards identified should be assigned a value commensurate with the extent of damage or severity of its effects and the frequency of its occurrence. The ranking should be judiciously made. This aspect is termed as 'ranking'.

12.4.2 Step 2: Risk Analysis

There is a necessity to identify and categorise risks to provide decision support concerning choice of arrangements and measures. The ability to define what may happen in the future, assess associated risks and uncertainties and choose among alternatives is the core of the risk management system. Alternative design solutions of an activity are formulated and evaluated both qualitatively and quantitatively. The best alternative is the one that gives the highest profitability, no fatal incidents and no damage to the environment. But it is impossible to know with certainty which alternative is the best, as there are uncertainties associated with each risk involved with any choice.

Determination of risk by a definite quantity is advantageous, as it is more clearly stated and can be understood by a large number of users. But such determination is only possible when sufficient data are available. When data are limited or when it is not possible to quantify a risk, expert decisions are relied on, which is called as qualitative assessment. Qualitative decisions depend on the expert's opinion and can change from person to person.

Quantitative risk analysis requires precise and quality data to establish a value to a hazard. A combination of qualitative and quantitative analysis is also used in risk assessment. An example of making risk estimates is shown in the Table 12.6 as a risk index (RI) with the help of Tables 12.2 through 12.5 showing a stepwise estimation procedure. Table 12.2 shows the severity of different types of risk with a quantitative value in the last column. Table 12.3 shows the frequency (probability) of occurrence of such risks shown as P. The calculation of the SI is shown in Table 12.4 as the logarithm of the severity value multiplied with a constant, in this case, 1000. In Table 12.5 the frequency index (FI) calculation is shown in a similar manner, the constant being 1,000,000. Finally, the RI is calculated and the risk matrix is tabulated in Table 12.6 by summing the RI and FI since these are logarithmic values. Data from IMO have been used in generating Tables 12.2 and 12.3. The risk rank gives a quantified value using the RI matrix, and in the present case, the risk above an index of 8 has been shown as 'high risk' with main and standby controls, between 6 and 8 as 'medium risk' with one main control and others as 'low risk' with no control. When a new circumstance is observed and the associated risk needs to be determined without sufficient data, an expert can classify this new risk as high, medium or low risk. This grouping is the qualitative determination of the risk. Qualitative risk should be improved throughout its life cycle, using day-to-day data of its likelihood and consequence.

As an example of risk analysis applications, let us take a cargo pipeline on a parcel chemical tanker ship which is to be designed. The identified risks are spill and fire.

The statistics maintained by the design firm indicates that the likelihood of a pipeline failing is remote ($FI = 3$). However, if such a failure occurs, it will spill hazardous cargo and may cause at least one or more fatalities (cargo watch crew on deck). The severity risk attached is therefore severe ($SI = 3$):

Risk during operation = $FI + SI = 3 + 3 = 6$ based on indexes shown in Table 12.2–12.6.

From the RI matrix maintained, the risk of rank '6' falls under the category that requires at least one control. So the designer refers to standards and recommendations in the International Safety Guide for Oil Tankers and Terminals (ISGOTT) and Tanker Safety

TABLE 12.2

Severity Index (*SI*)

SI	Severity	Effects on Human Safety	Effects on Ship	*S*
1	Minor	Single or minor injuries	Local equipment damage	0.01
2	Significant	Multiple or severe injuries	Non-severe ship damage	0.1
3	Severe	Single fatality or multiple severe injuries	Severe damage	1
4	Catastrophic	Multiple fatalities	Total loss	10

Source: International Maritime Organisation (IMO), Reduced/ No B.W. Ship Designs, GloBallast Monograph 20, IMO, London, U.K., 2011.

TABLE 12.3

Frequency Index (*FI*)

FI	Frequency	Definition	P
7	Frequent	Likely to occur per month on one ship	10
5	Reasonably probable	Likely to occur once per year in a fleet of 10 ships, i.e. likely to occur a few times during the ship's life	0.1
3	Remote	Likely to occur once per year in a fleet of 1000 ships, i.e. likely to occur in the total life of several similar ships	10^{-3}
1	Extremely remote	Likely to occur once in the lifetime of a world fleet of 5000 ships	10^{-5}

Source: International Maritime Organisation (IMO), Reduced/ No B.W. Ship Designs, GloBallast Monograph 20, IMO, London, U.K., 2011.

TABLE 12.4

Calculation of Severity Index (*SI*)

Severity	Calculation	Assigned Value (*SI*)
Minor	$Log_{10}(0.01) = -2$	$-2 + 3 = 1$
Significant	$Log_{10}(0.1) = -1$	$-1 + 3 = 2$
Severe	$Log_{10}(1) = 0$	$0 + 3 = 3$
Catastrophic	$Log_{10}(10) = 1$	$1 + 3 = 4$

TABLE 12.5

Calculation of Frequency Index (*FI*)

Frequency	Calculation	Assigned Value (*FI*)
Frequent	$Log_{10}(10) = 1$	$1 + 6 = 7$
Reasonably probable	$Log_{10}(0.1) = -1$	$-1 + 6 = 5$
Remote	$Log_{10}(10^{-3}) = -3$	$-3 + 6 = 3$
Extremely remote	$Log_{10}(10^{-5}) = -5$	$-5 + 6 = 1$

TABLE 12.6

Matrix of Risk Index and Risk Ranking

		\multicolumn{5}{c}{Risk Index (*RI*)}

		\multicolumn{5}{c}{Severity (*SI*)}

FI	Frequency	1 Minor	2 Significant	3 Severe	4 Catastrophic	
7	Frequent	8	9	10	11	
6		7	8	9	10	
5	Reasonably probable	6	7	8	9	At least 2 independent controls
4		5	6	7	8	
3	Remote	4	5	6	7	At least one control
2		3	4 No control are required	5	6	
1	Extremely remote	2	3	4	5	

Guide (TSG) and provides equipment accordingly and inserts a remark in the design drawing that the entire system, after completion of fabrication, should be subjected to a hydraulic test at a pressure equal to 1.5 times the design pressure. Similarly for minimisation of fire risk, the designer ensures that pipeline electrical continuity should be maintained by bonding cables across each flange used.

Three risk analysis principles are used for risk analysis.

Fault Tree Analysis

This is a top-down approach in which each hazard with causes or events below it, together with a relationship between the events, constitutes an understanding in constructing a 'working model' of hazardous occurrence.

The use of fault tree analysis (FTA) for 'cargo spill' as discussed earlier in the example of a cargo piping system for a chemical tanker is illustrated later. The development of such drawings will effectively give the designer an indication to target the item that requires attention, and he or she can be able to write down some comments for fabricators. FTA should be produced for each hazard identified (Figure 12.5).

Event Tree Analysis

ETA is an inductive analysis technique for identifying and evaluating a sequence of events leading to a hazardous occurrence following an initiating event. Unlike FTA, this a is bottom-up analysis. We first start up with an initiating event and look at the consequences.

Determination of the seriousness, probability and frequency of incidents requires a large amount of data. The example shown in Figure 12.6 (Rausand and Arnljot 2003) would help understand the ETA of an initial event 'explosion'. The analysis produces data required for qualitative risk analysis.

Risk Contribution Tree

If we were to combine both FTA and ETA for an accident incident, we get a risk contribution tree (RCT), a conceptual model of the risk. FTA details out what all the individual contributing factors to an accident are, while ETA applies the laws of probability with the help of sufficient data, to predict the outcome of the accident. Various techniques for risk analysis have been developed for different scenarios using the one or more of the aforementioned principles (some of these are discussed later):

- Hazard and operability (HazOp)
- Hazard analysis (HazAn)

The IMO has issued 'Guidelines for Formal Safety Assessment (FSA) for use in the IMO Rule-Making Process' as a guidance on measures and tolerability of risks. The best practices mentioned are on a three-level scale – 'Intolerable', 'As Low As Reasonably Practical' (ALARP) and 'Negligible', which form a guidance on setting the safety target(s) or consequence of unattended hazards.

The risks which are identified during an FSA process are listed and ranked. The level of risks that should be eliminated at all costs is assessed for the system. Further risk reduction

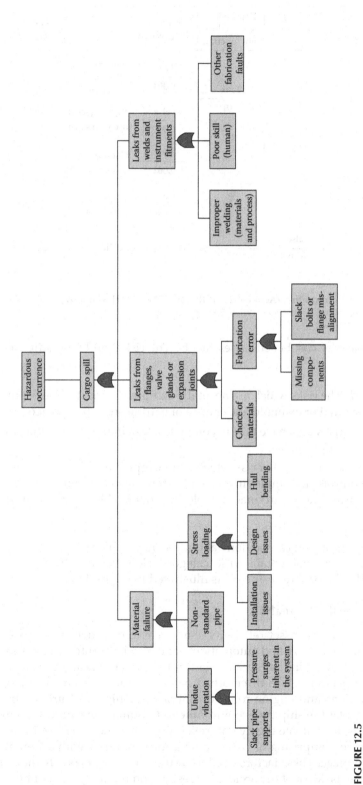

FIGURE 12.5
Fault tree analysis of cargo spill hazard in a chemical tanker.

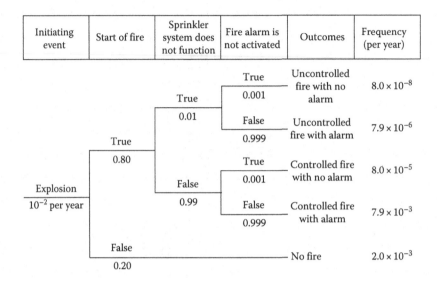

Initiating event	Start of fire	Sprinkler system does not function	Fire alarm is not activated	Outcomes	Frequency (per year)
			True 0.001	Uncontrolled fire with no alarm	8.0×10^{-8}
		True 0.01	False 0.999	Uncontrolled fire with alarm	7.9×10^{-6}
	True 0.80		True 0.001	Controlled fire with no alarm	8.0×10^{-5}
Explosion 10^{-2} per year		False 0.99	False 0.999	Controlled fire with alarm	7.9×10^{-3}
	False 0.20			No fire	2.0×10^{-3}

FIGURE 12.6
Event tree analysis in a typical fire. (Adapted from Rausand, M. and Hoyland, *A., System Reliability Theory: Models, Statistical Methods and Applications,* Wiley, 2004, p. 210, fig. 3.23.)

is done on the basis of cost-benefit assessments. The risks can be broadly categorised as follows:

1. *Negligible risk*: The risks which are normally accepted by most of the people in their daily endeavour. For example, the chance of getting stuck by a meteor.

2. *Tolerable risk*: Risks which we take to complete a task in the process that are considered essential or profitable.

3. *Unacceptable risk*: The level of risk which is unacceptable, for which we do not have sufficient controls. By employing suitable steps, a responsible design/management team brings down an unacceptable risk to tolerable risk levels or negligible risk levels.

There is nothing much that can be done about the 'negligible risk' category and nothing needs be done for the 'unacceptable risk' category. However, all other risks can be controlled to ALARP. The ALARP principle is illustrated in Figure 12.7.

12.4.3 Human Reliability Analysis

Assessment of a maritime asset reliability shall have to consider its hardware – design, structure, equipment, control and automation – and also importantly, operations by crew. Ships such as container ships, oil tankers, gas tankers and bulk carriers have a number of human effected operations. The three broad day-to-day activities on a ship, viz. navigation, cargo operations and engine room operations, are highly controlled by crew. It is important to consider the ship type (environment) to human–machine–automation interaction for analysis, to evaluate the overall reliability. Also, the maritime industry is truly global in workforce, comprising of crew from various countries with different traditions, expertise and language. These factors emphasise the need to incorporate the human factor in safety assessments. Most of the accident investigation reports point out human error as

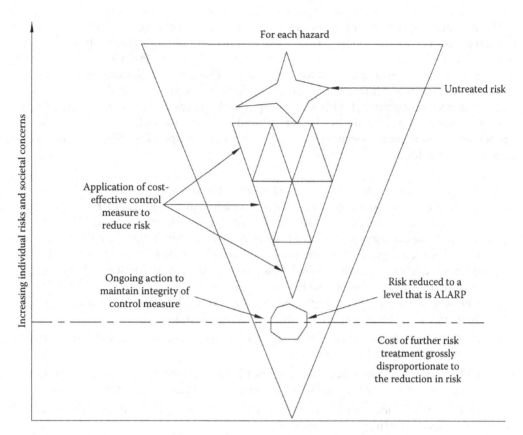

For each hazard

Untreated risk

Application of cost-effective control measure to reduce risk

Ongoing action to maintain integrity of control measure

Risk reduced to a level that is ALARP

Cost of further risk treatment grossly disproportionate to the reduction in risk

Increasing individual risks and societal concerns

FIGURE 12.7
The ALARP principle.

a cause. Justifiable statistics are available in the maritime sector to reason human error as a major concern to minimise or eliminate accidents and failures. Human reliability analysis (HRA) is a developing science to mitigate the human error from critical operations. Human reliability can be defined as the successful performance of human activity necessary in a healthy-system operation. HRA is a method of estimating human reliability and is based on the idea that human error is not a random event, but humans can be pushed to a faulty action due to the context, environment or situation under which they are operating. A safety assessment and its incorporation into a marine system design are detailed here using an example of a fire extinguishing system on a ship.

The ultimate aim of FSA is safety and cost-effectiveness; therefore, managing human element for increased reliability and cost-effectiveness is important. Some systems such as fire detection or a fire extinguishing system should have high operating reliability. It is common in such systems design to opt for complete automation to avoid human errors. Still, some of the fire extinguishing systems can harm the personnel or passengers in the spaces by suffocating them; examples of such types are CO_2 extinguishing systems. Hence, we have a safety barrier in these systems by way of incorporating a responsible manual release, after people have completely evacuated these spaces. The human element is currently indispensable in taking critical decisions in maritime operations. Therefore, employing HRA in such system designs yields certain important conclusions, which are incorporated to make the system highly reliable.

These operating stages and steps are a fallout of the safety assessment including the human element. The overall objective of ensuring high reliability of the system is achieved with such consideration of the human element at the design state. HRA is a process employing varied analysis techniques for the evaluation of risk due to human error in an FSA. Quantitative risk due to human error can be evaluated in terms of probabilities (referred to as human error probability [HEP]) and incorporated into the ETA models for probabilistic risk assessment. The IMO states that the present database for HRA risk quantification is limited and most benefit is derived by an early qualitative approach. HRA inclusion in FSA is done with the following objectives:

1. *Identify key human tasks for hazard identification*: The objective of this stage is to analyse the human–human interaction and human–equipment interaction to understand human error that could lead to system failure. Human hazard is identified by analysing the ways by which a human error can contribute to accidents during normal and emergency operations. Standard techniques such as HazOp and failure mode and effect analysis (FMEA) are recommended in this stage. A high-level functional task analysis is also highly recommended to gain a broad but shallow overview of the main functions that are performed by humans to accomplish a particular task. The steps of implementation are as detailed:

   a. Modelling the system (being investigated) using task analysis to identify main human tasks and its sub-tasks.

   b. For each task identified, techniques like HazOp and HazAn are applied to identify the contributing factors to human errors and its associated hazards.

   c. Each hazard identified and the consequence or scenario from its occurrence is ranked according to its criticality.

   d. More critical hazards and consequences are considered for further risk assessment, while lesser critical hazards, which are acceptable within the set safety targets, are left unattended.

2. *Risk analysis – task analysis*: Risk analysis in HRA identifies the probable zones in the process that are vulnerable to the human element. The analysis also brings into focus the factors affecting the risk. The aim is implemented in the following steps:

   a. Key tasks are analysed in detail using techniques such as horizontal task analysis or cognitive task analysis.

   b. The detailed analysis is carried out exhaustively to cover all identified sub-tasks.

   c. Likely human errors, which can lead to undesirable events, are identified.

   d. The human errors are classified based on

      i. Cause of the error.

      ii. Likely measures to recover error, i.e. to restore the system back to the stage it was before the human error was committed.

      iii. The consequence of the error to the system, i.e. monetary loss, process interruption and damage to the system or environment.

   e. Human errors are quantified in the HEP, which can be further used in FSA analysis. The HEP quantification is optional and the IMO has offered some guidance on the methodology that could be adopted to quantify human errors. Using direct measurement, utilising expert opinion and using historic data by employing techniques such as the human error assessment and reduction technique (HEART) or technique for human error rate prediction (THERP), the IMO emphasises the need for such quantification and, at the same time, cautions to heed the FSA objective.

3. *Risk control options*: To minimise or subvert the consequences from human error, risk control options are framed. Cost-benefit analysis for the risk control options is carried out and decisions are made judiciously.

12.4.4 Step 3: Risk Control Options

The purpose of risk control options (RCOs) is to provide effective and practical risk control options for risk analysis. It has three principal stages:

1. Focus on the risk areas that need controls: Identifying the risk control measures (RCMs).
2. Evaluating the effectiveness of the RCMs. Both historical and newly identified risks should be addressed here.
3. Grouping RCMs into practical regulatory options.

Identification of Potential RCMs

This stage groups RCMs into well-thought-out pragmatic regulatory options. Often such grouping is achieved by the following:

1. General Approach: Provides risk control by mitigating the likelihood of initiating incidents. They are likely to be effective in preventing several different accident sequences.
2. Distributed Approach: Provides control of escalation of accidents and/or later stages of escalation of other related and unrelated incidents.

Table 12.7 explains the common risk control options exercised on ships to bring the risks to an acceptable limit.

12.4.5 Step 4: Cost-Benefit Analysis

Cost-benefit analysis is a practical management decision-making qualitative technique which weighs the consequences of investment decisions and facilitates in identifying the projects that secure profits. Cost-benefit analysis can be applied for small projects such as selection of system components of a small system or large projects such as overall ship-building with regard to higher safety level in either case.

The steps of a typical analysis are discussed here to limit the subject to a macro-level (interested students can refer the economic texts on cost-benefit analysis for more intricacies):

1. *Determine the scope and objectives* of the proposed investment (project). It is called statement of the problem, which is very general in formulation.
2. *Identify all feasible alternatives* for the investment.

TABLE 12.7

Common Risk Control Options

Elimination	If the activity is redesigned or the substance concerned is eliminated so as to remove the hazard, then the redesigned method should not prove less effective or cause unacceptable result from the activity. Then this is a risk control option.
Substitution	If some material or process is substituted with alternative means, which results in a lesser hazards, then this means becomes a risk control option.
Engineering controls	If employed additional automation or machinary or separating and enclosing dangerous items results in mitigation of the hazards, then exercising these options is from an engineering risk control option.
Administrative controls	If some rules are framed such as avoiding smoking and limiting the workers' continuous exposure time to the hazard to reduce hazards, then these are the administrative risk control options.
Use of PPE	If the hazards associated with the activity are minimised and if the personnel involved use appropriate personal protective equipment (PPE), then providing personnel with such gear constitutes a risk control option. Examples of PPE are safety helmet, coveralls, visor or goggles, gloves, safety shoes, etc.

3. *Identify what the costs and benefits are* of each of the alternatives.

4. *Quantify the costs and benefits.* The costs and benefits are purely financial and include all the expenditure in setting up the project – price of equipment at the location, installation, testing, approvals, commissioning and running and maintenance cost over the lifetime.

5. *Discount the future benefits and costs.* Discounted cash flow analysis at present worth.

6. *Apply a check* of the calculations accounting for uncertainty.

7. *Report the findings.* List out the cost-benefit analysis for all the alternative options for the project. Also include the intangibles and threats in the findings.

12.4.6 Step 5: Decision-Making

Decisions are then made using net present value (NPV) over the life of the project as discussed in Chapter 4.

12.4.7 Overall Design Application

The loss of Panamax bulk carrier M.V. Derbyshire in 1980 was primarily due to damage of ventilators and air pipes on the upper deck in heavy seas which led to shipping of green seas into forward holds causing forward trim, more green seas, progressive flooding and finally sinkage. More recently, the loss of cruise ship Costa Concordia off the coast of Tuscany, Italy, was due to complete flooding caused by a 70 m tear in the engine room bottom due to grounding over a rock on a deviated path from the planned course. The loss of Korean passenger ferry M.V. Sewol in April 2014 off the South Korean coast was perhaps due to faults incurred during modifications of an 18-year-old vessel increasing its top weight, compromising its stability, its manoeuvring qualities and, in general, its seaworthiness. Frequently there are reports of loss of boats and lives in inland, port and coastal waters in Asian countries such as India, Bangladesh, China and Indonesia. It is difficult to prescribe rules for every hazardous event to be taken care of at the design stage for the safety of vessels and property. Perhaps an RBD based on safety assessment and steps based on economic viability at the design stage could lead to a more practical design with adequate safety.

An example of Formal Safety Assessment of Bulk Carrier Safety was presented to IMO vide report no. MSC 74/5/4. It includes risk assessment with a view to forming guidelines for safety during design and operation, but does not include cost-benefit analysis. A complete case study of RBD of a RO-PAX vessel and a cruise liner has been presented by Vassalos (2009) which does the overall (formal) safety assessment of either of these vessels including flooding, fire and evacuation.

While designing new platforms, ships or submersibles or installing new systems such as liquified natural gas (LNG) fired ships and nuclear ships, it is necessary to do FSA studies at the design stage to decide design features at the concept design stage and selection of equipment, materials and facilities later so that only negligible risks exist during operation and maintenance of the product or system.

13

Design for Sustainability

The World Commission on Environment and Development (WCED), commonly known as the Brundtland Commission, gave the definition of sustainability in its report in 1987 as 'Development that meets the needs of the present generation without compromising the ability of future generations to meet their own needs'. Sustainability, therefore, includes all technologies and their effects on the environment. Technology whose use is intended to mitigate or reverse the effects of human activity on the environment is known as green technology. Needless to say that transportation is a key element of technological development since transport is keenly related to trade and movement of goods and people. Furthermore, 90% of the world's trade is done through the seaborne trade. Therefore, marine transportation should have a major effect on the environment. The technological development that reduces environmental effects due to seaborne transportation is commonly known as green ship technology. Besides, there are a number of industrial activities at sea such as oil exploration and production, ocean mining, energy extraction from the oceans, and so on. Such activities should not also cause environmental degradation. And the rate at which shipping tonnage is increasing across the world (Figure 13.1) gives further impetus towards green technology application in shipping. Whereas care during construction, operation and dismantling must be taken to reduce the effects of environmental degradation; all this has its origin in the design of the artefact and therefore, sustainable design or green ship technology applications have become very important at the design stage.

The marine environment gets affected by pollution of various kinds – gas emissions, cargo and fuel oil discharge due to oily bilge discharge, accidents and damage, ship waste discharge due to garbage, black and grey water, ballast water movement across the seas, discharge of underwater coatings and underwater emissions affecting marine biological habitat and ship dismantling. Even though it may produce some undesirable environmental effects, shipbuilding activity is non-marine in nature and, therefore, is not included as a part of marine sustainability. The United Nations, the representative body of the world's nations, has entrusted the International Maritime Organisation (IMO) as the body responsible for monitoring and control of marine environment. The Marine Environment Protection Committee (MEPC) under the aegis of IMO deliberates on various environmental issues and recommends guidelines for marine environment protection from time to time which becomes mandatory on signatory nations. The MARPOL regulations prescribe regulations for pollution control to be adopted by ships as discussed in Section 12.3. IMO also deals with efficient utilisation of energy for transportation at sea and prevention of environmental changes due to ballast water movement. Although these rules cover pollution generally at sea, IMO also designates special areas and emission control areas (ECAs) which are ecologically and oceanographically sensitive areas, particularly with respect to shipping activity, which must follow a stricter environmental requirement. IMO also defines particularly sensitive sea areas (PSSAs) which must give special protection to ecological, scientific and socio-economic aspects by controlling maritime and shipping activities by strict the appliance of MARPOL discharge regulations. IMO circular MEPC.1/ Circ.778 Annex 1 gives the list of special areas

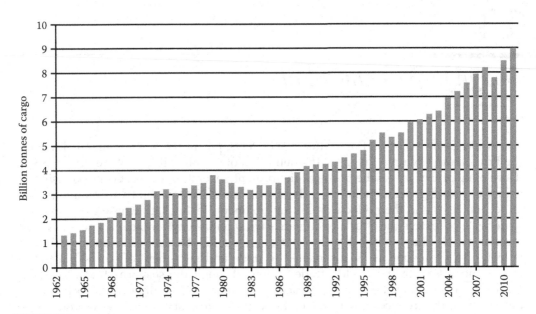

FIGURE 13.1
Increasing seaborne trade across the world.

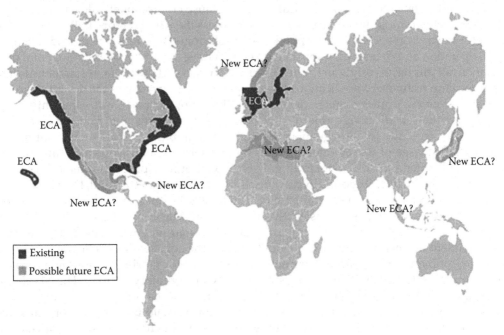

FIGURE 13.2
Emission control areas.

including ECA for air pollution and Annex 2 gives the PSSA. Figure 13.2 shows the ECA around the world as designated by IMO.

This chapter discusses the various maritime pollution issues and design methods and principles to reduce the same. It discusses the guidelines set by IMO for pollution reduction, but does not stop there and goes on to discuss measures and vision of a zero pollution environment. It does not discuss the IMO rules and regulations in detail. This leads to an exciting world of new design developments which includes a lot of innovations and creative thinking. To discuss these issues, it is necessary to understand the various machinery and ship systems on marine artefacts as are used conventionally using diesel oil today, which has already been discussed in Chapter 9.

13.1 Air Pollution

Air pollution from marine operations is due to exhaust gases, mainly carbon dioxide (CO_2); nitrogen oxides (NO_x), mainly nitric oxide (NO), nitrogen dioxide (NO_2) and nitrous oxide (N_2O); sulphur oxides (SO_x), mainly sulphur dioxide (SO_2) and sulphur trioxide (SO_3); and particulate matters (PMs) including volatile organic compound from oil burning main diesel engines, boilers, auxiliary machinery, combustion of wastes, chlorofluorocarbon (CFC) mainly consisting of freon and bromotrifluoromethane (halon) gases; and evaporations from cargo. Freon and halon gases come under the category of ozone-depleting substances (ODS) which deplete the protective ozone layer of earth from harmful ultraviolet radiation of the sun. The United Nations Environment Program, in its Montreal Protocol of 1987, has banned production of ODS and therefore, marine artefacts do not use CFC gases anymore.

13.1.1 Air Pollution from Diesel Oil Burning Engines

Sulphur oxides cause acid rain which affect coastal areas the most, which has a detrimental effect on the environment. SO_x emission must be controlled and if not, the coastal areas will have large deposits of sulphates causing environmental hazards. Figure 13.3 shows the sulphur concentration originating from international shipping on European coast in 2000 and what it is likely to be by 2020 if no action is taken to reduce SO_x emission from shipping activity (OSPAR Commission 2009). The use of SO_x scrubbers, either closed loop, open loop or hybrid systems, for treating exhaust gas using sea water and caustic soda has proved as a useful equipment for SO_x reduction. IMO has recommended that sulphur content in bunker

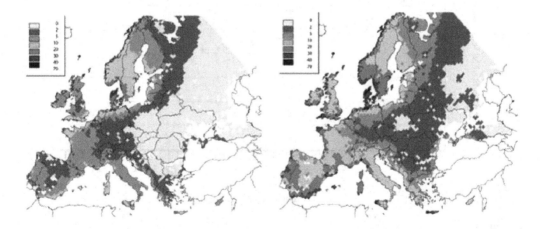

FIGURE 13.3
Sulphur deposit from international shipping if no action is taken.

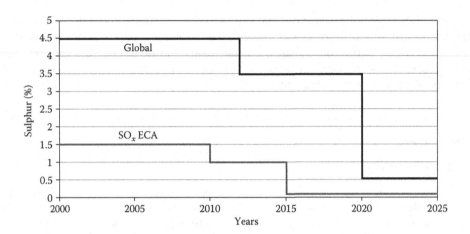

FIGURE 13.4
Maximum sulphur content in bunker fuel for ships (Annex IV, MEPC, IMO).

fuel must be reduced drastically by the year 2020 as suggested by MEPC. This is shown diagrammatically in Figure 13.4. The diagram indicates that use of lighter diesel oil with reduced sulphur content will increase in future increasing the cost of fuel used in ships.

Nitrogen oxides also cause acid rain causing deterioration of air quality in coastal areas. Therefore, it is also necessary to have a regulatory mechanism to control NO_x emission. MEPC-suggested measures have been adopted by IMO as per Annex IV for NO_x control in three tiers: tier I should have been applicable from 1 January 2000, tier II is to be applicable from 1 January 2011 and tier III from 2016. This is shown diagrammatically in Figure 13.5. The various methods for NO_x reduction, as mentioned in the following, require some improvement in diesel engine and auxiliary equipment design: (1) water addition to reduce combustion temperature

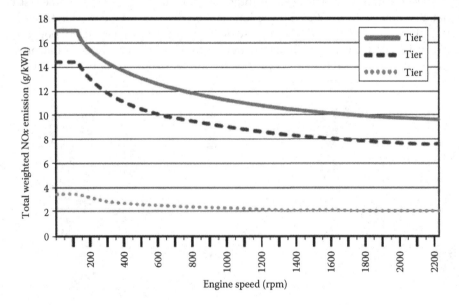

FIGURE 13.5
Maximum NO_x content in exhaust gases of ships (Annex IV, MEPC, IMO).

introducing humid air motors or fuel water emulsification; (2) exhaust air treatment such as selective catalytic reduction, selective non-catalytic reduction, exhaust gas recirculation or humidifying scavenging air; and (3) engine modification with common rail control with delayed fuel injection and control of exhaust for optimised NO_x emission.

PMs contain different matters, of which the individual components, size of particles or their properties have not been clearly studied. Also, the relationship of PM with smoke opacity has not been established. However, it is known that PM affects the health of local populace and therefore, these should be reduced. Fortunately, if sulphur content is reduced, PM automatically reduces. There need not be any separate requirement for PM reduction.

Carbon dioxide (CO_2) is generated by burning carbon-based fuels, typically fossil fuels such as coal, oil, natural or petroleum gas and biofuels containing methane in power plants and transport vehicles. CO_2 concentration in air varied between 180 and 280 ppm for the last 400,000 years, but has already increased to 396.81 ppm in the past few years, as shown in Figure 13.6. Increase of carbon dioxide in the atmosphere causes global warming or increase of earth's temperature. The Intergovernmental Panel on Climate Change (IPCC) report on climate change states that the earth's temperature has increased by about 0.85°C between 1880 and 2012. If carbon dioxide emission is not checked, the temperature will rise fast by anywhere between 2°C and 4°C in the next 100 years. If the carbon dioxide concentration is limited to 450 ppm, the temperature rise may be limited to 2°C. The effects of global warming leads to melting of ice caps in the Atlantic and Arctic causing sea level rise leading to immersion of large land masses. This would mean displacement of millions of coastal inhabitants creating a large number of 'climate change refugees' across the world. Climate change also means severe heat waves and draughts across the world. This also means change of weather pattern, severe cyclones, floods and similar natural calamities and drying up of natural resources in the biosphere. In the last 40 years, global warming has caused warming

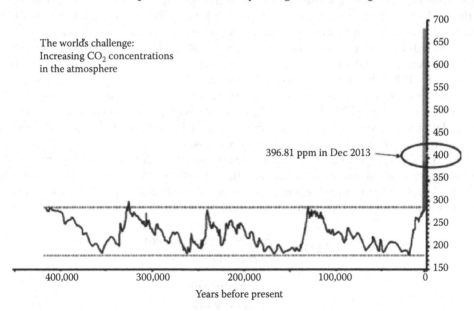

FIGURE 13.6
CO_2 concentration in the atmosphere. (From Schneider, T., *How We Know Global Warming is Real—The Science Behind Human-induced Climate Change*, Skeptic Magazine, Vol. 14, No. 1, pp 31–37, The Skeptics Society, Altadena, CA, 2008. With permission.)

of ocean waters up to 750 m depth by 0.09°C–0.13°C. If this temperature rise is not contained, it will affect ocean currents, atmospheric patterns and accelerate climate change. The temperature rise in the oceans will increase carbon content in the ocean causing ocean acidification leading to destruction of ocean planktons affecting ocean food chain, ultimately leading to reduction of ocean biosphere and elimination of many bio species. Also, global warming may lead to gasification of frozen methane hydrates in the earth's crust emitting methane gas accelerating global warming. Although IPCC has recommended that global warming should be limited to 2°C rise in this century, experts opine that it should be much less than that to avoid catastrophic climate change. Therefore, it is essential that all human activities must be directed towards reduction of carbon dioxide emission.

Marine use of conventional diesel engines and boilers produces CO_2 by burning heavy or light diesel oil. In the short term, the endeavour should be to reduce carbon dioxide emission by more energy efficient ships burning diesel oil. Towards this effort, IMO has suggested an energy efficiency design index (EEDI) applicable to newly designed ships. The next stage should be to use alternate fuels, such as natural gas (NG) or petroleum gas, containing carbon which may burn efficiently so that CO_2 emission may be reduced. But effort should be on finding renewable and non-polluting energy sources to propel ships across the oceans, in ports as well as in rivers for inland transportation. These aspects are discussed in the following sections.

13.1.2 Energy Efficiency Design Index

Roughly 50,000 ships carry 90% of the world's traded cargo every year and these ships tend to run on heavily polluting oil known as bunker fuel. According to the IMO, shipping was responsible for 2.7% of global carbon emissions in 2007, but that could double or even triple by the mid-century if no action is taken now. IMO's MEPC has concluded that all ships built in the future must reduce pollution from today's averages. The levels of emissions reduction will be based on an efficiency index for ships of varying sizes and types. The new rules mandate that ships contracted in the first 5 years after 2015 must improve fuel efficiency by 10%. The standards are to be tightened every subsequent 5 years. By 2030, a 30% reduction rate would be set for most types of ships, based on the average of those built between 1999 and 2009. Based on this, a new Chapter 4 has been added to Annex VI on regulations on energy efficiency for ships to make it mandatory that the attained EEDI must be lower than the required EEDI for new ships.

The EEDI formulation has been developed initially for the largest and most energy-intensive segments of the world merchant fleet, thus embracing 72% of emissions from new ships and covering the following ship types: oil and gas tankers, bulk carriers, general cargo ships, refrigerated cargo carriers and container ships as

$$\text{EEDI} = \frac{\text{Impact to environment}}{\text{Benefit to society}}$$

$$= \frac{CO_2 \text{ emmission}}{\text{Transport work}}$$

The CO_2 emission represents total CO_2 emission from combustion of fuel, including propulsion and auxiliary engines and boilers, taking into account the carbon content of the fuels in question. If energy-efficient mechanical or electrical technologies are incorporated on board a ship, their effects are deducted from the total CO_2 emission. The energy saved by the use of wind or solar energy is also deducted from the total CO_2 emissions, based on actual efficiency of the systems. The transport work is calculated by multiplying the ship's capacity (dwt), as designed, with the ship's design speed measured at the maximum design load condition and at 75% of the rated installed shaft power. Then

$$\text{EEDI} = \frac{\left(\prod_{j=1}^{M} f_j\right)\left(\sum_{i=1}^{nME} C_{FMEi} SFC_{MEi} P_{MEi}\right) + P_{AE} C_{FAE} SFC_{AE} + \left(\sum_{i=1}^{nPTI} P_{PTIi} - \sum_{i=1}^{nWHR} f_{effi} P_{WHRi}\right) C_{FAE} SFC_{AE} - \left(\sum_{i=1}^{neff} f_{effi} P_{effi} C_{FMEi} SFC_{MEi}\right)}{f_i \, \text{Capacity} \, V_{ref} f_W}$$

1. C_F is a non-dimensional conversion factor between fuel consumption and CO_2 emission based on carbon content as shown in Table 13.1. The suffices MEi and AE indicate values for each main engine and for all auxiliary engines taken together respectively.
2. V_{ref} is the ship speed, measured in nautical miles per hour (knot), on deep water in the maximum design load condition (capacity).
3. Capacity is defined as follows:
   a. For dry cargo carriers, tankers, gas tankers, containerships, ro-ro cargo and general cargo ships, deadweight should be used as capacity.
   b. For passenger ships and ro-ro passenger ships, gross tonnage should be used as capacity.
   c. For container ships, the capacity parameter should be established at 65% of the deadweight.
4. Deadweight means the difference in tonnes between the displacement of a ship in water with a relative density of 1025 kg/m³ at the deepest operational draught and the lightweight of the ship.
5. P is the power of each main engine (suffix MEi), all auxiliary engines (suffix AE), input to each power take-off system (suffix PTi) and innovative technologies or waste heat recovery systems for auxiliary power (suffix WHRi) and innovative technologies for main propulsion power (suffix effi), measured in kW.
6. V_{ref}, capacity and P should be consistent with each other.

TABLE 13.1

Carbon Content of Different Fuels

Type of Fuel	Reference	Carbon Content	C_F (t-CO_2/t-Fuel)
Diesel/gas oil	ISO 8217 Grades DMX through DMC	0.875	3.206000
Light fuel oil (LFO)	ISO 8217 Grade RMA through RMD	0.86	3.151040
Heavy fuel oil (HFO)	ISO 8217 Grade RME through RMK	0.85	3.114400
Liquefied petroleum gas (LPG)	Propane	0.819	3.000000
	Butane	0.827	3.030000
LNG		0.75	2.750000

7. *SFC* is the certified specific fuel consumption, measured in g/kWh, of the engines with suffices MEi and AE as used before.

8. f_j is a correction factor to account for ship-specific design elements.

9. f_w is a non-dimensional coefficient indicating the decrease of speed in representative sea conditions of wave height, wave frequency and wind speed (e.g. Beaufort scale 6).

10. $f_{eff(i)}$ is the availability factor of each innovative energy efficiency technology. $f_{eff(i)}$ for waste energy recovery system should be one (1.0).

11. f_i is the capacity factor for any technical/regulatory limitation on capacity and can be assumed one (1.0) if no necessity of the factor is granted.

The aforementioned formulation is flexible and can be modified for some other ship types such as general cargo ships and chemical tankers, a number of main propulsion units, power take-offs, innovating technologies etc. This is the attained EEDI for a new ship which must be lower than the required EEDI, which is shown diagrammatically in Figure 13.7. There are a number of limitations to this formulation of EEDI which may be listed as follows:

- The formulation of EEDI is not really an energy efficiency formulation since it does not confirm to the standard definition discussed in the previous section. Whereas energy efficiency should increase with improved hull form and propeller design, the present EEDI formulation should reduce with improved design. This formulation can best be described as an emission efficiency design index.

- This formulation is based on CO_2 emission only, whereas it should include all green house gas (GHG) emissions, particularly methane dispersed through methane slip in liquefied natural gas (LNG) fired ships, discussed in the next section.

- As stated by MEPC, the present EEDI formulation is applicable to about 72% of all transport vessels and it will be extended to the remaining transport vessels with suitable corrections in due course.

- Even then it leaves out a large number of smaller vessels, port craft, inland vessels, non-cargo carrying vessels and a vast number of naval vessels. These vessels do not do any transport work as defined in the formulation, i.e. cargo carried over a distance. Perhaps a more general denominator could have been 'useful work' instead of 'transport work'. 'Useful work' has to be defined in more general terms which should include 'transport work'.

- One important observation that can be made from Figure 13.7 is that smaller vessels emit higher levels of carbon dioxide and not much can be done in terms of design improvement and these vessels will have a higher EEDI even in the future. Such vessels include port craft, trawlers, coastal vessels, a large percentage of naval vessels and all inland vessels across the world. Since there are a very large number of small vessels around the world which operate within limited distances, emissions from such vessels might have been affecting the local environment apart from contributing to global warming. This needs to be addressed. Perhaps these vessels require alternate non-polluting energy source to drive them urgently.

The following could be some of the design measures to improve EEDI:

- Arrangements of machinery could be such that total emissions can be reduced.
- Waste heat recovery system could be installed for emission reduction.

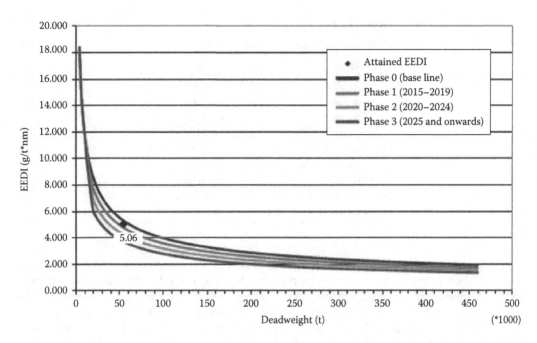

FIGURE 13.7
Energy efficiency design index.

- Use of shaft driven generators is to be investigated from EEDI angle.
- Diesel electric propulsion system, particularly in vessels having varied load demand and podded propulsion systems could be investigated for reducing EEDI.
- Use of non-polluting energy sources could be used as auxiliary power sources to reduce emission.
- Hull form and propeller design can be improved to reduce energy consumption.
- Most importantly, designing ships at reduced speed reduces energy consumption substantially. Thus, EEDI application points at generally low speed design and operation of ships.

13.1.3 Natural Gas as Marine Transportation Fuel

Improvements or reductions in polluting gases can be achieved only as a slow process. It is now understood that the use of gaseous fuel such as LNG reduces emissions to some extent and does not require much technology change for its shipboard use. Other energy sources must be fully investigated before they replace carbon burning fuels completely. Therefore, LNG and its use as shipboard fuel are discussed in the following as it seems to be the most practical alternative fuel.

13.1.3.1 Physical Properties of Natural Gas

NG is a gaseous mixture of hydrocarbons occurring in the earth's crust. It is often found together with petroleum and coal deposits and as a hydrate on sea bed. NG is also generated during decomposition of organic matter such as human and animal wastes and biomass in marshy areas. NG is a colourless, odourless and tasteless gas, and is lighter than air.

TABLE 13.2

Properties of Processed Natural Gas

Property	Value	Remarks
Boiling point	−161.5°C	At 1 atm
Freezing point	−182.6°C	At 1 atm
LNG-specific gravity	0.43–0.47	Rel to water = 1
Gas density	0.7–0.9	kg/m³ @ STP
Flammability limits	4–15	By volume in air
Ignition temperature	538°C	At 1 atm
Carbon content	73	By weight
Hydrogen content	24	By weight
Oxygen content	0.4	By weight

At atmospheric pressure, it is in gaseous state above −161°C temperature. NG composition varies at each point of origin. Its properties can be determined only after assessment of its composition and a popular technique is by gas chromatography. NG contains mainly methane, about 70%–90%, other hydrocarbons such as ethane, propane, butane between 0% and 10%, pentane and other higher hydrocarbons between 0% and 10%, carbon dioxide 0%–8%, oxygen 0%–0.2%, nitrogen 0%–5%, hydrogen sulphide 0%–5% and traces of some rare gases like helium. The impurities that have no heat value are removed by applying different processes, leaving mostly a hydrocarbon mixture of gases. NG is further refined and the recoverable hydrocarbon gases like propane and butane are extracted based on the properties of weight, boiling point or vapour pressure of constituents. Table 13.2 shows some properties of processed NG. If more quantity of recoverable higher hydrocarbon gases is present in the NG, it is termed as 'Rich', otherwise 'Lean'. Sometimes the NG is termed 'Wet' if liquefiable ethane, propane or butane is present and 'Dry' if these are absent. The energy content of a specific volume of NG is variable, depending on the composition. The physical properties of NG vary depending on the temperature and pressure. The condition that refers to standard temperature and pressure (STP) is a dry bulb temperature of 15°C and pressure of 101.325 kPa and the normal temperature and pressure (NTP) condition refers to a temperature of 0°C and pressure of 101.325 kPa.

The heating value or calorific value of NG varies depending on its constituent gases. Gross or higher heating value (GHV or HHV) is the total amount of heat recoverable by complete combustion of a unit volume of NG with stoichiometrically correct amount of air. On the other hand, if the same fuel is combusted in engines, boilers or turbines, hydrogen is oxidised to water, which is present as vapour or steam above 100°C and evaporates with exhaust. This reduces the heat released. The available heat is the lower heat value or LHV. Table 13.3 shows energy density in NG and marine fuel.

13.1.3.2 Storage of Natural Gas

Gases are made up of a number of atoms or molecules, which tend to occupy whole of the volume they are confined to. The gas density is quite low compared to solids or liquids. This is a major drawback for use of NG in gaseous state on ships. For fuels, the energy density follows the following order: solids > liquids > compressed gases > gases. The ability to store more energy in a given volume increases endurance of the vessel and reduces bunkering intervals, associated costs and interruption to the ship's activity. To increase the

TABLE 13.3

Energy Density of Natural Gas and Marine Fuel

Fuel	LHV mJ/kg	Density kg/m³	Energy Density gJ/m³	Tank Weight-Volume (Approx. Relation for Vol. in Litres, Weight in kg)
LNG	48.632	428.22	20.825	0.23 × vol.
CNG	47.141	182.14 at 200 barg	8.586	0.67 × vol. (fibre)
				0.43 × vol. (carbon)
ANG (190 V/V)	47.141	148 at 35 barg	7	0.43 × vol.
NG	47.141	0.7769	0.0366	NA
MDO/MGO	40.612	846.94	36.089	0.072 × vol.
HFO	40–42.5	989.1	39.564–42.036	0.073 × vol.
Ice gas oil	42.79	836.63	35.80	0.072 × vol.

energy density, NG can be compressed, cooled or adsorbed or dissolved in media. LNG is stored in cryogenic temperature of –162°C at normal pressure. CNG is a compressed NG stored or transported as a pressurized gas at about 250 bars. Storage of NG at pressures at or in excess of 200 bars increases the stored energy and is a potential risk. Increasing potential energy of a flammable gas by way of increasing pressure (compression) for storage increases risk of explosion hazard. Adsorbed natural gas (ANG) was developed to reduce this risk of stored energy and at the same time achieve CNG equivalent energy density. ANG has an advantage over LNG, as it does not require energy for cooling. ANG is based on principles of reducing kinetic energy of constituent gas particles by keeping a highly porous medium inside the tank. The porous medium adsorbs (penetration of methane gas into solid porous structure) methane and reduces gas-to-gas and gas-to-cylinder wall collisions. Thus, ANG has been developed as an alternative to CNG to reduce stored pressure which is about 35 bars with equivalent energy density as that of CNG. The compression of volume from gaseous stage at NTP/STP to liquefied or pressurised form is expressed in terms of volumetric efficiency which is the ratio of maximum volume of gas at NTP/STP to the net storage volume of the container. This is 615 for LNG, 200 for CNG and 180 for ANG achieved in the laboratory which can go up to 270 ideally. Table 13.4 gives the various properties including energy content of liquid and gas fuels.

Liquid fuel is stored in tanks which can be a part of the ship and do not contribute much additional weight. LNG, on the other hand, is stored in separate tanks made of material suitable for carriage of cryogenic cargo. CNG and ANG are carried in separate tanks suitable as pressure vessels made of fibre re-enforced plastic which could even be carbon fibre of high strength. Taking the data of Table 13.3, the mass and volume and tank weight for different fuels can be calculated and Table 13.4 shows these values equivalent

TABLE 13.4

Comparative Storage Requirement of Fuels

Fuel Type	Capacity (m³)	Energy (gJ/m³)	Mass (kg·wt)	Tank Weight (kg·wt)
HFO	1	40	989.1	0.073
LNG	1.92	40	929.23	0.442
CNG	4.66	40	848.77 (at 200 bars)	3.12 (fibre)
				2.0 (carbon)
ANG	5.71	40	845.08 (at 35 bars)	2.46

for 1 m³ of HFO. The last column of Table 13.4 gives an approximate estimate of tank weight for carrying fuel having same energy content equivalent to 1m³ of fuel oil.

13.1.3.3 Emissions from Natural Gas

NG does not contain any sulphur and so, SO_x emission and, along with it, emission of PMs is completely eliminated. The nitrogen oxide emission is drastically reduced. Similarly, carbon dioxide emission is also reduced. Table 13.5 shows the comparison of emissions between marine diesel oil (MDO) with 1% sulphur and NG. Using NG, there could be knocking in the internal combustion engines causing some methane to escape with the exhaust gas and this phenomenon is known as methane slip. Methane has an impact on global warming, the specific impact being more than that of carbon dioxide. This methane slip must be taken into consideration while accounting for total emission effects. Table 13.5 gives approximate numerical values of emission gases by weight for MDO (1% sulphur) and NG burning engines.

13.1.3.4 NG Engines: Design Implications

NG has been used as ship fuel in LNG-carrying ships. Dual fuel engines have been developed such that the IC engines could use both diesel and NG as fuel. The dual fuel engines have been equipped with auxiliaries such that boil-off gas from the cargo tanks could be fed to the engine in gaseous form as and when required. Dual fuel engines can be designed to work on HFO and gas or MDO and gas or MGO (marine gas oil) and gas. Alternately, boil-off gas could fire a boiler generating steam running a steam turbine which could be used to provide additional power to propulsion machinery. Boil-off gas could also fire an air boiler producing hot air running a gas turbine which could be used similarly.

NG-fired internal combustion engine has been developed and it is projected as the engine of the future primarily due to their superior emission qualities which will satisfy the stringent EEDI requirements in the coming years. NG requires temperature higher than that required by diesel engines for auto-ignition which is difficult to achieve in the IC engine through compression alone. Therefore, a pilot diesel oil injection may be required to start NG firing. Alternatively, a spark ignition system can be provided in NG-fired engines.

From Table 13.4, it can be observed that LNG requires nearly double the volumetric capacity required for HFO. But CNG and ANG require 4.5–5.5 times the volume of HFO. This volumetric capacity is to be provided by the ship's internal volume if the tanks are housed inside the hull. Of course, tanks could be placed on deck if this arrangement does not conflict with the ship's operational requirement. The weights of the tanks of LNG or CNG have to be provided for, which means slight loss of deadweight. Eswara et al. (2013)

TABLE 13.5

Comparison of Emissions from MDO and NG

Emission Type	MDO (1% Sulphur) (gm/kWh)	NG (gm/kWh)
CO_2	600	450
SO_x	4.1	—
NO_x	11	2
PM	0.2	—
CH_4	—	3.46

have studied feasibility of using NG as fuel for harbour tugs and they have considered alternatives of carrying LNG, CNG or ANG as fuel for the vessel and in all cases emissions are reduced compared to that of diesel-fired tugs.

13.1.4 Alternative Energy Sources for Ship Operation

Conventional marine engines for ships and for installation on offshore platforms have been discussed and improvements have been made in the engine design to reduce air pollution. However, none of these technologies indicate a drastic reduction of pollution levels which can control global warming. It is necessary to find alternative energy sources for ship operation which reduces pollution drastically. Most of the technologies in this regard, mentioned below, are in the process of development and experiments on ship board applications are being carried out. However, the hazards these devices may cause are not fully visualised and it is necessary that, before a full shipboard application, a formal safety assessment must be carried out.

13.1.4.1 Biofuel

Biofuels are fuels derived from biomass consisting of living organisms consisting of animals and plants and their products or from other fossil fuels such as coal, residual oil and sludge. The biofuels developed so far are biodiesel (chemically known as fatty acid methyl ester) and bioethanol derived from animal and plant-based biomass reacting with other chemicals and alcohols. The more recent biofuel, dimethyl ether (DME), is derived from NG, coal, oil residues and biomass. Biofuels have the potential to be an alternative to natural fuel, but technology needs to be developed for maritime use. Perhaps a more appropriate use can be the use of biofuel in conjunction with normal fuel oil. Although the use of biofuel can reduce demand on fossil fuels, it is not clear whether it leads to emission reduction. Perhaps more laboratory research is necessary to establish emission norms of biofuels. One other disadvantage of bofuel is that production of this fuel in large scale for marine use needs large land mass for production which may hinder the use of this fuel in the future.

13.1.4.2 Nuclear Power

Existing power generation systems on marine vehicles depend on technology that releases energy from fossil fuels by chemical reactions. Nuclear power, on the other hand, is generated by nuclear fission reaction where large, heavy nuclei are broken into smaller nuclei and large amount of heat is released in the process. This heat energy can be used to generate steam (or heat air) in a boiler which in turn can run a steam turbine (or gas turbine) to generate rotational motive power. The nuclear material is generally uranium or, more commonly available, thorium. Nuclear fuel does not require air for its operation and so it has been successfully used in underwater submarine propulsion. Furthermore, unlike battery-powered submersibles which require frequent surfacing for charging the batteries, nuclear powered submarines can stay underwater for a very long period provided other conditions are met. Nuclear fuel emits no CO_2, SO_x or NO_x and it is totally pollution-free. Experience exists with regard to safe design and risk-free operation of nuclear reactors and nuclear reactors have been found suitable for marine operation and merchant ship applications. Furthermore, unlike diesel engines, nuclear-powered ships have the freedom of choice of hull form and other design parameters. Technological development may lead to production of small modular reactors suitable for shipboard operation. Although initial cost of nuclear fuel and reactor is high at present, the fuel cost in operating condition is very small compared to ships requiring to bunker frequently. In spite of all these advantages, nuclear fuel has not found much use in merchant ship applications. It must be understood that nuclear reactors require heavy shielding which demands space and adds weight to the vehicle in which it is installed. Also, nuclear waste disposal is a problem which has not been solved at the international level. Nuclear installation requires an entirely new system planning for installation, maintenance and operation. Training of personnel is completely different from that for other installations. Public perception against nuclear applications plays a major role in the installation of nuclear reactors. Furthermore, insurance issues for nuclear installation also require to be addressed. It is necessary to address issues of nuclear peril such as reactor protection, leakage, radiation protection and accidents and their mitigation. Similarly, issues related to machinery failure risks and predictive preventive maintenance including equipment reliability and condition monitoring have to be studied. Fire protection including compartmentalisation, plant segregation and control of other hazardous operations has to be addressed. On the whole, a safety and reliability analysis through formal safety assessment has to be done before installation of any nuclear plant on board ships.

13.1.4.3 Batteries

Batteries are stored electrochemical energy which can be made available as and when required. Conventional batteries are the 'wet' lead acid batteries or the dry zinc–carbon or nickel–cadmium cells. Batteries have been conventionally used in submarines as a means of propulsion underwater when the air breathing diesel engines cannot be operated. At present, battery technology development includes metal–sulphur batteries where metal can be magnesium, sodium or lithium and metal–oxygen batteries where metal can be zinc, lithium or sodium. Of these, lithium–air battery development is most promising. Also promising is the lithium–ion battery development. Batteries suffer mainly from weight and size limitations on board ships. Also, a matter of concern is its shelf life which means replacement after a certain number of charging and discharging. Batteries also have the distinct advantage of being totally pollution-free with regard to harmful emission; these do not generate noise and can operate underwater. With more sturdy and reliable battery development, it is possible to develop and install a battery pack on board a vessel to be an auxiliary power source. It could also be the main power pack on small- and medium-sized vessels provided the economics can be satisfactory.

13.1.4.4 Fuel Cell

The science of fuel cells (FCs), i.e. extraction of electrical energy from electrochemical process, has been known for a long time; the technology of FC development is of recent origin. Although a few demonstration vessels have been built with FCs, its extensive use in marine application awaits further technology development with cost benefit. Two reactants such as hydrogen and oxygen combine to form water-releasing electrical energy and small amount of thermal energy. Unlike batteries, which work on similar principle of electrochemical reactions, the reactants are stored externally and not inside the FC. There are no moving parts in an FC and it does not emit any polluting gases such as carbon dioxide, sulphur oxide or nitrogen oxide. If hydrogen and oxygen can be supplied directly, the efficiency of the FC is the highest. But then hydrogen and oxygen have to be stored in liquid form in cylinders (like in spacecraft) which is an expensive process. On the other hand, oxygen is freely available in air which can be used directly at the expense of small efficiency loss. Hydrogen, on the other hand, is difficult to manufacture and store in large quantities to be useful in marine applications. Hydrogen can be generated on board from distillate diesel oil, NG or methanol and fed to the FC on a continuous basis. The use of HFO for hydrogen generation is also possible. The Nordic countries, particularly Iceland, are developing 'green hydrogen' using power source of hydro- and geothermal energy source. The type of FC used in marine application using hydrogen is normally the proton exchange membrane type (PEMFC). One can also use methanol and air in an FC which can be either molten carbonate type (MCFC) or solid oxide type (SOFC). Even if the FC itself does not have any moving parts, the externally stored reactants require auxiliary equipment such as pumps, fans and humidifiers. Furthermore, each individual FC produces very little electricity and large stacks of FCs are required to get reasonable voltage and current.

The FC generates DC power which can run an electric motor for propulsion drive. FC technology has advanced to a stage when a PEMFC can generate up to 1 MW power for boats. Higher power up to 10 MW is reported to be developed by MCFC or SOFC. FCs do not emit polluting gases if 'green hydrogen' is used. Studies must be conducted to estimate emission quantity in an FC installation where reactants are produced on board. FC has less number of moving parts and is easy to maintain. Perhaps FCs offer the most attractive alternative for elimination of GHG emission.

13.1.4.5 Wind Energy

The principle used in a Flettner rotor is that wind passing across a rotating cylinder produces a lift force (known as Magnus effect) which can be directed in the direction of ship's forward motion. A number of such long rotating cylinders can be installed on the deck of the ship to provide forward force augmenting the thrust produced by the propulsion mechanism. Soft sails are the age-old technique of sail boats. Imitating the skill of the boat man of the sail boat, it has been possible to provide automatic control system for soft sails (flexible or metal sails) on board merchant vessels experimentally to provide auxiliary thrust similarly. Wing sails and kites are similar devices to provide auxiliary thrust to the ship. If such devices are installed on board, then the thrust to be generated by the propeller must be estimated and the propeller must be designed accordingly for optimum efficiency. It must be realised that the auxiliary power from wind is dependent on the wind speed and wind direction. Also it must be noted that the super structure for wind energy extraction must be lowered or protected in case of adverse wind or rough weather. Wind turbines, if installed, generate electrical power which must be fed to the main switchboard augmenting electrical power supply from generator sets. But wind turbines also generate large amount of gyroscopic force and heeling moment which make such devices unsuitable for mobile structures.

13.1.4.6 Solar Energy

Photovoltaic technology to extract solar energy and convert to electrical power has been well developed. If deck area on the top structure of the ship or other marine structure is available, solar panels can be installed to generate electrical power. But this power is likely to be small and can only be used along with main electrical power as auxiliary power. However, solar power has been successfully used as power for electrical drive of small boats.

13.1.4.7 Other Devices

Technology development has led to many new devices and systems and designers for marine applications should utilise these devices for power systems for a clean environment. Supercapacitors which can store large amount of energy can be utilised for driving small boats provided the capacitors can be charged regularly. Superconductivity is a mechanism by which electrical power transmission can be done almost without any loss. The use of superconductivity in ship operation is a distinct possibility. Magnetohydrodynamics is a process by which hydrodynamic action in the presence of large magnets can produce forward thrust. At least one boat has been manufactured based on magnetohydrodynamic principles. These are some examples of power generation using modern technologies.

13.1.5 Emission Reduction by Increasing Energy Efficiency

It has already been observed that pollution due to the use of LNG as fuel is much less compared to the use of diesel. But use of LNG is dependent on LNG supply at bunkering ports. All ships, particularly small vessels such as port and harbour craft, trawlers, tugs and inland vessels, may not be able to call at LNG bunkering ports. In such cases, the use of CNG may be considered as a feasible alternative. Hybrid systems combining diesel/LNG machinery systems and renewable energy sources for main propulsion as well as auxiliary power can reduce pollution by reducing consumption of fossil fuels. The use of diesel electric propulsion and shaft generators can integrate total power consumption, both propulsive and auxiliary power, thus optimising total power system. Consumption of fossil fuel can be further reduced by using intelligent control for main engine such that the engine can run at its optimum efficiency even at partial load conditions.

As has been discussed in previous sections, hydrodynamic design of the hull form can be done by designing a fore body for minimum wave-making resistance. Stern design can be optimised for minimum viscous drag and improving flow into the propeller increasing its efficiency. By designing the shoulders and bilges properly, viscous form drag can be minimised. Similarly, proper appendage design can reduce appendage drag.

The treatment of external surface of the hull with extremely low friction self-polishing paint can reduce frictional drag by restricting corrosion and not allowing fouling on the hull and affect a GHG emission reduction. The recently developed system of air lubrication of the bottom of the ship by permanent installation on board can reduce GHG emission further (Class NK 2013).

The choice of suitable propulsion system and the use of waste heat for power generation (Figure 9.10) have been discussed in Section 9.2. Furthermore, if there is a variable load demand on the engine, one could consider the use of controllable pitch propellers for improved fuel consumption compared to fixed pitch propellers which is normally fitted on merchant vessels operating mostly in single load conditions. Propulsion systems other than conventional screw propulsion can be considered for optimal use based on operational requirements. Similarly, propellers can be designed with some modifications, additions or alterations for improved flow around the propeller increasing its efficiency. Figure 13.8, taken from Ghose and Gokarn (2004), shows the different propulsion alternatives diagrammatically.

Waterjet propulsion system (Figure 13.8a)

Surface piercing propulsion system – used when water depth is low (Figure 13.8b)

Podded propeller and steering duct suitable for propulsion and steering, sometimes known as azimuthing propellers (Figure 13.8c and d)

Tandem propellers, fore propellers, two propellers, one above the other and overlapping propellers – thrust distributed on two shafts (Figure 13.8e through h)

Contra-rotating propellers – two propellers rotating in opposite directions on two concentric shafts to improve efficiency (Figure 13.8i)

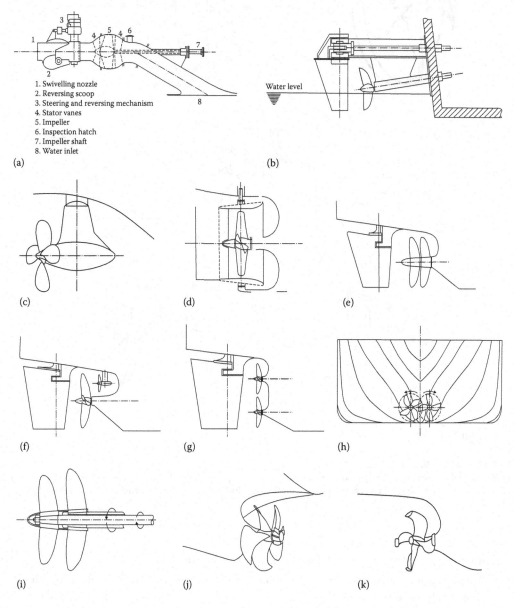

1. Swivelling nozzle
2. Reversing scoop
3. Steering and reversing mechanism
4. Stator vanes
5. Impeller
6. Inspection hatch
7. Impeller shaft
8. Water inlet

FIGURE 13.8
Different propulsion devices. (a) Waterjet propulsion system; (b) surface piercing propulsion system; (c) podded propeller; (d) steering duct; (e) tandem propellers; (f) fore propellers; (g) two propellers, one above the other; (h) overlapping propellers; (i) contra-rotating propellers; (j) hub vortex suppression; (k) trailing vortex suppression.
(Continued)

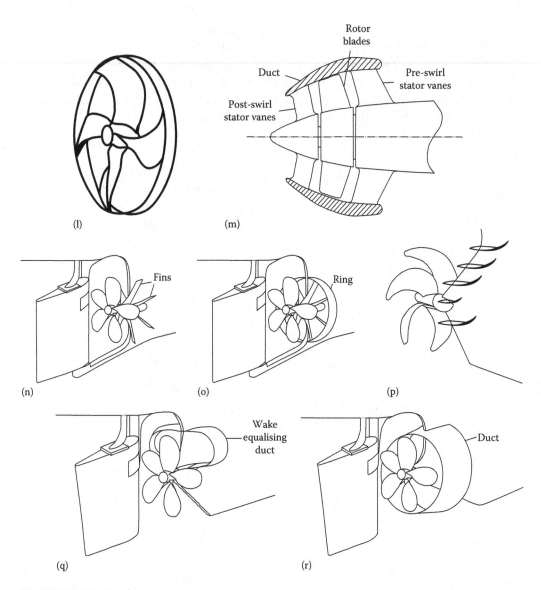

FIGURE 13.8 (*Continued*)
Different propulsion devices. (l) ring propellers; (m) pump jet propellers; (n) Mitsubishi reaction fin; (o) Mitsubishi reaction ring; (p) Grothues spoilers; (q) Schneekluth wake equalising duct; and (r) Mitsui integrated duct.

Hub vortex suppression – propeller boss cap fin (Figure 13.8j)

Trailing vortex suppression – propeller with end plates and ring propellers (Figure 13.8k and l)

Improvement of flow on to the propeller – pumpjet propeller, Mitsubishi reaction fin, Mitsubishi reaction ring, Grothues spoilers, Schneekluth wake equalising duct and Mitsui integrated duct (Figure 13.8m through r)

13.2 Ocean Pollution

As already discussed, oceans are already getting affected by global warming. The operation of marine vehicles and structures should not further pollute the oceans. Since oil is carried across the seas as cargo as well as fuel in vehicles, it is a primary source of pollution due to accidental or voluntary discharge. All these artefacts house a large number of personnel as crew, as workmen or as passengers. Human beings generate wastes such as sewage, garbage and sludge apart from oil. If discharged to the sea, these can cause pollution of coastal areas affecting health and livelihood of coastal people and marine bio species. The design issues related to such pollution are discussed in the following section.

13.2.1 Pollution due to Oil

About 2.4 billion tonnes of oil cargo is transported across the seas of the world quietly and mostly safely every year. Furthermore, there is oil exploration and production at the sea. Of this amount, about 500,000 tonnes are discharged to the sea intentionally or unintentionally every year. Data collected over the years indicate that about 45% of this oil is due to normal ship operation in the form of fuel and bilge oil. Tanker accidents causing cargo oil outflow and other ship accidents causing fuel oil outflow account for about 26% of total discharge. Tankers cargo operation and offshore platform operation such as tank cleaning account for about 28% of this discharge. Statistics also indicate that most of the oil spillage due to accidental damages occurs near the shore affecting coastal habitation at land, coastal sea areas and shore based industrial activity.

Since the Torrey Canyon disaster (see Section 11.1.3.1), IMO has been rather active in considering ways and means of restricting oil pollution at the sea from merchant ships and platforms. The developments have been rather gradual, but effective. Of the total amount of oil spill between 1970 and 2009, the 1970s account for 55.7%, the 1980s 20.7%, the 1990s 20% and the 2000s 3.7%, indicating a drastic decline in oil pollution. One other factor that has been recorded is that, of the total number of ships causing pollution, a large percentage is small vessels of less than 700 tonnes capacity though the quantity of pollution may be much less. This is primarily because it is difficult to implement all the design guidelines including installation of equipment on small vessels due to operational and economic constraints. Furthermore, small vessels operate in coastal and harbour waters where the probability of grounding and collision is much higher. IMO has looked into the three aspects of pollution and developed design guidelines to tackle them. These are bilge oil discharge, accidental and damage discharge and discharges due to oil operations such as tank cleaning.

The residue coming out of treatment in separators in ship's engine room is called 'sludge' which is a mixture of oil and water. This cannot be used for any useful purpose and a large ship can generate a few tonnes of sludge per day. This sludge can settle in a sludge tank from which water that settles at the bottom can be removed and sent to an oily water separator and the separated water containing less than 15 ppm of water can be discharged to the sea as clean ballast. Sludge can be carried in a sludge tank to be discharged to a shore reception facility at a port or burnt in a ship board incinerator provided certain conditions are met. Depending on the capacity of oil-fired boiler on board, certain types of sludge can be burnt in the boiler. Some types of ship waste can be burnt in an incinerator installed on board the vessel. These include domestic waste such as garbage and sewage residue, cargo-associated waste which are non-polluting, maintenance waste, operational waste, cargo residues and fishing gear. However, certain types of material, such as polychlorinated biphenyls, garbage containing more than a trace of heavy metals, refined petroleum containing halogen compounds and residues of exhaust gas cleaning systems, are prohibited from burning in the incinerator since it produces polluting gases. Other types of bilge collection which is a mixture of oil and water can be passed through an oily water separator, and water having less than 15 ppm oil can be directly discharged to the sea and oil can be used on board.

As far as practicable, oil and water should not be allowed to mix due to ship operation. For this purpose, fuel oil tanks and water ballast tanks must be completely segregated. Water ballast should not be carried in any oil tank, either fuel or cargo oil. The oil and ballast or sea water pipe lines and pumps should be completely separated with no cross connection with only emergency provisions. One type of liquid pipe line should not pass through a tank carrying another liquid. The designer has to take care in design of tank layout in this regard and see that number of tanks is minimum reducing piping and number of pumps so as to ensure minimum pipe line length.

For oil tank cleaning in tankers and platforms, crude oil washing (COW) system has been recommended in which the sludge or residue generated in oil tanks are to be washed with oil itself. This can be done by laying COW pipelines on the underside of the cargo tank top with COW electric rotating motors with jets which eject oil to different corners of the cargo hold. A number of motors should be installed such that the entire hold, including the corners, is washed of residue. The oily residue mixture that is generated must be removed from the hold space by installation of a stripping system consisting of stripping pipe lines and stripping pumps such that not only the cargo hold but also the cargo pipe line and pump can be stripped of the dirty oil. This oil can go through settling tanks removing the cleaner oil from the residue (or sludge) which can be fed back to the cargo tank or used as fuel. COW system is used in an empty oil tank when the tank is full of oil–air mixture, essentially an explosive mixture. Also, the use of COW motors means generation of static electricity at the mouth of the jets due to friction which is likely to cause explosion. To avoid such eventuality, the tank must be filled with inert gas (IG) replacing oxygen-rich air. The IG must have less than 8% oxygen. Therefore, the vessel must carry an IG system including an IG generating plant and associated pipe line. Sometimes it is necessary to prepare cargo tanks for carriage of ballast water in emergency situations. Therefore, the cargo tanks may be required to be cleaned with water after COW in a crude oil carrier or in a product oil carrier in empty condition. The oil film, when washed with water, generates oily water which cannot be discharged directly to the sea. This must be stored in a slop tank which must carry the oily water mixture to be delivered to a port reception facility or treated through an oily water separator when the clean water can be discharged to the sea and oil can be reused. Therefore, tankers must have slop tanks of adequate volume, typically between 2% and 3% of the volume of cargo-carrying capacity.

To reduce discharge of cargo oil from accidental damage to tankers, there has been a gradual improvement in tanker design led by IMO guidelines. Oil tankers used to be very safe ships since they carried oil of nearly equal density as that of water having a continuous top deck without any opening and only single bottom. There was no danger to the ship in case of damage with regard to strength or stability. But subsequent to the Torrey Canyon disaster, in the 1970s, it was decided to provide protective locations to cargo in a tanker which meant providing ballast tanks in the double bottom and side tanks to a limited extent at selected locations. Subsequently, these days tankers have to be completely double hull construction with full double bottom and side tanks over the entire length of

FIGURE 13.9
Double-hull tanker. (a) Plan and section with two additional longitudinal bulkheads. (b) Block under construction with one longitudinal bulkhead.

cargo holds. Figure 13.9a shows the plan and section of a typical double hull tanker with two additional longitudinal bulkheads and Figure 13.9b shows a block under construction of a double hull tanker under construction with one additional centre line longitudinal bulkhead. The statutory requirement of side tank width or double bottom height should be at least 0.76 m for ships less than 5000 tonnes deadweight. For bigger vessels, this value is between 1.0 and 2.0 m subject to some formulations in the rules.

To reduce discharge due to damage from fuel oil tanks, it is necessary that these must also be protected. Barring small vessels and small oil tanks for which it may be impossible to provide protective locations, fuel tanks should not be provided next to shell side or next to fresh water tanks. However, the fuel oil tanks can be provided on the shell side above load water line. Figure 13.10 shows two alternatives of probable fuel tank arrangements.

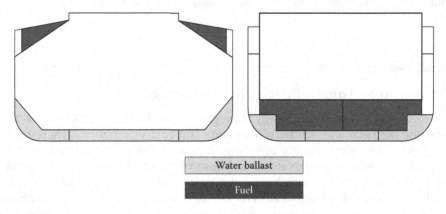

| Water ballast |
| Fuel |

FIGURE 13.10
Protective location of fuel oil tanks.

13.2.2 Pollution due to Garbage

Domestic wastes including food wastes and operational wastes due to cargo operations, such as dunnage, wood pieces and ropes, are known as garbage. These items are required to be discharged continuously or at regular intervals. It is commonly believed that the sea is a great assimilator of all rubbish material and garbage can be suitably absorbed by the sea. But all items are not biodegradable and remain as such for a long period. When discharged to sea, 70% of garbage sinks to the bottom of the sea. If it contains broken glass or metallic material or plastic material, it can harm fishes and other bio species. About 15% of discharged garbage floats ashore and affects the beaches. The remaining 15% float on or just below the sea surface. Sometimes these items collect together due to oceanic activity and float away to distant places forming an 'island'. Sometimes these may float to shore having sensitive ecosystems such as the Atlantic shores. This must be avoided and the IMO has provided regulatory mechanism for the same.

Generally, synthetic materials such as plastics, nettings and riggings cannot be discharged to the sea at all. All other materials can be discharged to the sea at a distance from the shore while the vessel is moving forward with a speed such that the probability of these materials floating back to the shore line or collecting together is minimised. Garbage disposal in special maritime areas is highly restricted. If the vessel has an incinerator, garbage can also be burnt in it.

13.2.3 Pollution due to Sewage

Sewage is termed as black water which contains faecal matter and urine (water from toilets) which are human and animal wastes. Grey water means water generated from domestic activities from dish washing, laundry etc. If sewage is discharged directly to the sea, it can cause hygienic problems by releasing harmful germs and bacteria into the water which can spread over a long area including the shore line. In addition, in sheltered coastal areas, spread of harmful nutrients, detergents and such other material can harm the environment. Therefore, sewage disposal has to be done with caution. The ship can have a sewage treatment plant and the solid waste could be disposed to the sea as per garbage disposal provisions or can be burnt in an incinerator. Sewage can be stored in a tank and disinfectant can be added to disinfect it so that it can be discharged to the sea at a distance from the shore line while the vessel is moving at a speed. Of course, sewage can be stored in a designated sewage tank to be discharged to a shore reception facility.

13.3 Dispersal of Aquatic Species due to Shipping

The oceans in tropical areas have nearly homogeneous and uniform biodiversity. This area separates the northern and southern temperate and cold water zones where bio species have evolved quite independently resulting in quite different biodiversity between the northern and southern waters. Natural dispersal of aquatic species occurs due to ocean currents, waves and winds which have not altered the biodiversity of marine ecosystem over millions of years. But human activity related to shipping has caused much unnatural dispersal of aquatic species across the oceans. One of the causes of dispersal due to garbage dispersal from ships has already been discussed in the last section where it was mentioned that aquatic species ride over garbage flotsams and float to distant places. The other two methods of dispersal are due to ballast water movement and fouling on ship surfaces which are discussed in the following section.

13.3.1 Ballast Water

Modern shipping cannot operate without ballast water which is required to maintain adequate draught and trim for safe shipping operation including stability and structural strength. After a fully laden voyage at a discharge port, cargo is unloaded and ballast water is loaded. After the vessel reaches the destination port, ballast water is discharged and cargo is loaded. This operation is most important for tankers and bulk carriers which have a ballast leg of voyage in each round trip, i.e. about 50% of the total number of voyages. Other ship types also carry ballast water, but to a lesser extent. Maximum volume of seaborne trade is carried out in bulk cargo and oil. In modern tankers, to avoid pollution and contamination, cargo and ballast water are carried in separate compartments and ballast water capacity is roughly 30% of the cargo deadweight capacity. Thus, a tanker of 200,000 tonnes dwt may require to carry 60,000 tonnes of ballast water in its ballast voyage and a bulk carrier of 40,000 tonnes dwt will require to carry 12,000 tonnes of ballast water in its ballast voyage. It is estimated that about 3–5 billion tonnes of ballast water are transported over large distances in a year and a similar amount is transported over smaller distances over a geographical region. So total ballast water transported across seas can be anywhere between 7 and 10 billion tonnes annually. Thus, aquatic species that can pass through the ballast water ports and pumps get carried by ships to distant seas with a different ecosystem where these are discharged. Most of the aquatic species go through a life cycle including the planktonic stage. So even though large aquatic species cannot get through ship's ports, eggs, spores, seeds, cysts and larvae of various aquatic species along with microbes, micro-algae, small invertebrates etc. get carried over to different locations.

The ecological impact of large-scale movement of bio species is large and varied. The newly introduced species may grow in numbers competing with native species for space and food. They may even prey over native species. This may lead to reduction of native species volume and even extinction of some species. A typical example is the transport of the American comb jelly from North American waters to the Black Sea where these grew to large volume reducing the local fish production drastically including extinction of some species. If such growth is uncontrolled, it may alter the food web and also the ecosystem.

Economic impact of such biodiversity changes is heavy. Reduction in fisheries production and also effects on aquaculture may affect economic environment of coastal regions. This may sometimes lead to closure of recreational facilities and tourism infrastructure. One of the major impacts of growth of invasive species in the new environment is growth of fouling on coastal infrastructure such as inlets of cold water pipes and on heat exchangers of industrial infrastructure on the coast. Zebra mussels transported from the Black Sea to the North American coast and its uncontrolled growth have caused an expenditure of nearly a billion U.S. dollars between 1989 and 2000 in cleaning the coastal infrastructure of fouling due to these Zebra mussels.

To avoid this migration of aquatic species to new and uncharted destinations, the ballast water has to be managed in such a manner that no aquatic species are introduced in a new environment through movement of ballast water or sediments deposited through carriage of ballast water. Until today no perfect solution in this regard has been achievable. Therefore, IMO has introduced a ballast water management programme to reduce the likelihood of transfer of harmful aquatic organisms to an acceptable level without compromising the ship safety until better feasible alternative is available. This is based on the fact that ballast water can be exchanged in the open ocean during the voyage killing fresh water organisms and flushing out coastal organisms taken at the source. Ballast water exchange should be such that flushing, rinsing and cleaning of ballast water tanks and removal of sediments accumulated through ballast water carriage is affected and at least 95% of the ballast water carried at the origin port is exchanged with mid-ocean water. Two methods of ballast water exchange have been suggested:

1. Sequential method which involves emptying the ship's ballast tanks at sea and refilling them with clean ballast water in a sequential manner.
2. Flow through method which involves pumping clean ballast water into the bottom of each ballast tank via the suction head and allowing overflow water to exit through the air pipes or access hatches such that at least three times the volume of the tank is exchanged.

Either of these methods has some severe limitations. Removal of water and sediment is affected only up to 95% of the tank volume and the remaining 5% is carried to a new location. Sediment transport is similarly not eliminated and so, the biological effectiveness is incomplete. Ballast water exchange at high seas is dangerous with regard to stability and strength, particularly in rough weather.

Some attempts have been made with regard to ballast water treatment, but further developments are required for implementation:

- Use of biocides for killing all bio species which has limitations that when the treated ballast water is discharged, it may have toxic impacts on the ocean environment. Furthermore, handling of chemicals can be dangerous on board. It has not been established if all bio species are inactivated.
- Filtration process can be used with different mesh sizes for separating aquatic species. This method also has the limitation of not filtering micro-sized species. Filter choking can affect ballasting and discharging rates. Large filters are required for filtering large amounts of ballast water and this requires design considerations for implementation.
- Heating of ballast water using ultraviolet rays can kill all bio species. But this requires large amount of ship board energy and also a lot of time which may not be available unless the voyage length is long. Some microbes and organisms may escape elimination.
- Deoxygenation can be implemented to kill oxygen breathing species which means that cysts and anaerobic bacteria can escape destruction. This method also requires design considerations for implementation.

All these methods require further technology development for implementation and are expensive both for procurement and operation.

Most attractive research and development to eliminate aquatic species migration is the no-ballast concept with regard to design evolution. Ship hull form is so designed that it can achieve the forward draught requirement of the Classification Societies with full propeller immersion in the aft without carrying any ballast in light load condition. Three novel concepts taken from the Globallast Monograph 20 of IMO (2011) are listed as follows:

1. Zero ballast concept development is as follows:

   a. Monomaran Hull (DUT) – Widened beam with catamaran hull underside of a broad single hull.

   b. Tri-Hull or Cathedral Hull (DNV) – Widened beam with three hulls underside of a broad single hull.

   c. No-Ballast Ship-NOBS (SRI, Japan) – V-shaped wide-beam hull form.

   These concepts generally generate a much bigger vessel with larger wetted surface resulting in higher drag though some of the penalty due to drag increase can be offset by obtaining higher efficiency from twin podded propulsion system. This solution may be appropriate for small high value vessels.

2. Ships carrying ballast other than sea water have been developed as follows:

   a. Arrangement to carry solid ballast in containers in the lowest tier in the hold of containerships can be made. The containers need not be removed in full load condition. Since container ships have sufficient margin in full load condition, carrying solid ballast is not a problem (DSME, Korea).

   b. Fresh water can be carried as ballast in ships which need not be discharged at ports, but shifted from tank to tank for trim control and stability. This concept can be used in container ships and high volume carriers.

   c. Fresh water can be produced in fresh water generators which can be used as ballast as and when required. This is a power consuming and slow process and can only be used in yachts, patrol vessels, naval vessels and vessels with less than 3000 GT.

   These concepts are not suitable for deadweight carriers such as tanker and bulk carriers since there is a permanent loss of deadweight. Also these require extra equipment and extra power consumption.

3. Continuous flow though methods have been proposed in double bottom of vessels such that water keeps flowing continuously from forward to aft end of the vessel in light load condition as follows:

   a. The University of Michigan proposal includes forward and aft plenum chambers in the double bottom tanks which are joined by continuous pipes from forward to aft. The plenum chambers are open to sea such that water automatically moves from forward to aft when the vessel has a forward speed.

   b. There are a number of continuous tank flushing systems consisting of adjacent and multiple tanks such as Auba Flow/LoBE system/Dyna Ballast systems.

   c. The proposal of the Indian Institute of Technology, Kharagpur, has continuous flow through pipes forward to aft of the ship in its double bottom. The pipes remain closed and empty in full load condition with water tight forward and aft valves of the flow through pipe closed water tight. In ballast or light draught condition, valves remain open allowing water to flow through them continuously (Figure 13.11).

FIGURE 13.11
Flow through continuous pipes from forward to aft.

This method has the inherent advantage that since there is continuous flow of water in pipes, there is no accumulation or stagnation of ballast water or sediment. But since the effective wetted surface is increased, there is an increase of frictional drag. This method, though attractive, require a lot of research with regard to all design and performance characteristics and behaviour in the sea.

13.3.2 Paints

The marine environment is one of the harshest environments and causes extensive deterioration of ship's materials and surfaces. The outer hull surface that is below water is subject to very severe actions – corrosion and fouling. Corrosion is an electrochemical process causing formation of iron oxide, commonly known as rust, in the presence of moist and saline climate. This causes wasting away of steel surface material in a manner which increases surface roughness apart from reducing thickness. Effects of corrosion include reduction of plate thickness affecting strength properties of the hull structure and increasing drag of the ship to forward motion due to increased frictional resistance. Fouling is the growth and accumulation of marine organisms on vessels and other movable marine structures which affect their performance and can lead to spread of invasive species. There are two types of fouling: microfouling which refers to a layer of microorganisms such as microbes, microalgae, cysts, larvae and spores eggs, and the slimy substances they produce; and macrofouling which refers to large, distinct multicellular organisms visible to the human eye, such as barnacles, tubeworms, mussels, fronds of algae and other large attached or mobile organisms. Due to attachment of fouling, surface roughness increases drastically causing a heavy drain of ship's energy to counter the increase in frictional drag. Furthermore, aquatic species get carried over long distances causing ecological problems discussed earlier. Surface coatings, commonly known as paints, are used to protect structures from deterioration from corrosion in an aggressive environment as well as to protect the underwater hull surface from fouling.

Generally, a ship's underwater hull is coated with a layer of primer followed by a layer of anticorrosive paint and finally, on the top surfaces a layer of antifouling paint. Whereas anticorrosive paint prevents corrosion mainly by forming a barrier between the steel surface and sea water, antifouling paint slowly releases biocides to water which prevents aquatic organisms from attaching themselves to the surface. This leads to reduction and removal of paint surfaces and then corrosion starts. The deterioration of paint surface is not only due to slow bleaching of antifouling paint, but also due to viscous fluid flow on the hull surface as shown in Figure 12.2. Painting schemes including painting procedures are important parameters for preservation of painted surface and these are generally provided by the paint manufacturer in association with the designer. With degradation of painted surface, the ship must be drydocked and the hull surface must be prepared by sand/shot blasting and repainted to a new ship finish.

In the 1970s, tributyltin (TBT)-based paint was developed which was self-polishing in nature, i.e. the smooth surface is maintained by not allowing fouling to grow on the surface. This increased drydocking period to 5 years. However, tin-based paints have been found harmful to aquatic species in sea water and these have now been phased out due to environmental concerns. Tin-free antifouling paints have now been developed though these are not as effective, but are generally satisfactory in the control of fouling as far as routine operation of the vessel is concerned. However, even the best maintained vessels are fouled to the extent of at least 5% of the total surface area. This fouled area, although a small fraction of the entire vessel surface, is the primary vector for transmigration of invasive species. The more recently developed antifouling paints or surface coatings can be briefly summarised in the following:

- Self-polishing copolymer paints free of TBT have been developed which include TBT replacement with copper acrylate or zinc acrylate or silyl acrylate. Copper-based biocides have been found effective and are widely used nowadays. But copper-based biocides are ineffective against soft fouling. Also heavy barrier coats are required to insulate the toxicants from metal surface to avoid corrosion. Furthermore, it has now been reported that copper concentrations in water and sediments are environmentally harmful particularly to aquatic species such as mussels and to damage larval stages of aquatic invertebrates and fish species. The use of copper in paints has been banned in some countries and is likely to be completely banned in the near future.

- Controlled depletion polymer based on rosin which consists of 90% abietic acid and is extracted from trees. It is only slightly soluble in water and can form insoluble matrix antifouling.

- Hybrid self-polishing copolymers – This is a hybrid of SPC acrylic copolymers and a certain amount of rosin.

- Foul release coating is based on polydimethylsiloxane/fluoropolymer backbone polymer which is extremely flexible allowing the polymer chain to readily adapt to the lowest surface energy configuration. These coatings provide ultra-smooth finished surface and low surface energy minimizes the strength of adhesion of fouling. Also fouling can be easily removed. Furthermore, this paint does not contain any biocides, is chemically durable and is suitable for vessels moving at speeds higher than 15 knots. Furthermore, it can be applied over existing SPC system. Thus, it provides a good means of fuel saving and is environmentally friendly. But it is a very expensive painting system requiring utmost care in its application and is susceptible to abrasion and shear.

- Nano-coatings are coatings that are produced by usage of some components at nanoscale to obtain the desired properties. Some of these properties that are imparted to the paints by using nanomaterials are much greater surface area per unit mass reducing total paint requirement, enhanced internal mechanical properties, high wear and scratch resistance, resistance to ultraviolet radiation, can be fire retardant, anticorrosive, antibacterial and antifouling. The self-polishing and self-smoothening capabilities offer a low hydrodynamic drag coefficient, which results in a reduction of fuel consumption.

- Hydrogels – Some fishes increase their speed in short time due to skin secretions (mucus) which are slightly soluble in water. In some areas where strong micro-turbulence exists, this substance is locally dissolved and this reduces friction in the boundary layer. Hydrogels are prepared to mimic this slick skin which could respond to external stimuli, including pH, changes in temperature, humidity, salt concentration and even specific molecules. These compounds trap a large amount of water on their gel surfaces, acting as a lubricating film similar to the mucus on fish skin. Natural and synthetic hydrogels have been developed due to their high water absorption capacity, long service life and wide varieties of raw chemical resources and excellent antifouling capability against barnacles both in laboratory assays and in the marine environment. But the most important aspect of hydro-gels is the capability to reduce frictional drag saving fuel consumption.

13.4 Underwater Noise

Noise travels much faster in water and over much longer distance compared to that in air. This disturbs the marine ecosystem to a large extent. Aquatic life is extremely sensitive to noise pollution since it is dependent on sound for communication, food and searching for a mate. Noise can be very painful to marine animals due to haemorrhages, affection of auditory systems of fishes as well as mammals, can even cause death and cause mass stranding of animals on beaches leading to death. Noise can lead to changed diving patterns, migration to newer places, distorted communications and even mass extinction. Sources of noise in water could be due to flow around ships, flow due to propellers, low frequency sonar sounds for submarine detection, seismic air guns for oil and gas exploration and even coastal tourism and sports activities. In the near future, design of marine structures and vehicles will have to take into account noise emission from these products so that the environment and ecosystem are not affected.

13.5 Ship Recycling

Recycling of any man-made product is an essential component of sustainable development. Ships and other maritime structures have an operating life span of 20–25 years. Recycling of parts and equipment should be considered to be the best option for disposal.

Steady withdrawal of old ships and their replacement with products should be the natural process of development. Steel is the main component of a ship's structure and its retrieval and recycling from an old ship is both economic and environmentally friendly since the resources required in steel making from iron ore and the environmental problems in steel production are of a much larger magnitude compared to ship breaking. Similarly, individual equipment including all machinery items, auxiliary items, fittings, furnishings and similar items have a life span of more than 25 years and therefore these items can be disposed off as used items to other users. But ship breaking and subsequent recycling has problems associated with release of hazardous material during ship dismantling to the environment and occupational safety hazards in difficult working conditions associated with ship breaking.

Ship breaking can be undertaken in three alternative facilities: (1) in floating condition outside the beach area where complete dismantling is not possible, (2) by beaching method at locations where tidal variation is high and one end of the ship can be dragged to the beach during high tide, a portion can be removed by gas cutting and the ship can be dragged further into the beach during the next high tide and so on and (3) in human-controlled environment in a drydock. Hazardous material disposed during dismantling can be retrieved to some extent based on dismantling procedure, but the rest is released to the environment. Waste materials include hazardous material in ship structure and equipment installed during construction, wastes generated during ship's operation and those in ship's stores. Such materials include asbestos, glass wool, oil sludge and contaminated material, plastics and cables, paint chips, rubber, fibre glass, cardboard and packing material, iron chips, chicken mesh and residues of lead, arsenic, antimony, mercury and rarely nuclear material.

Occupational hazards during ship breaking include working at height, in enclosed oxygen deficient areas, in flammable environment, in toxic, corrosive, irritant or fumigated atmosphere or residues, with insufficient light, unsafe material handling by manual or semi-automatic means, etc.

IMO has been discussing the problems related with ship breaking and it has stipulated that at the design stage itself, a list of hazardous material used in ship structure or in machinery and equipment must be prepared and submitted to the ship owner and authorities so that properly planned dismantling procedure can be worked out. But naval vessels, pleasure and small craft, costal vessels, port craft, etc. are not covered under this. Time will come when dismantling procedure will be considered at the design stage itself which should include reduction of, if not elimination of, occupational hazards.

14

Design for Production

In Chapter 3, it has been mentioned that production follows detailed design. In the later part of detailed design, production drawings and information are generated based on which production planning and production scheduling are carried out. Production is done based on information provided in these design documents. Preparation of these design documents is known as production design or manufacturing design. Manufacturing design has to take into account the production process in the builder's yard such as materials availability and flow, human resource and skill level in the yard, lifting capacity of assembly blocks and production processes. Therefore, even if an external design consultant provides the design up to basic design stage to a builder, owner or any other client, the builder must have his own design office to prepare the manufacturing drawings and information. The production departments and the design office must work in close coordination to ensure quality, cost control and maintenance of building schedule.

A shipbuilder's constant endeavour is to reduce building cost and reduce time of construction. This, more or less, defines shipbuilding efficiency which primarily depends on the production process. Has design a role to play in this as separate from manufacturing design? It is possible to identify areas where design evolution can play a major role in reducing both cost and time of production. For this, it is necessary to integrate design with production process from the concept design stage itself. This is known as design for production. These two aspects, manufacturing design and design for production, are discussed in this chapter.

14.1 Manufacturing Design

On obtaining the contract, the design department of the construction facility proceeds with the manufacturing design. Till this stage the designer had based his calculations on the functional system of the ship or the basic design. Design and drawings are produced to satisfy requirements of the owner, classification society and statutory bodies. Then there is a transition state where information generated for functional groups is mapped in terms of constructional groups based on the facilities, resources, production practices and constraints of the particular shipyard. Resource requirements in terms of materials, manpower and facilities and due dates for completion are calculated for each constructional group which can, in turn, be related to the total material requirement and the building cost of the product.

Ships are built by forming and joining plates and sections to sub-assembly and assembly blocks (constructional groups) with a reasonable content of pre-outfitting. Overview of the ship production process can be stated as steel work such as straightening, cleaning, cutting, forming and various assembly stages, transportation of outfitting and machinery components (purchased or manufactured) and assembly of the same into steel works and associated activities such as materials handling, painting, inspection and approvals,

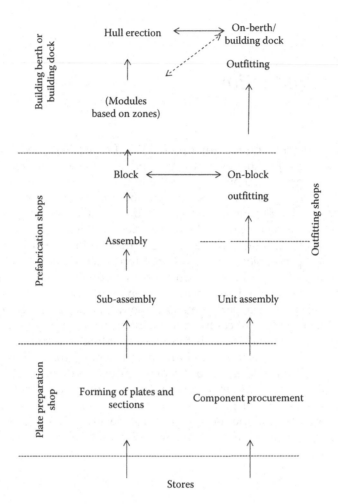

FIGURE 14.1
Modern ship production process.

transfer to water, trials and delivery. Modern developments include advanced cutting and welding technologies, computer-controlled machines, robotics and computers in management of such activities. Ship production process has been very well described in Storch (1988). Figure 14.1 shows a much shortened marine construction procedure. Based on the yard facilities and practices, the functional groups must be further subdivided and distributed to the individual constructional blocks. In all these activities, the design department has a key role to play. The role of the design department can be termed as 'designing is the beginning and the end of production engineering'.

The design department has the following four tasks:

1. Determine the shape of the ship with defined functions and acceptable performance. This work is mostly completed at the basic design stage though there is some overlap of this functional design activity with manufacturing design at subsequent stages.

2. Preparation of constructional information at block, assembly and element levels while examining with what materials, equipment and methods a ship can be built

TABLE 14.1

Differences of Generated Information between Basic and Manufacturing Design

Item →	Information for Product Development from Design (Basic Design)	Information for Production from Manufacturing Design
1.	Defines the complete ship	Defines intermediate products
2.	Arranged according to functional groups and subgroups	Arranged according to production stages with intermediate products consisting of elements from different functional groups
3.	Emphasis is on the operation of the complete ship, not on the production processes involved	Emphasis is on production processes, materials lists, work content and other production parameters associated with intermediate products
4.	Generally independent of the shipyard in which the ship will be built	Largely governed by the facilities, standards and working practices of the shipyard

at a reasonable cost while at the same time, satisfying the desired functions and performance specifications. The design department has to express their results in terms of engineering documents and drawings including materials listing and production sequence and procedure based on yard practices and facilities. Table 14.1 shows the differences between the information generated by the basic design group and the manufacturing design group.

3. The design department and the production and planning department must work in close collaboration so as to decide on the materials and equipment procurement dates. The design department has to provide information to the materials procurement department on specifications, quantities and delivery dates for various materials and equipment required.

4. The design department should make a systematic analysis of the differences between estimates and actual figures in terms of costs, quality and performance both during the building process and at the completion of the ship. The department should endeavour to incorporate improvements on the basis of this experience in future building strategy.

The design engineers cannot contribute to cost reduction as long as they consider their job as simply producing design drawings. The design department must take the responsibility towards cost reductions by various means, some of which are mentioned in the following:

- The cost of materials can be as much as 65% of a ship's cost. It is therefore vital that the design department plays a key role in the reduction of material costs. The design department should not only try to reduce the quantity of materials but also make an effort towards selecting quality materials that are affordable and easily available. If early procurement leads to lower price, an inventory of such materials may be generated.

- The timely procurement of material is of vital importance for maintaining time schedule of production. The design department must provide information to the procurement department such that sufficient lead time is available for procurement. However, search should be on to procure materials on just-in-time (JIT) basis to reduce inventory cost.

- The total work content at the assembly and block levels should be minimised. This can be achieved at design stage by
  - Reducing total weld length in a panel, assembly or block.
  - Reducing materials handling requirement.
  - Reducing plate and section forming activities by reducing three-dimensional curvature.
- Effort must be made for reducing rework by providing proper design information such as
  - Proper and complete welding information so as to reduce the possibility of development of locked-up stresses or generate areas of stress concentration.
  - Proper and complete information on line heating if it is used for plate forming.
  - Joining sequence of plates and sections based on yard facilities so that no deformation takes place and accuracy is fully controlled.
  - During outfitting at various stages, minimising interference between equipment and materials.
- Many of the previously stated requirements may be conflicting in nature. Design is always a process of compromise between such requirements to reduce cost and time of production.

The manufacturing design must provide information to satisfy production requirements as stated earlier for computer-aided manufacturing. The information should include the following. However, if computer use is limited, then such information must be provided in graphical form or drawings.

- Digital information on element preparation such as plate cutting, edge preparation, etc.
- Digital information on nesting of plates for minimisation of scrap.
- Digital information on rolling lines for machine-based plate and section bending and line heating information for plate forming.
- Working drawings from element level to assembly level to block level and finally at berth or building dock. The working drawings must include materials and equipment list and the production sequence including welding details.
- For clear understanding of the structure, three-dimensional isometric views of structures should be provided by the design office.
- For reducing interference between materials, equipment, cabling, ducting, piping and structural elements during installation, it is necessary to provide three-dimensional composite drawings showing the layout of these items together in an assembly or block.
- The design department must ensure a continuous dialogue with the production department with regard to consumption of materials and progress of production to provide quick solutions to on-site problems and also to check on technical matters in case of deviations from design information, cost and delivery schedule. A good solution to ensure dialogue is to adopt 'product modelling' software techniques linking design and other relevant departments on a single platform.

14.2 Design for Production

Design means definition of the product and its systems, whereas production engineering means definition of production processes to build the ship. The goal of design and production together is reducing production cost without sacrificing design performance or product quality, and such design practice is known as 'design for production'. Design and construction of a marine platform is a technology-intensive complex process, and it is necessary to understand the features of this process for doing design for production.

Concurrent design, already discussed in Section 3.4.2, can be effective in design for production where teams of specialists work together continuously to consider design, production, purchasing and other factors to improve design and make production more efficient. The general principles for this are using common sense, planning and definition of the production process, basing the design work on available resources, simplification of work and standardisation. Some of these aspects which may be implemented in the design are mentioned in the following:

- Simplicity in design
  - Minimum number of parts
  - Minimum number of parts to be formed
  - Reduction in part variability
  - Reduction in welding joint length
  - Standardisation of parts
  - Minimum fitting and fairing of erection joints
  - Elimination of the need for very accurate fitting
  - Integration of structure and outfit
  - Minimisation of staging
  - Consideration for access
- Design based on shipyard facilities
  - Limits on ship dimensions
  - Maximum weight and size of blocks
  - Maximum size of panels—panel line turning and rotating capabilities
  - Maximum berth loading
  - Launching limitations
- Other production considerations
  - Simplified hull forms
  - Avoidance of double curvature and large single curvature
  - Developable surfaces
  - Use of maximum plate sizes
  - Transom stern and cruiser stern
  - Bulbous bow with chine
  - Constant hold or tank length, constant hatch size

- Underhung rudder instead of horn rudder
- Corrugated bulkheads and stiffened bulkheads
- Deckhouse shape: aesthetics and producibility
- Access for workers and equipment

To implement these concepts of design for production effectively and to achieve good producibility, it is necessary to understand the features of marine construction process, to understand and improve producibility in the context of marine construction.

14.2.1 Features of Marine Construction Process

The marine construction process is significantly different from a mass production-based industry including large transport sector industries such as aircraft industry or car industry. The features of this industry are perhaps unique to this industry and can be listed as follows.

- The marine platform to be constructed is generally 'one of a kind'. Even if a repeat order for a few similar platforms is placed on the builder, each sister vessel differs from the other in details based on owner's requirements and fancies. Even then the repeat order is for a few vessels only. Therefore, mass production is not possible.
- Though ships have been transporting goods across the seas and do so even now, and it would seem that offshore oil production platforms are frequently in demand considering our appetite for fossil fuel, the demand for marine platform construction is not constant and fluctuates based on temporary global market fluctuations and supply of ships (see Chapter 4).
- Marine platforms are highly technology intensive, varying in intensity with the type of ship such as commercial ships, naval vessels, special-purpose ships, high-performance vehicles, fixed and floating oil/gas production platforms, etc. These require different skill sets for system design and system integration.
- Such construction requires a high degree of expertise in different skill sets for different systems installation, testing and system integration in the final stage. Thus, it has a number of intermediate production stages that are dependent on each other.
- This construction is not suitable for a single flow line production and, therefore, has varying manufacturing principles at different stages of production.
- Different types of equipment are needed for the fabrication process.
- The design, planning and manufacturing processes have a high degree of overlap.
- The working environment is harsh.
- The builder gets to know the final definition of the product only after the contract is signed, and therefore, his inventory of materials and equipment is low.
- The sales contract is signed prior to the start of the production process and payments are made in stages throughout the production process.
- The builder has to take important decisions during the product definition (design) stage based on uncertain stochastic information.

- It takes a long time to complete and deliver a single product.
- Marine platforms come under the category of heavy industries and have a high product value.
- Ships have a long product life of around 20–25 years and oil production platforms have a still longer life span.
- The material cost is very high in these products ranging between 60% in merchant ships to 70%–75% in naval vessels.

This list, though not complete, gives some of the reasons why marine construction is not suitable for adopting the manufacturing principles of mass production. So the producibility aspects of such products have to be investigated separately.

14.2.2 Producibility

Producibility can be defined as the technical and managerial effort necessary to reduce the production (including acquisition) cost of a system without adversely affecting the system performance. Producibility can also be looked upon as the capability to manufacture, build or assemble goods in a cost-effective manner. In the context of marine platform manufacture, producibility can be defined as the reduction in costs as a result of the effort put in to make the product cheaper to construct without compromising on its desired capabilities like size, speed, displacement, payload, stability, strength etc.

The producibility framework is a systematic plan for considering producibility in the design and construction process and should include the following four steps:

1. Identify potential producibility concepts.
2. Evaluate the impact of producibility concepts on the ship along with cost estimates.
3. Integrate the desirable producibility concepts into the design.
4. Provide for a lesson-learned mechanism and feedback loop.

14.2.2.1 Producibility Concepts

It is necessary for the builder or manufacturer to understand the international demand for ships from market study and analysis, and it is also necessary for him or her to decide on which market segment he or she should be interested. Coupled with the yard's budget business and financial forecast, the yard should be able to define its product mix for the short term and also the medium term. Once the product mix is identified, one could implement methods to improve producibility of the product. Then it is possible for the design to develop designs of such products and for the marketing team to aggressively do the marketing.

One of the main problems in shipbuilding is JIT procurement of steel plates and sections. Very often, steel requires long lead time to procure and orders cannot be placed till building order is placed. To avoid such difficulties, a yard could maintain an inventory of steel plates and sections irrespective of the type of ship it is likely to construct. For this purpose, it is necessary for the yard to rationalise plates and sections to a limited number of plate thicknesses and sections shapes. This can be decided at the design stage based on the product mix such that about 80% of the steel material for any ship conforms to the shipyard's store of plates and sections.

Use of standardisations of material elements, equipment and production processes means repeated use of same parts and processes again and again, reducing design and production time and the risk of error. The following can be considered for standardisation based on the product mix and or even a single product:

- Standardisation of material elements such as brackets, beam knees, flats, etc. such that these can be cut to shape and stored in large numbers to be used when necessary. Manholes, lightening holes, access hatches and companion ways can be standardised so that cutting of these openings can be a standard process.

- Standardisation of small manufactured items such as foundations, seatings, flats, ladders, tables, chairs and other furniture, etc. These standardised items can be used in many products.

- It is not necessary to have a single standard for any item. One could have more than one standard for a single item.

- Standardised equipment such as bollards, fairleads, railings, scuppers, sea chests, valves, etc. go a long way in reducing procurement effort and time as well as installation time.

- Standardisation of work processes reduces time required for development of different processes and such processes can be incorporated as instructions for production from the design office. These include welding processes, fabrication processes of flat panels and singly curved panels as well as erection of panels.

- Is it possible to standardise a major portion of the marine platform including the outer shape covering a large portion of the shipyard's identified products? A large module of a ship can be identified to be standardised for a number of products conforming to the technical and production-related issues. The next section discusses an example of such a modularisation attempt.

- All standards of materials, equipment or processes should be made into booklets by the design office for internal use in design as well as distributed to other departments to be used on the production floor.

- Standards can be used over and over again in many products. Therefore, a standardised design can be bigger than that required by the product going beyond satisfying the minimum requirements. There could be an increase in weight and volume of standardised item over and above the minimum. It must be ensured that there is no compromise on technical performance due to this.

It is necessary for the shipyard to decide on the product mix based on its business plan. This defines the range of products for which marketing efforts are to be made to get orders. Once the product mix is defined, it is possible to identify the components such as steel plates and sections; structural elements such as brackets and beam knees; equipment such as bollards and fairleads; work procedures such as welding, assembly and painting; and similar other components and procedures, and decide on single standards or more than one standard for each item. Then standards can be finalised. So it is imperative that standards have to be integrated with the overall product manufacture of the entire range.

To improve producibility, one should avoid difficult manufacturing processes. Forming plates with double curvature or three-dimensional shapes means application of difficult processes and such curvatures should be reduced in the design stage itself. To reduce handling and welding, the number of elements in a structure should be reduced without compromising strength. To reduce overall work content during erection, care should be taken at the design stage in block breakdown and sequencing of outfitting.

14.2.2.2 Evaluation of Producibility Concepts on Cost

It has already been noted that the cost of material forms the largest portion of the product cost. One way to reduce material cost is to order material in advance at its lowest price and order in large quantities/numbers so that rebate on price can be demanded by the buyer and obtained. This can be achieved by the process of standardisation.

If components are known a priori, the design efforts towards their installation are repetitive in nature and require much less design time. It is well known that while doing a job for the first time, there is a process of learning which takes time, and even with utmost care, there is a possibility of mistakes and, then, rework. But in case of the installation of a standard item, the procedure is repetitive or standardised and, therefore, the learning process is already complete. The time of installation should be the minimum and rework is not required. Thus, the cost and time of installation is the basic minimum. It has already been discussed in Chapter 4 that reduction in construction time reduces not only labour cost but also overhead expenses. But this reduction in construction time must reflect in a full-order-book position realised by the marketing efforts of the organisation.

14.2.2.3 Integration of Producibility Concepts into Design

Design oriented towards production leads to improved producibility, reduced time of construction and reduced building cost. Marketing a product necessarily means that the design information (may be up to preliminary or basic design stage) on the product is available. Therefore, if a product is to be marketed on a competitive basis, then lowest cost at the design stage itself must be ensured. Storch et al. (1988) have suggested an integrated hull outfit production process (IHOP) incorporating the producibility concepts.

It has already been discussed in Chapter 4 that large automation does not necessarily mean reduced cost of production compared with semi-automatic or manual process of production. So standardisation and reduction of rework must be considered keeping in view these aspects in connection with the building facility.

Modularisation is the next step of standardisation where a number of components can be put together to form a module which can be a unit of a mechanical, hydraulic or pneumatic system or a geographical unit containing the components of all systems in that unit or a portion of the hull form which can be repeated in different ships. Modularisation, if planned properly, can reduce the time of construction, complexity of construction and, therefore, cost of construction. Section 14.3 discusses this concept in some more detail.

14.2.2.4 Feedback for Improvement

Standards must be documented and used during the design activity. As has been mentioned earlier, standards do lead to selection of material and equipment above the minimum requirement, in terms of weight, capacity and size. It is possible that if standards are kept for too long a period, technological developments and advances may be bypassed leading to rejection. Therefore, with development of technology and knowledge, the standards are likely to change. So, a self-analysing process of modifying standards, after getting feedback from the shop floor and clients, also must be implemented as a part of design activity. If standard component manufacture is done by outside vendors, it is necessary that their views also be considered while modifying standards through a vendor feedback system.

It has been suggested that in the manufacturing industry if a ship was built with only straight plates without incurring penalty on speed and power, it would be cheaper to build. However, the improved producibility with this concept is questionable since work on plate cutting and joining increases, whereas there is idle time in the plate and section forming shops. Such concepts require analysis from technical as well as production point of view by getting feedback from the shop floor.

14.3 Modularisation

Modularisation in ship production has been studied and the various methods suggested across the globe have been described, including an exhaustive list of references, by Bertram (2005). Bertram has quoted the Booz-Allen report (1968) to define modularity as 'constructed of modules or unit packaging schemes, usually having all major dimensions in accordance with a prescribed series of dimensions; capable of being easily joined to or detached, as an entity, from other components, units, or next higher assemblies'. Rommel et al. (1995) have suggested that modularisation leads to simplicity and, therefore, makes it easy to produce where product varieties can be created during production itself (by simply changing modules). Such concepts have been successfully utilised in automobile, aerospace and appliances industries. In shipbuilding or offshore construction industry, modularisation has been incorporated only on a piecemeal basis.

The general objectives of the modular ship design concept may be stated as follows:

- *To reduce design and construction cost*: Once modular concept is accepted, even to a limited extent, the learning process of implementing it in either design or production is limited to the first few applications, and subsequently, cost of doing the job reduces with almost no rework being necessary. Also if modular units are to be installed on a large number of products, vendors can be identified and a cost reduction may accrue due to placement of assured orders in large quantity.
- *To reduce design and construction time*: With no rework being necessary, quality of product can be assured easily reducing design and construction time. This leads to reduction of overhead cost of production and delivery on time.
- *Greater flexibility for updates later in the ship's life*: In modular concept, certain modules can be removed and replaced with new modules even at a later stage of ship's life giving a lot of flexibility.
- *Shorter and cheaper maintenance periods*: Since the components of modules are installed on a large number of products, it is possible to store spares on a regular basis for JIT repairs. Also, since the modules are very well studied prior to installation, preventive maintenance can be incorporated making shorter and cheaper maintenance possible.
- *Reduced maintenance cost*: Since the time period of maintenance is reduced, the maintenance cost as well as down time of the product is reduced.

Design effort is necessary to study specific modularisation concepts and their implication on the particular construction industry. If the concept can be applied successfully, reduction in cost and time of production can be achieved; otherwise, just the opposite may happen. As an example, sometimes, to utilize the design already available, a ship of different length can be manufactured from an existing design by adding a midship module to lengthen the ship or removing a similar module to shorten the parallel middle body. Sometimes, the bulbous bow forward body is altered to provide a better speed performance to an existing vessel. Though these changes are incorporated into an existing design, these are basically adhoc processes that come into effect after the design of the original ship is over and are generally need based. The shipbuilder who tries to alter the design at a later stage in this respect has to do a considerable amount of re-work on design as well as on production without taking advantage of any standardisation which may lead to an overall delay of the production process.

Modularisation also means that design effort should be constrained by the established modules, thus reducing the designer's freedom for new designs. Care should be taken to see that modularisation does not lead to avoidance of new or emerging technologies.

Generally, while designing modules, the geometry of the module itself as well as the components is chosen for the requirement of the entire product range for which this module would be implemented. Thus, it may be of higher size, capacity and weight than necessary giving a product of slightly higher weight, consuming a little more power and having a little less space for carriage of payload.

Design is a compromise of many conflicting requirements and the designer has to take appropriate decision at each step. In the case of modularisation, the designer has to weigh the losses mentioned earlier with the gains in terms of reduction of overall building cost and time. Normally, modularisation, designed and implemented properly, is a profitable concept.

Modularisation can be conceived for many different kinds of modules as stated in Bertram (2005), Gallin (1977), Jacobi (2003) and many others. Some of the modularisation concepts are stated in the following:

- A system can be modularised such as the propulsion system, the HVAC system, the steering system for merchant vessels or a weapon system for naval vessels. It means that the system should consist of components and their connecting mechanisms which should be standardised for implementation in a range of products. While designing such a module, care should be taken to mention what is the range of components that can be accepted by the module. For example, in a propulsion module, the shaft length can have some variations and so also the number of bearings. Similarly, in a module of HVAC system, the duct length can have limited variations in different ships or platforms.

- Accommodation installation on board a marine platform is a long and tedious process consisting of numerous components integrating with a number of ship systems such as HVAC, electrical system, fire fighting and access arrangement. If it is possible to modularise rooms complete with their components such as furniture, ducting, cabling and materials, the modules can be sourced to outside parties (vendors) for supply which can be just fitted when ready. This can reduce the complexity of installation process and reduce installation time and cost. The standard toilet module is a fine example. This module is now being installed on small as well as large vessels. It is possible to build standard modules of rooms of sleeping accommodation with some flexibility of room size. The module size must be integrated with the overall ship system at the design stage itself.

- Most of the ship structure and the location and sizing of equipment are based on the hull shape. So, if the hull shape or portions of it can be modularised, the systems in it could also be modularised. Hull form modularisation would mean that portions of the shape of the hull form can be modularised with the standard shape so that systems in it could be standardised or modularised. If such hull modules can have variations, it is possible to generate different overall hull forms for multiple products. Misra et al. (2002) and Sha et al. (2004) have suggested such a hull form modularisation process.

14.3.1 Hull Form Modularisation

The entire hull form can consist of a number of modules, some of which may have variations so that by connecting different modules different hull forms may be generated. To do this successfully, the following steps need to be followed.

- Based on the identified product mix of the yard, the preliminary product specification for all products of the mix can be generated. This should be based on market study, identification of technical requirements/constraints and client requirement for all these products. The product mix of a shipyard should be so defined that a large amount of standardised material, equipment and production techniques can be utilised in each product.
- Features of the different ships conforming to modularisation concept have to be identified.
- Design process has a significant role to play in developing a series of ship designs or a design mix suitable for the identified product mix incorporating standardised hull modules. Conceptually the hull form has to be split into different parts, each part being modularised separately.
- Identification of functional and geometric requirements of each modular portion of the hull form has to be done. Typical example can be the three regions of a ship – aft, midship and forward – as shown in the following table. It can be observed that the functional requirements of the three regions are different with overall geometric requirement as shown in the last column. If a separate deck module is considered, it would have a functional requirement of deck space. Similarly, a forecastle module may require to have adequate deck area as well as enough under-deck space and behaviour in a sea way.

Region/Zone	Requirements	
	Functions	Geometry
Aft body	1. Hydrodynamics of the stern	Main dimensions
	2. Propulsion	Deck area
	3. Steering	C_B and LCB location
	4. Accommodation	
Mid body	1. Cargo	
	2. Cargo volume	
	3. Production kindliness	
Fore body	1. Hydrodynamic of the fore body	
	2. Production kindliness	

- Identification of construction requirement of each modular portion of the hull form. If the module structure consists of flat plates or plates with single curvature, it is easy to construct, and if it has curvature in two directions, it has to go through a difficult construction process.
- Generation of shapes of each portion satisfying both functional and construction requirements.
- Merging of different zones to make a whole ship hull form.
- Generation of number of hull forms using different portions satisfying the functional requirements identified earlier.

Once the shape modularisation is achieved, it is possible to modularise further, for example certain geographical extents such as accommodation area, forecastle area, etc. and certain ship systems such as the main propulsion system, the steering system and the anchoring system. It is understood that standardisation of systems is to be based on the highest demand made by any of the products and, thus, there is to be a penalty on equipment on other products. Before standardisation to this extent is to be implemented, it is necessary to estimate the cost of all the penalties due to standardisation and ensure that the cost benefits outweigh the penalties.

It is easy to explain the working of a modularised hull form concept through an example with three modules – aft body, midship and fore body.

14.3.1.1 An Example

Product mix for a medium-sized shipyard was decided to consist of (1) a 500–700 TEU container feeder vessel for short sea operation, (2) a products tanker of about 10,000 dwt for coastal and short sea voyages, (3) a 10,000 dwt bulk carrier and (4) a multipurpose cargo vessel. A detailed economic analysis indicated that the most suitable speed for the container vessel of 120 m length would be 14–15 knots, whereas that for the bulk carrier or tanker would be about 13 knots. The product could be designed in a modular fashion, particularly the hull form, in three parts – fore body, mid body and aft body, a combination of different such units making a different product. The functional and construction requirements were identified as follows:

- *Aft body*: The aft body is perhaps the most complex portion of the ship having the maximum content in terms of construction and assembly of equipments. It is also the most demanding in terms of cost and time. Therefore, a single modularised/standardised aft body has been designed to suit a range of product mix of ships by combining with different fore body and mid body modules. Both cost and time of construction can be brought down. This could lead to standardisation of machinery, equipment and production process of the aft body to a large extent. This is subject to the constraints of functional requirements such as:
  a. *Hydrodynamics*: Good flow characteristics around the stern, propeller disc and rudder
  b. *Propulsion*: Adequate internal volume to house a range of main machinery of the propulsion system subject to various requirements of the product mix
  c. *Steering*: Proper aft body shape with constant position of aft perpendicular and stern aperture

  d. *Accommodation*: Standardised accommodation matching with aft body shape to serve the varying requirements of the product mix

- *Mid body*: The mid body of the ship is the biggest portion in terms of volume and length and is the freight-earning portion of the ship. It is structurally simpler than the other two zones of the ship but contains the maximum steel work in terms of weight. The variation in this area can include length, bilge radius (midship area) and prismatic coefficient. This would be subject to constraints of function and requirements such as:

  a. *Cargo*: Different internal arrangement for varying product mix, such as containers, liquid cargo (POL), bulk cargo and general cargo, etc.

  b. *Cargo volume*: Cargo volume can be adjusted to suit product mix by adjusting the following geometric characteristics in this region:

     i. Length (change in parallel middle body length).

     ii. Block coefficient (by changing sectional area curve)

  c. *Production kindliness*: This can be incorporated in the product mix by keeping the same internal of volume but varying the bilge radius and thereby altering the length of parallel middle body.

  d. Two bilge radii, 2.2 and 3.3 m, have been considered with three lengths, 113, 120 and 127 m, by adding 7 and 14 m lengths of parallel middle body to the lowest length vessel.

- *Fore body*: The fore body of a ship is a highly curved three-dimensional structure, complex and difficult to fabricate. Therefore, a standardised fore body can reduce production cost and time substantially. The fundamental requirements of the fore body design include consideration of the following aspects:

  a. *Hydrodynamics*: To suit the varying requirements of product mix particularly with regard to design draught (container ship design draught is lower than that for a tanker) and ballast draught (important for tankers and bulk carriers). The hydrodynamic design of the bulb should also take into account the range of speeds for the varying product mix.

  b. Design draught of the container ship is 6.6 m, whereas that of the bulk carrier or tanker is 7.8 m.

  c. *Production kindliness*: The fore body shape is to be designed so that it can be produced easily and is able to house standardised anchoring and mooring equipment for the product mix.

The modularised design of the three zones satisfying the separate functional requirements described earlier are able to satisfy the overall geometric constraints as well.

Figure 14.2 shows the outline of the two bulb shapes, F1 and F2, conforming to the draughts 6.6 and 7.8 m, respectively, and Figure 14.3 shows the outline of the profile where lines L1 and L2 indicate the demarcation of fore body and mid body and mid body and aft body, respectively. It also shows the centre-line buttock indicating the two alternative bilge radii, 2.2 and 3.5 m. Figure 14.4 shows the models of the three modules with their alternatives. Other modules such as accommodation module, the steering module, etc. can also be standardised following the same logic of modularisation. Figure 14.5 shows the general arrangement drawings of three types of vessels with modularised hull forms.

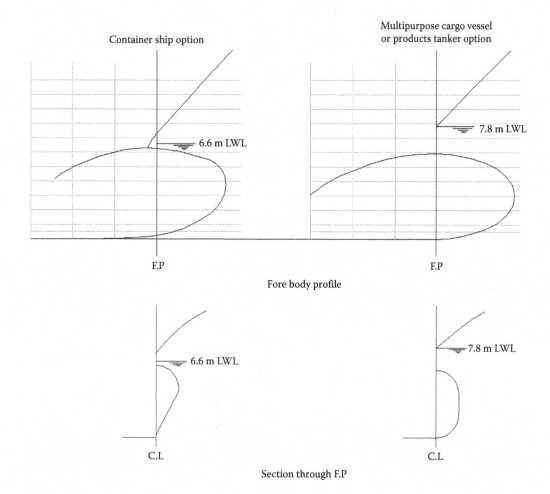

FIGURE 14.2
Two fore body shapes for two draught conditions.

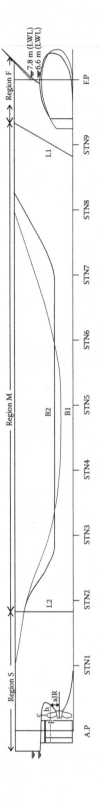

FIGURE 14.3

Profile showing the merger of Stern and two mid body and two fore body alternatives.

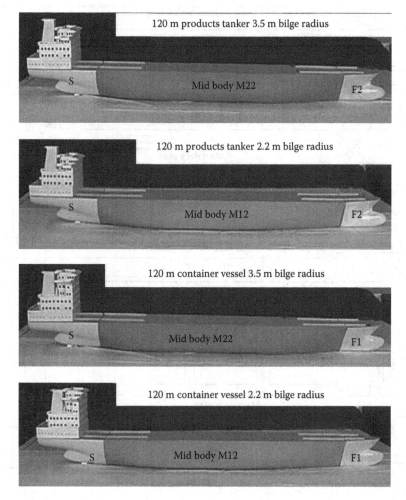

FIGURE 14.4
Four alternative ship hull forms generated from modularisation.

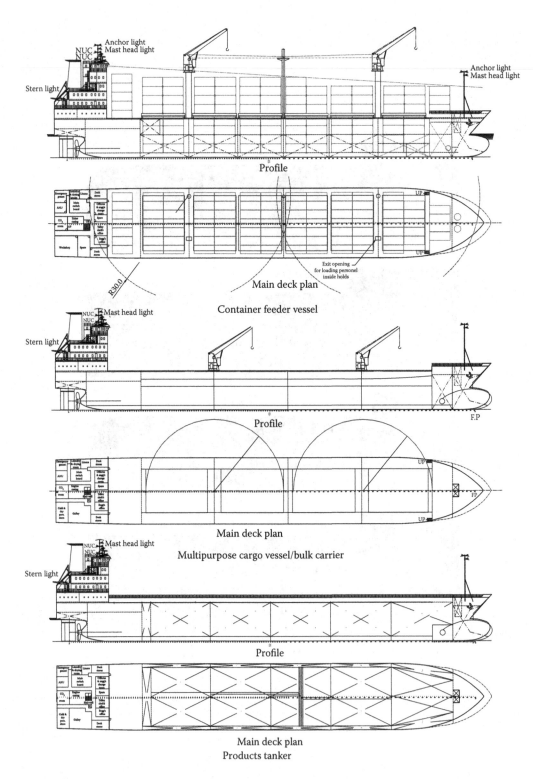

FIGURE 14.5
General arrangement drawings of three product alternatives.

15

Decision-Making Process

Marine product design process consists of designing individual components, designing sub-systems consisting of a number of components, system designing integrating a number of sub-systems and finally designing the whole product. At each stage and level of design, it is necessary to find a solution which must satisfy a certain performance level or a number of performance levels where the level is not absolute, but a maximum or minimum level such that performance achieved or predicted is below or above the pre-determined level, respectively. The design must also be subject to a number of pre-defined constraints. A standard overall ship design problem could be that the main dimensions and form of the ship need to be determined subject to constraints consisting of dimensional limits, stability and freeboard requirements, etc. with the performance or merit function which could be the ship's first cost or net present value (NPV) of ship operation. Identification of the design variables and determination of complete specifications of the merit function(s) and constraints are required to define the design problem. One can observe that there could be a number of design solutions to the same problem and the designer has to take a decision as to which solution must be accepted. Thus, the design problem reduces to a decision-making process. The design decision process can be clearly divided into three stages: the 'idea' stage in which the designer does the concept formulation for the design, the 'calculation' stage where all required calculations are carried out with (or without) the help of computers and finally the 'evaluation' stage where the designer takes the required decisions for development of the product, as shown in Figure 15.1, which can be with or without the use of computers. Figure 15.1 also shows the iterative design process already discussed in Chapter 1 and shown in figure 1.1. The final design solution can be arrived at based on one of the three methods specified in the following:

1. Feasibility study where the aim is to find a feasible design solution which does not violate any constraint. The performance or merit function is not necessary to be defined. The example is selection of the ship's main dimensions and form parameters which do not violate any design constraint such as weight, stability and other hydrodynamic and structural requirements. Merit functions such as NPV over ship lifetime or building cost, etc. are not considered. In case a merit function such as NPV over ship lifetime is to be the design basis, then the merit function value is estimated for all feasible design solutions individually and the best one is chosen. Such an approach is known as parametric study.

2. Algorithmical optimal design which is the mathematical optimisation process where the merit function as well as the constraints are mathematically defined as functions of a number of free or design variables. The design variables must be chosen for either maximisation or minimisation of the merit function, both processes being the same. Whereas a positive value of the objective function can be maximised, the negative value of the objective function can be minimised as shown in Figure 15.2. The next sections list the various mathematical optimisation methods that could be used for finding the optimum solution.

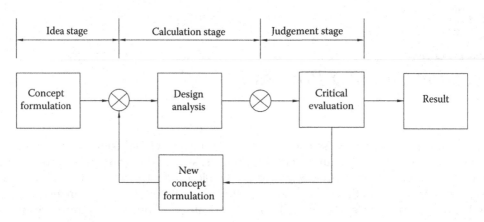

FIGURE 15.1
Design decision process.

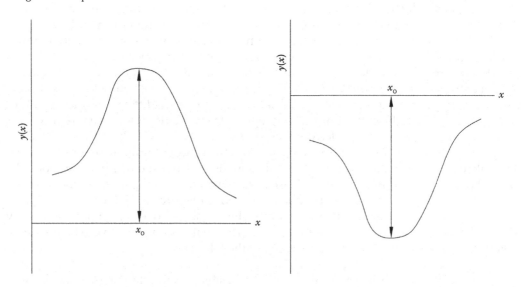

FIGURE 15.2
Conversion of a maximisation problem to a minimisation problem.

3. In many everyday design applications, it is customary to use a measure of merit for comparison of designs as mentioned in (1). Changes to design are made without resorting to formal optimisation strategies to improve the merit function. This type of design decision making is referred to as heuristic optimal design.

Ship design traditionally has been based on a sequential and iterative approach to find a feasible solution. Even the concurrent approach highlighted in Chapter 3 leads to a feasible solution only through an iterative process. With the advent of computers and sophisticated optimisation tools, researchers have been attempting to address the ship design problem using different optimisation techniques. The subsequent sections in this chapter highlight various optimisation techniques but do not elaborate on mathematical techniques involved. There are a number of textbooks on operations research and optimisation to which the reader may refer, some of which are Taha (1971), Fletcher (1987), Birk et al. (2009) and Lavy (2009).

15.1 Modelling the Optimisation Problem

15.1.1 Problem Formulation

Optimisation seeks the best solution from a proposed set of alternatives based on a defined criterion satisfying a set of proposed constraints. For this purpose, it is necessary to pose the problem in such a manner that a solution exists which is either unique or one of several acceptable ones and that a solution algorithm exists. Optimisation does not relieve the designer of his or her responsibilities to submit an optimum design solution based on relevant design criteria and constraints and accuracy of the numerical algorithm chosen. Modelling of the optimisation problem includes

Enumeration of n design variables x_i with $i = 1, 2, \ldots, n$ that are required to be evaluated

Definition of m constraints, C_j with $j = 1, 2, \ldots, m$ as functions of the variables

Choice of the measure of merit or the objective function O as a function of design variables. This could be a single objective or multiple objectives.

Generally speaking, an objective function O is to be minimised (or maximised) and is expressed as

$$O = fn(x_1, x_2, \ldots, x_n)$$

Subject to constraints $C_j = f_j(x_1, x_2, \ldots, x_n), j = 1, 2, \ldots, m$
  where there are n free design variables subject to m constraints defined by the functions C.

15.1.2 Problem Characteristics

The nature of the optimisation problem can be identified by a number of characteristics of the functions involved in its formulation. Some of these are listed as follows:

- *Dimensionality*: The number of dimensions or degrees of freedom n is defined by the number of independent parameters or design variables in the formulation. In general, it can be stated that the greater the number of variables, the greater is the complexity and time required to find a solution. In particular, single-dimensional problems are amenable to a number of very efficient optimisation techniques which are not applicable to higher dimensions.
- *Linearity*: Problems are normally divided into two categories – linear and non-linear – depending on whether all the functions are linear in all the parameters. A purely linear programme can be solved by the special technique of linear programming. If all functions are quadratic in nature, then quadratic programming can be used. Normally, ship design problems are highly non-linear.
- *Continuity*: Most solution methods take advantage of the assumption that the functions are continuous in position. However, in practice, the functions may have discrete values. For example, selection of power of the main engine or steel plate thickness depends on discrete values. This problem may be overcome by approximating discrete functions by continuous ones and then selecting the discrete

values close to optimum. One has to be careful about the quantum of error due to this discretisation.

- *Convexity and modality*: A function $y = f(x)$ is said to be convex if, given any two points x_1 and x_2, the value of the function at any intermediate point $ax_1 + (1 - a)x_2$ is less than the weighted arithmetic average of $y(x_1)$ and $y(x_2)$, that is,

$$y[ax_1 + (1-a)x_2] \le ay(x_1) + (1-a)y(x_2) \quad \text{where } 0 \le a \le 1.$$

A function is said to be strictly concave if the relationship is an equality. Similarly, a function is said to be concave if

$$y[ax_1 + (1-a)x_2] > ay(x_1) + (1-a)y(x_2) \quad \text{where } 0 \le a \le 1.$$

Figure 15.3a shows a convex function and Figure 15.3b shows a concave function. A merit function is said to be uni-modal if there is one optimum (maximum or minimum) only in the design space and multi-modal if there are more than one optimum point. The modalities are shown diagrammatically in Figure 15.4a and b. If there are

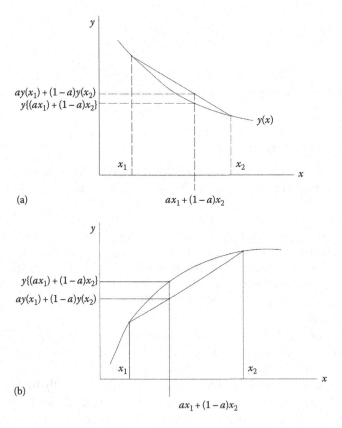

FIGURE 15.3
Convex and concave functions: (a) convex function and (b) concave function.

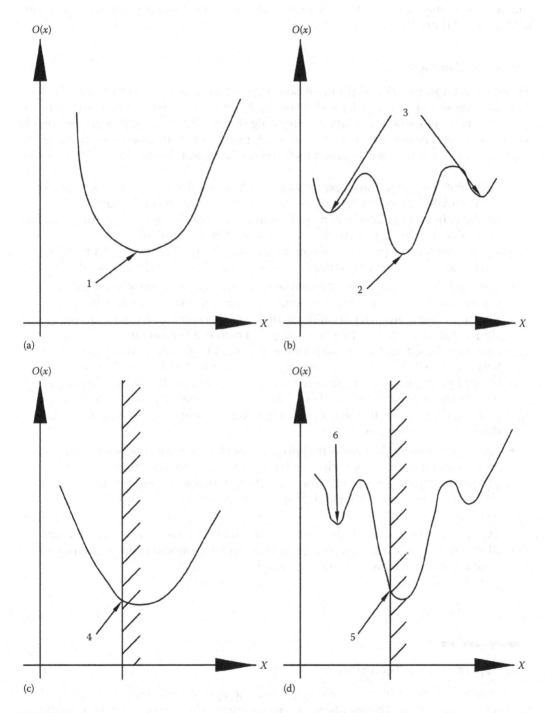

FIGURE 15.4
Uni-modal and mul-timodal functions: (a) unconstrained uni-modal, (b) unconstrained mul-timodal, (c) constrained uni-modal and (d) constrained multi-modal. *Notes:* (1) Minimum, (2) global minimum, (3) local minimum, (4) feasible minimum, (5) feasible global minimum and (6) feasible local minimum.

constraints on the function, then the optimum point shifts according to the constraint as shown in Figure 15.4c and d.

15.1.3 Solution Methods

There are a large number of different solution methods to the optimisation problem. Normally, these can be characterised by the problem characteristics where attempts are made to gain in efficiency of solution by limiting the class of problem that can be solved. The more general or complex a solution method, the less efficient it becomes. Some of the characteristics for selection of solution methods are listed as follows:

- *Analytical or numerical*: An analytical solution is possible where the problem can be formulated by an analytical merit function $y(x)$ which can be differentiated. At the extremum (maximum or minimum) the slope $dy/dx = 0$ and the second derivative d^2y/dx^2, if greater than 0, confirms a minimum value. However, this solution method is particular to the function $y(x)$ only. Marine design problems can be rarely described analytically. Numerical methods are more general.

- *Constrained or unconstrained*: Sometimes design variables are not entirely free within the design space, but are bound by some constraints or functional relationships. These auxiliary conditions are known as constraints. An example of a simple constraint can be that beam $B \leq$ St. Lawrence Seaway limit and that of a constraint based on functional relationship could be that transverse metacentric height $GM_T > 0.15$ m. When all constraints are inequalities, the number of degrees of freedom of the problem equals the number of free variables. However, an equality constraint reduces the number of degrees of freedom by one. Unconstrained problems are easier to solve than constrained problems. However, most ship design problems are constrained.

- *Linear or non-linear*: If there is non-linearity in the objective function or the constraints and if this non-linearity can be approximated to a linear function, linear programming can be applied for a solution. Otherwise, non-linear optimisation techniques must be applied. Ship design problems are highly non-linear in nature.

- *One or N-dimensional*: Highly efficient techniques exist for uni-modal one-dimensional problems. As the problem becomes multi-modal and N-dimensional, solutions become more difficult and complex.

15.2 Optimisation Techniques

As has been discussed in the previous section, the optimisation techniques depend on the type of the problem. The details of various techniques have been given in a number of reference books, some of which have been mentioned before. A gist of various techniques of optimisation is given in the following sections to acquaint the reader of the relevance of the techniques.

15.2.1 Unconstrained Optimisation

Unconstrained optimisation can be accomplished by direct search of the value of the merit function and selecting the best. One could also track the function by its value as well as the gradient of the function to reach the optimum. In addition, there are methods which use the second derivative of the merit function for reaching the optimum.

15.2.1.1 Unconstrained One-Dimensional Search

1. *Exhaustive search*: The objective function $O(x)$ need not be uni-modal, continuous or differentiable. The algorithm of this technique is evaluation of $O(x)$ at n equally spaced x locations finding out the minimum of these values. Then the minimum of the function O lies in the interval on either side of the minimum $O(x)$ already obtained. If more accuracy of evaluation is required, this identified interval can be subdivided further and the process repeated. But if n is too large, the process becomes inefficient and, on the other hand, if n is too small, a global optimum may be incorrectly evaluated or missed.

2. *Bisection search*: The objective function $O(x)$ must be uni-modal, but need not be continuous or differentiable. The algorithm of this technique is evaluation of $O(x)$ at two points $\pm\varepsilon$ on either side of the centre of the range, discarding the side with the highest result and repeating the process with the remaining half and continuation of this calculation till desired result is obtained.

3. *Golden search*: The objective function $O(x)$ must be uni-modal, but need not be continuous or differentiable. Similar to bisection search method, two points are chosen, one on either side of a chosen value of free variable x; the function $O(x)$ is evaluated at these two internal points and the segment beyond the higher $O(x)$ is discarded. This process is repeated till an extremum is reached. The chosen value of x is a function of the range of x such that the number of calculation cycles is reduced.

15.2.1.2 Unconstrained N-Dimensional Search

1. *Random search*: The objective function O is evaluated at randomly selected values of variables. This process is repeated again and again and the function values are compared to get the minimum. This is the simplest method of finding the optimum by exhaustive random search which can be aided by a random number generation program to enter the value of each variable. This method is easy to implement and can be used for discontinuous and non-differentiable multi-modal functions. But this method is computationally inefficient and is independent of the behaviour of the underlying function which may lead to large inaccuracies.

2. *Univariate search*: In this method one-dimensional search is made to get the minimum in the first dimension keeping the values of the other variables constant. Once the value of the variable (say x_1) is obtained, the process is repeated for other variables x_i with $i = 2, 3, \ldots, N$. To obtain the required accuracy in the final solution, the process may have to be iterated a number of times. This method is useful for uni-modal functions which may be discontinuous and non-differentiable. This method is very simple and the efficiency of the method reduces as the number of dimensions increase. Sometimes, this method may fail to arrive at the minimum.

3. *Hooke and Jeeves method of pattern search*: This method is used for uni-modal multi-variate optimisation problems which need not be continuous or differentiable. In this technique, first, a base point is established and local search is made around the base point to find reduction of the function value in all dimensions and a pattern is established. Then the pattern indicates the direction of search and a global search is carried out in the direction indicated. If the global move does not indicate improvement, a new base point is selected for the previous local search and a new pattern is established. The method continues till an optimum is located. This is a simple, efficient and reasonably accurate method except in cases of functions with sharp ridges. The number of trials increases linearly with the number of dimensions.

4. *Nedler and Mead simplex search*: The concept of a simplex, which is a multi-hedron of $n + 1$ vertices in n dimensions, is used in this method. It approximates a local optimum of a problem with n dimensions when the objective function varies smoothly and is uni-modal. The behaviour of the objective function is evaluated at each of the test points which are the corners of the simplex. Then one of the test points is replaced by a new test point. The simplest case is to replace the worst point with a point reflected through the centroid of the remaining n points. If this point is better than the best point of the simplex, then this point can be stretched exponentially in the same direction. If this point is not much better, the new point can shrink the simplex towards a better point.

15.2.2 Constrained Optimisation

15.2.2.1 Linear Programming

Linear programming is the most important special case of constrained optimisation. It deals with the optimisation of linear measure of merit function subject to linear inequality constraints. Mathematically,

$$\text{Maximize (or minimize) } O(x) = \sum_{i=1}^{n} c_i X_i$$

$$= c_1 x_1 + c_2 x_2 + \cdots + c_n x_n$$

Subject to m constraints,

$$C_k(x) = \sum_{i=1}^{n} a_{ik} \times x_i - b_k \leq 0, \quad \text{for } k = 1, 2, \ldots m$$

that is,

$$a_{11} x_1 + a_{12} x_{12} + \cdots + a_{1n} x_n \leq b_1$$

$$a_{21} x_1 + a_{22} x_2 + \cdots + a_{2n} x_n \leq b_2$$

$$a_{m1} x_1 + a_{m2} x_2 + \cdots + a_{mn} x_n \leq b_m$$

and $x_i \geq 0$, that is, $x_1 \geq 0$, $x_2 \geq 0$, $\ldots$,$x_n \geq 0$, that is, all design variable are non-negative.

15.2.2.2 Integer Programming

Integer programming is a special case of linear programming with added restriction that only integer values of the variable x_i are feasible.

15.2.2.3 Constrained Non-Linear Optimisation

- *Variational calculus formulation*: The complete non-linear programming problem can have equality and inequality constraints. Classical variational calculus deals with extreme value problems in the presence of equality constraints. It is possible to remove inequalities from the non-linear-programming formulation by introducing non-negative unknowns called slack variables. In variational calculus the extreme value of $O(x_i)$ subject to equality constraints is determined by introducing further auxiliary unknowns and solving for the unconstrained minimum of the modified function.

- *Penalty function method*: The objective of transforming the constrained non-linear-programming problem into an unconstrained problem leading to the same solution, which was one of the basic principles in the variational formulation, can also be achieved without any additional unknowns. This is the basis of the so-called penalty function methods. The principle of these methods is to create an artificial objective function by adding a penalty term to the natural objective function of the given problem. The penalty term, which may be constructed in one form or another, allows for the presence of constraints by raising the objective function when a constraint is violated or nearly violated. There are two principal variants of this method, the interior and the exterior penalty function techniques.

  The interior penalty function technique is commonly referred to as sequential unconstrained minimisation technique (SUMT). This method operates by adding a function in the feasible search area bound by the constraints. As the boundary is approached, the modified function assumes very high values.

  The external penalty function operates by adding a function only in the infeasible region, that is, external to the constraints. Thus, the penalty takes effect only when a constraint is violated.

15.2.3 Dynamic Programming

Very often an optimisation problem is encountered which cannot be solved by a single technique, but if broken down to a number of optimisation problems, the problem can be tackled effectively. Dynamic programming is an optimisation framework that transforms a complex problem into a sequence of simpler problems and their solutions. Its essential feature is the multi-stage nature of the optimisation procedure. Dynamic programming can provide a general framework for analysing many problem types. Within this framework a variety of optimisation techniques can be employed to solve particular aspects of a more general formulation. A thorough insight of the problem and ingenuity in finding a solution is necessary to recognise that a particular problem can be cast effectively as a dynamic program and it may be necessary to restructure the formulation so that it can be solved effectively.

15.3 Heuristic Methods for Decision Support Systems

A heuristic is a technique which seeks good (i.e. near optimal) solutions at a reasonable computational cost without being able to guarantee either feasibility or optimality or, even in many cases, to state how close to optimum a particularly feasible solution is.

Presently, there is great deal of growing interest in the application of heuristics for optimisation problems. On one hand, the development of the concept of computational complexity has provided a rational basis for exploring heuristic techniques rather than pursuing absolute global optimum. At the same time, there has been significant increase in the power and efficiency of either approach with more modern techniques.

What is actually being optimised is a model of a real-world problem. There is no guarantee that the best solution to the model is also the best solution to the underlying real-world problem. In a different form can one choose an exact solution of an approximate real-world model or an approximate solution of an exact model? Of course, one cannot hope for a true exact model, but heuristics are rather more flexible and are capable of coping with more complicated (more realistic) objective functions and/or constraints than exact algorithms.

Methods like simulated annealing and genetic algorithm are classes of those methods where objective functions need no simplifying assumptions of linearity. Thus, it may be possible to model the real-world problem rather more accurately than is possible if an exact algorithm is used.

15.3.1 Simulated Annealing

As already identified, the main disadvantage of local search is its likelihood of finding a local rather than global optimum. By allowing some uphill moves in a controlled manner, simulated annealing offers a better chance of reaching the global optimum. The annealing algorithm is similar to the random descent method in that the neighbourhood is sampled at random. It differs in that a neighbour giving rise to an increase in the objective function may be accepted and this acceptance will depend on the control parameters (temperature in an annealing problem) and the magnitude of the increase. The algorithm for a minimisation problem with solution space S, objective function O and neighbourhood structure N can be stated through an iterative random search.

If the randomly selected initial point is close to the global minimum, then simulated annealing is likely to attain the global minimum regardless of the objective function terrain or the annealing schedule. However, if the initial point is at a large distance from the global minimum and both the annealing schedule and the objective function terrain are relatively flat, then the global minimum is likely to be found only at the expense of a large number of function evaluations. However, if the temperature decreases fast as in the annealing process and the objective function terrain is hilly, simulated annealing is most likely to fail. A slow decrease in the temperature or start from different initial points or both may be necessary to ascertain the global optimum. Simulated annealing has been applied for the global solution of the ship design optimisation problem treating all the variables as continuous by Ray et al. (1995).

15.3.2 Genetic Algorithm

Genetic algorithm is a class of iterative procedures that simulate the evolution process of a population of structures subject to Darwin's survival of the fittest principle. The process of evolution is random yet guided by a selection mechanism based on the fitness of the individual structures. An application of the genetic algorithm exhibits a behaviour similar to that described in Darwin's theory of evolution, that is, larger fitness individuals will have a higher chance of survival and produce even higher fitness offsprings.

While most of the optimisation techniques employ a point search process, the genetic algorithm develops a population to be explored. Thus, the genetic algorithm goes beyond a mere sequential search procedure and identifies the building blocks of good structures within the population. These building blocks correspond to the regions of space where good solutions are likely to be found. The genetic algorithm uses a set of genetic operators to select, recombine and alter existing structures according to design heuristics to direct the search towards the global optimal solution.

15.4 Multiple Criteria Decision-Making

The basic departure from precise to imprecise decision situation gives rise to a formal decision-making approach of multiple criteria decision making (MCDM). MCDM refers to making decisions in the presence of multiple, usually conflicting, criteria. The terms generally used in the MCDM process are as follows:

- *Criteria*: A criterion is a measure of effectiveness. It is the basis for evaluation. Criteria emerge as a form of attributes or objectives in the actual problem set.

- *Goals/constraints*: Goals, synonymous with targets, indicate levels of aspiration. These are to be either achieved or surpassed or not exceeded. Often these are referred as constraints since these are defined to limit or restrict alternate sets.

- *Attributes*: An attribute provides a means of evaluating the level of an objective. Performance parameters, components, factors, characteristics and properties are synonyms for attributes. Each alternative can be characterised by a number of attributes (chosen by the decision-making process concept of criteria).

- *Objectives*: An objective is something to be maximised or minimised over the entire solution space. An objective generally indicates the direction of change desired in the continuous or discrete solution space.

- *Multiple objectives or attributes*: The design problem has multiple attributes or objectives from which the selection is to be made. A decision maker must generate relevant objectives or attributes for each problem setting.

- *Conflict among criteria*: Multiple criteria usually conflict with each other. For example, requirement of good resistance characteristics of a ship (high length-to-breadth ratio) conflicts with good stability requirement (high breadth).

- *Incommensurable units*: Normally, each objective or attribute has a different unit of measurement. For example, speed and stability have different units of measurement.

- *Design or selection*: Solution to the design decision problem is either to design the best alternative or to select the best amongst the previously specified alternatives. The MCDM process involves designing or searching for an alternative that is most attractive over all criteria or dimensions.

15.4.1 Multi-Attribute and Multi-Objective Decision-Making

The MCDM problem involves the selection of an alternative from a set of decision alternatives. Identification of the sets of alternatives is by no means a trivial task. There are two alternative sets due to different problem settings:

1. One set containing a finite number of alternatives
2. Another set containing infinite number of alternatives

In this respect the MCDM problem can be broadly classified into two categories:

- *Multiple attribute decision making (MADM)*: Decision making based on alternative sets defined explicitly by a finite list of alternatives.
- *Multiple objective decision making (MODM)*: Decision making based on alternative sets defined implicitly by a mathematical programming structure, theoretically giving infinite number of solutions.

The distinguishing feature of MADM is that there are usually a limited number of predetermined alternatives. The alternatives have associated with them a level of achievement of attributes, which may not necessarily be quantifiable, based on which the final decision is to be taken. The final selection of the alternatives is made with the help of inter- or intra-attribute comparisons. The comparison may involve explicit or implicit trade-offs.

MODM is not associated with the problem where alternatives are pre-determined. The thrust of an MODM model is to determine the best alternative by considering the various interactions within the design constraints which best satisfy the decision maker by attaining some acceptable levels of a set of some quantifiable objectives. The common characteristics of MODM methods are that they possess

- A set of quantifiable objectives
- A set of well-defined constraints
- A process of obtaining some trade-off information, implicit or explicit, between the stated quantifiable objectives and also between stated or unstated non-quantifiable objectives

Thus, MODM is associated with design solutions rather than selection process as in MADM. Table 15.1 compares the MADM problem with MODM process.

TABLE 15.1

MADM vs. MODM

	MADM	MODM
Definition of criteria	Attributes	Objectives
Objectives	Implicit (ill defined)	Explicit
Attributes	Explicit	Implicit
Constraints	Inactive (incorporated into attributes)	Active
Alternatives	Finite and discrete (prescribed)	Infinite and continuous
Interaction with DM	Not much	Mostly
Usage	Selection/evaluation	Design

15.5 Decision Support Applications in Ship Design

Any engineering design activity can be carried out in either of the following two ways:

1. The basic idea behind intuitive or indirect design in engineering is the memory of past experiences, undefined motives, approximate empirical relationships, statistical data sets or sometimes mere intuition. This, in general, need not lead to optimum merit of the designed product.

2. This can be overcome by adopting a direct or optimal design procedure. The most important feature of such a design process is that it consists of only logical decisions. Such decisions are arrived at by maximising or minimising a well-defined objective function subject to a number of constraints. Thus, optimisation techniques lead to finding a solution so that a designer or decision maker can derive maximum benefit from available resources. The techniques can be mathematical or heuristic or, in the event of multiple criteria, consisting of MODM/MADM procedures.

As has been mentioned before, a decision system can be arrived at for designing a single component or element, a sub-system, a system or an entire product such as ship or platform. An example of a component design could be the design of a propeller for a given engine power (break power) and RPM where, normally, propeller is designed for optimum efficiency with constraints on blade area ratio based on cavitation and diameter. Propeller efficiency, in such a case, can be written as

$$\text{Optimisation of } \eta = \text{fn}(D, RPM, z, P/D, A_E/A_0, V_s, \ldots)$$

$$\text{Constraints: } D_{\min} \leq D \leq D_{\max}$$

$$RPM_{\min} \leq RPM \leq RPM_{\max}$$

$$A_E/A_0 \geq \text{Minimum for cavitation}$$

$$\text{BHP: Discrete as per engine availability}$$

$$\text{Model for calculation: Series propeller } (B/\text{Gawn}/\text{Kaplan etc.})$$

$$\text{charts or lifting line/surface design}$$

where
D is the propeller diameter
z is the number of blades
P/D is the pitch ratio
A_E/A_0 is the blade area ratio
V_s is the ship speed

This is a non-linear constrained optimisation problem which should lead to an optimum propeller. Similarly, structural design of a pillar or girder could be optimised provided the maximum load on the structure is known a priori.

A system design problem would include design of components in the system and also the integrated system behaviour. An example could be the propulsion system design. In this system design, selection of equipment itself is a decision process. The propulsion mechanism could be single/multiple screws, fixed/controllable pitch propellers, normal or podded propulsion, propeller or water jet propulsion system and so on. The propulsion machinery could be single/multiple diesel engines, direct/geared drive, diesel/diesel-electric/boiler-steam turbine propulsion and so on. The shafting could be direct drive, z-drive or electrical pods. In such a system problem, the variables can be seen to be discontinuous and vary in a discrete manner. The optimisation could be to maximise overall propulsive efficiency, minimisation of installation cost or minimisation of running cost or fuel consumption. It can be observed that the technique of decision process involved in the choice of various components will be varied involving mathematical as well as heuristic optimisation.

Optimum design of a structural problem, whether a simple beam or complex three-dimensional ship structure, depends on the design philosophy employed. Ship or offshore structures are normally designed on the basis of permissible stress criterion. Knowing the maximum load on the structure and formulating the procedure for stress estimation, either by analytical method or sophisticated finite element techniques, structural design can be carried out for optimum weight, cost or some similar objective function. However, one could design a structure for a specified factor of safety against ultimate failure of the structure. In such cases, both the resistance of the material (strength) and load on structure are probabilistic in nature. The design approach in such cases is based on identifying various types of events leading to local or global failure associated with a probability of occurrence of the event. The objective function can be stated to be the sum of the effect of each event multiplied by the corresponding probability and suitable optimising procedure can be selected for this purpose.

Leaving out integration of propulsion system, hydrodynamic design normally means designing the hull form. The hull form is represented at a macro-level in terms of main dimensions such as length, breadth, draught, depth and location of centre of buoyancy and form coefficients such as mid-ship area coefficient, water plane area coefficient and prismatic coefficient. This can be a multi-objective optimisation problem of hydrodynamic performance with regard to resistance, propulsion, sea keeping and manoeuvring. One could also convert it to a single-objective problem by taking a suitable weighted sum of performance in all these four hydrodynamic aspects. Integrating the CAD and CFD analysis modules with the optimiser module, one can generate an optimised hull form as well.

At the concept design stage, knowing the design requirements, it is possible to define an objective function which can be maximised such as the net present value (NPV) or yield over the life cycle of the ship, or an objective function which can be minimised such as required freight rata (RFR), building cost or power consumption. The objective function, so chosen, can be expressed as a functional relationship of variables such as main dimensions and form coefficients. The constraints of such optimisation could be those due to restrictions on parameters, stability requirement, power limitations, strength and vibration-related constraints, etc. At concept design stage when details of external shape and internal arrangements are not known, optimisation can be carried out using approximate relationships. A possible brief problem definition could be as follows:

$$\text{Objective function } O = \text{fn}(L, B, T, D, C_B, V, \ldots)$$

$$\text{Constraints: dwt within predefined deadweight} \pm 10\% \text{ (say)}$$

$$\text{Volumetric capacity} = \text{Required capacity} \pm 10\% \text{ (say)}$$

$$GM_T \Rightarrow \text{Statutory } GM_T$$

$$D - T \Rightarrow \text{Statutory freeboard and so on.}$$

Measure of merit (O): First cost/lightweight/RFR (minimisation) or NPV (maximisation)

Model: Design and cost relationships, operating pattern, cash flow calculations, etc.

But as design progresses, optimisation can be continued supported by detailed scientific calculations and CAD modules to generate shape and internal arrangements. Sophisticated shape generation software is now available with the designer which can not only generate three-dimensional wire mesh diagrams and rendered models to give the visual impact but also change shape very quickly for new/altered local or global parameters. These can also generate numerical data to be the input to any other analysis solver. The solvers, be it FEM or CFD, are equipped with a pre-processor to model the data from the CAD software as required, the solver for FEM or CFD and a post-processor which gives the user both numerical and graphical output. This graphical simulation post-processor module provides the visual effect required for decision making at the design stage from different alternatives. Figure 15.5 shows diagrammatically this optimisation procedure integrating CAD module, the analysis module and the decision-making process.

It has already been shown in the form of a flow chart in figure 7.8 how hydrodynamic design can be done using CFD and EFD techniques and an example has been shown in figure 7.7. The underlying process may have involved investigating a number of hull forms and use of MADM/MODM techniques to decide on the final hull form as is shown in Figure 15.5. Ayob et al. (2012) have proposed a multi-criteria design framework for planning craft generating its shape for optimised hydrodynamic behaviour. Saha and Ray (2011) have proposed a robust design optimisation (RDO) where the results are not very sensitive to uncontrollable variations of design variables and Alam et al. (2012) have designed an unmanned underwater vehicle using this technique.

Any marine product design is a very complex process involving different scientific and engineering fields. The decision-making processes of different aspects of design as well as the whole design involve different numerical and heuristic optimisation processes and multi-objective decision-making methods. It may be convenient to develop a decision-making framework or dynamic programming framework involving all the techniques mentioned before so that the design process can be easily addressed. Currently, a number of research workers are involved in developing advance decision systems for ship design. Such an algorithmic framework is shown in Figure 15.6.

However, in the real world of marine design, a computerised framework for decision making involving optimisation is rarely used. An owner normally prefers a design with proven track records rather than a numerically optimised design. Since a marine product is a very high value item, the owner is normally reluctant to take any risk which is much reduced if he or she buys a proven product. Even if he or she procures a new product or new type of product, he or she is satisfied if all statutory requirements are met and it gives the required performance which can be reliably estimated using standard calculation procedures and model testing. The design solution thus arrived at generally works well, and therefore, this solution could be a near-optimum design. Further, the optimum

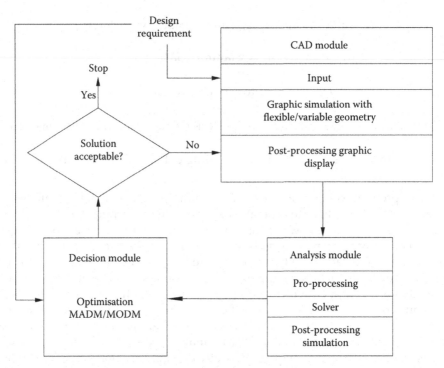

FIGURE 15.5
Interactive design optimisation process.

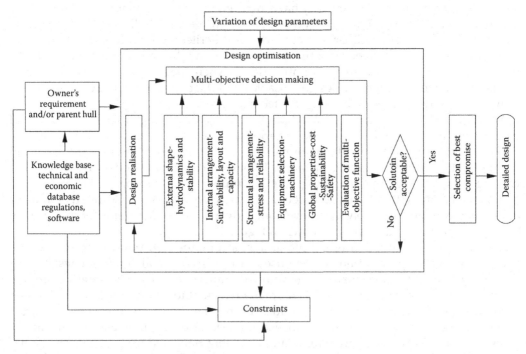

FIGURE 15.6
Generic ship design optimisation problem.

point should lie on a plateau in the design space such that slight variation in the variable values does not alter the performance. Perhaps the designer uses his or her own methods of using maximum attribute decision-making process to arrive at the solution without realising that he or she has done so.

In case of a marine design, the objective function is dependent on values of variables during the future life of the ship. As an example, if NPV is to be maximised, the economic quantities could only be predicted values with large uncertainties. Therefore, an optimised design with assumed economic values would not be the best solution when the economic environment changes. The optimisation framework has to include probabilistic values of economic parameters such as fuel oil price, freight rate, etc. which may lead to a prediction with a reliability factor. Use of optimisation techniques in marine design, for components, systems as well as overall design, is on the rise, which may be very important for a new design concept or a design with unconventional parameters.

16

Design Management

While marketing a design, the designer is frequently asked a question by the client: 'Is this a proven design?' Also very often the tender document for a ship has a clause asking if the designer has designed similar vessels earlier. If the designer fulfils this condition along with other conditions of the tender, the design is selected for conversion into a product. Then the design can be termed as a successful design.

The design requirements include the essential or 'must-be' requirements which must be fulfilled for the vessel to operate satisfying all national and international statutory regulations and classification codes. If any of these requirements is not met, then the design is rejected. A 'proven design' means a design that has demonstrated satisfaction of the 'must-be' requirements. Sometimes, the client may demand satisfaction of some extra requirements and the designer may be able to fulfill all or some of these beyond the essential requirements. The customer satisfaction is proportional to the satisfaction of these requirements. Sometimes, the designer goes beyond customer-stated requirements by providing a much higher performance level in which case customer satisfaction is the highest. Of course, extra features required for providing this higher performance level must be available to the customer at an affordable price.

It is often said that producing a bad design is rather difficult since, for conventional vessels, there are a number of scientific software packages and experimental facilities to check performance criteria satisfied by the design. The statutory requirements and implementation of classification society codes ensure that the predicted performance level is satisfactory. Therefore, to produce a bad design which does not satisfy one or more of the essential requirements is rather difficult. Thus, it is easy to produce a mediocre or average design which satisfies these requirements only. Going beyond it to produce an attractive and good design is rather difficult since it must have certain novel features beyond normal requirements.

A good design must satisfy all technical requirements and other the functional requirements. As it is often said, the design must showcase a product which must feel right, smell right, sound right and feel right. But over and above everything else, it must appeal to the professional and emotional senses of the client such that a good design becomes great. Some of the attributes that make a great design are discussed in the following sections.

16.1 Creativity and Innovation

Creativity, in an engineering sense, means capability of producing new solutions in an intellectual way to solve problems. Creative people challenge assumptions, are not afraid to fail and strive to generate multiple solutions to problems. The basic ingredients of creative thinking in engineering field can be stated as follows:

- *Domain skill*: Creative person must have adequate domain skills so as to have the basic fundamentals clear in the field of discipline.

- *Creative thinking ability*: A creative person must be able to think 'out of the box', seek novelty and diversity from the stereotype and must be able to think independently.
- *Motivation*: Creative ideas must come from within and not because of external pressure or aspiration of rewards. Creative ideas must be pursued by the creative urge and passion.

When offshore platforms were required for the first time for drilling in deep waters, it was necessary to have some new and imaginative ideas and semi-submersible was one such creative idea which could be made into a reality. Following this concept, a number of new and imaginative ideas have been thought of and implemented such as tension leg platforms and articulated platforms. The last half century has seen many creative ideas being implemented in the marine field. Thus, creativity is a new and imaginative idea translated to reality. If the idea is not translated to reality, at best it can be said to be imaginative.

Innovation is the act of introducing something new which did not exist before. By its very definition, innovation includes accepting risks of failure of the new concept. Thus, innovation includes creativity and the associated risk involved. Innovation can be something different and new, or it can be small with a minor improvement over an existing product, process or system. Innovations need not be large nor be an entire product. Innovations can be incremental improvements and it may mean developing a competitive edge over the competitor's product and also diversification. Thus, innovation builds up capacity for change leading to modification of the product adapting to changed environment. In a business setup, this leads to being a business leader in the competitive environment and long-term survival. It can be easily visualised that any product innovation has to start from the design stage itself. Thus, creativity leading to innovative design has the capability of transforming a good design to a great design. The need for technology change is accelerated by business demand. However, today technology change is demanded not only from the business angle but also from sustainability and safety aspects as has been discussed in the previous chapters. There is need for small and large innovations at the design stage leading to novel designs, novelty being small in an existing design or large, leading to different marine craft or process or system.

Very often, development of a design involves constraints which limit the way design can evolve in the traditional way. It is very often said that necessity is the mother of invention. The constraints can sometimes necessitate change leading to innovative thinking and design development. Innovation can lead to new product bypassing the limitations imposed by the conventional product development. Innovations can be breakthrough designs leading to a new type of product or, even small innovations increasing the product value substantially. Typically, as an example, the lack of stability in a small vessel can be obviated by designing a multi-hull vessel.

Innovation means going away from existing comfortable routine and mastered skills. Generally, designers and technologists like to continue with existing systems for the comfort of the knowledge of its working. So any innovative concept or design to create purposeful and focussed change in the design or design process is fraught with risk of failure when the design is transferred to a product. Marine products are large and expensive and any failure after the product goes into service may mean large financial loss, and therefore, risks are perceived to be high. Ways and means of reducing risk must be included in the design process and the client must be informed about the gains in the innovation proposed in terms of business or functional objectives.

Risk taking means the ability to move new concepts ahead in the face of adversity or the difference between the uncharted future and the known projection. Risk taking is often linked with motivation and empathy for the concept. In spite of motivation, if one does not go ahead with taking risk, perhaps one regrets later that a golden opportunity of doing something original has been lost. Very often one comes across a person who regrets that if he or she had another chance in life, he or she would have led it the other way. Risk taking succeeds if:

- Risk taking means taking challenges which is exciting.
- The innovation can be thoroughly evaluated from all aspects through analysis using high skill levels.
- Encouraging an attitude of change leads to implementing change.
- One understands that success of the project or idea is reward itself.

Design is a group activity consisting of people having different types of skills and different skill levels. In such a scenario, innovative ideas from an individual are most likely to be discouraged. Since the demand on the design group is to produce a design in short time, it is convenient to carry out the activity in the traditional way of vertical process of sequential activity. In such a system, skill sets are used for each activity to the satisfaction of each performance criterion. Any innovation which means change from this traditional approach is discouraged since it means change. But imagine if one worked laterally between various groups where each skill set is applied without knowledge of output from other skill sets. In this event, it may be possible to have 'irrelevant' design developments which may turn out to be innovative. Then analysis of this new design may lead to something extraordinary. Thus, it is necessary to develop a design system with innovation attitude where each individual is encouraged to project innovative concepts.

Any innovative idea carries the risk of failure. It is possible that, even after thorough investigation, the idea fails. It has to be ensured that the innovator is not taken to task for this. It has to be understood that failure is an inherent part of innovation or creativity. It is commonly said that failure is the pillar of success. Thus, failure must lead to learning ultimately leading to success.

The requirements of implementation of innovative ideas can be summarised as follows:

- Association or connection with personnel with different skill sets
- Power to challenge or question existing systems
- Observation of existing systems with all details
- Experimentation and evaluation of risk involved
- Empathy, excitement and motivation for change and for something new
- Use of design constraints to an advantage

16.2 Design Integration

A marine product consists of different technical systems and other non-technical systems such as economic, business and similar other systems. Each system must be evolved to an optimised level and all the systems must be put together to make the whole product.

FIGURE 16.1
Renoir painting: (a) line diagram, (b) painting of the line diagram and (c) actual painting.

The integration of all systems to realise the product is known as design integration which has already been discussed in Chapter 3 dealing with design process. Design of multiple systems functioning cohesively in harmony should be such that the whole is greater than the multiple parts. Mohan Ram (2007) has discussed this matter with some details and examples. Figure 16.1 illustrates this aspect in a field outside the marine field. Figure 16.1a is a line drawing of the Renoir painting and Figure16.1b is the painting of this diagram. But the actual painting is shown in Figure 16.1c which shows the total effect of painting of parts in overall harmony.

16.3 Design Management

As discussed so far, design activity includes multiple skill sets with different kinds of human resource with varied knowledge sets and levels. It also requires good communication skills within the design group as well as outside parties. This may include information technology systems and data and information management. A modern marine design house must also have sophisticated design tools which include software and hardware tools for scientific analysis, statistical analysis and CAD activity and integration of all of these. Perhaps a product lifecycle modelling (PLM) may also have been installed in the design house. So a good design effort involves managing high-quality human resource as well as sophisticated software and hardware tools so that design completion at different stages can be done within stipulated time. Further it must be remembered that a successful design is one that is built and works satisfying mission requirements much more than just adequately, and then, repeat design orders may be obtained. To do this, a design house has not only to compete with other design houses but also to have a technology jump over the competitors. These calls for technology leadership which must encourage creativity establish an attitude of innovation so that new features may emerge in the product in terms of equipment, construction, operation, maintenance or economic performance.

References

Alam, K., Ray, T. and Anavatti, S.G., A new robust design optimization approach for unmanned underwater vehicle design, *Proceedings of the Institution of Mechanical Engineers, Part M: Journal of Engineering*, 226, 235 originally published online for the Maritime Environment 2012, DOI: 10.1177/1475090211435450, 2012.

Almeter, J.M., Resistance prediction of planing hulls: State of the art, *Marine Technology*, 30(4), 297–307, October 1993.

Ang, A.H.-S. and Tang, W.H., *Probability Concepts in Engineering Planning and Design*, Vol. II, *Decision, Risk and Reliability*, John Wiley & Sons, New York, 1990.

Annual Report 2013 – Class NK in Research, Class NK, Tokyo, 2013.

Assessment of Impact of Shipping on Marine Environment, OSPAR Commission Report No. 440.2009, London, 2009.

Ayob, A.F.M., Ray, T. and Smith, W.F., *Beyond Hydrodynamic Design Optimization of Planing Craft*, SNAME, Jersey City, NJ, 2012.

Ayyub, B.M., Assakkaf, J.A., Beach, J.E., Metton, W.M., Nappi, N. Jr. and Conley, J.A., Reliability based load and resistance factor design (LRFD) guidelines for ship structures, *Naval Engineers Journal*, 114(2), 23–42, April 2002.

Bailey, D., *The NPL High Speed Round Bilge Displacement Hull Series: Resistance, Propulsion, Manoeuvring and Seakeeping Data*, RINA Maritime Technology Monograph No. 4, RINA, London, U.K., 1976.

Barltrop, N.D.P., Ed., *Floating Structures: A Guide for Design and Analysis*, Vol. 1, CMPT Publication No. 101, Oilfields Publications Ltd., Herefordshire, U.K., 1998.

Bertram, V., Modularisation of ships – Report within framework of project 'intermodul' s/03/G Intermare C, in *ENSIETA Second Rue Francois Verney*, France, 2005.

Bhattacharya, R., *Dynamic of Marine Vehicles*, Wiley-Interscience, New York, 1978.

Birk, L. and Harries, S., Eds., *Optimistic Optimisation in Maine Design*, 2nd edn., Mensch and Buch Verlag, Berlin, Germany, 2009, ISBN 3-89820-514-2.

Blount, D.L., Small-craft power prediction, *Marine Technology*, 13(1), 14–45, January 1976.

Blount, D.L., Achievements with advanced craft, *Naval Engineers Journal*, 106(5), 49–59, September 1994.

Booz-Allen Report, *A study of feasibility, cost and benefits of expanded use of modular concepts in Dx/DXG program*, Booz-Allen Applied Research Inc., Final report Contract N00024-67-C-0368, 1968.

Brix, J., Ed., *Manoeuvring Technical Manual*, Seehafen Verlag GmbH, Hamburg, Germany, 1993.

Bruzzone, D., Ferrando, M. and Gualeni, P., Numerical and experimental investigation in to resistance characteristics of symmetrical and unsymmetrical catamaran hull forms, in *Proceedings of the Fifth International Symposium on High Speed Marine Vehicles, HSMV99*, Capri, Italy, pp 24–26, 1999.

Burrill, L.C., Developments in propeller design and manufacture for merchant ships, *Transactions on Institute of Marine Engineers*, 55, 148, 1943.

Buxton, I.L., *Engineering Economics in Ship Design*, BSRA, Wallsend, U.K., 1971.

Buxton, I.L., Daggitt, R.P. and King, J., *Cargo Access Equipment for Merchant Ships*, E&FN Spon, London, U.K., 1978.

Carlton, J., *Marine Propellers and Propulsion*, 2nd edn., Buttersworth-Heinemann, Oxford, U.K., 2007.

Carreyette, J., Preliminary ship cost estimation, naval architect, *Transactions on Royal Institution of Naval Architects*, 120, 235–258, July 1978.

Clement, E.P. and Blount, D.L., Resistance tests of a systematic series of planing hull forms, *Transactions of the Society of Naval Architects and Marine Engineers*, 71, 491–579, 1963.

Compton, R.H., The resistance of a systematic series of semi-planing transom hulls, *Marine Technology*, 23(4), October 1986.

Comstock, J.P., Ed., *Principles of Naval Architecture*, SNAME, Jersey City, NJ, 1967.

Couser, P.R., Molland, A.F., Armstrog, N.A. and Utama, I.K.A.P., Calm water powering prediction for high-speed catamarans, in *Fast'97*, Sydney, New South Wales, Australia, 1997.

Design Council, Attaining competence in engineering design, Engineering Working Party report, The Design Council, London, U.K., 1991.

Doust, D.J., *Optimised Trawler Forms*, Vol. 79, NECIES, London, U.K., 1962–1963.

Doust, D.J. and O'Brien, T.P., *Resistance and Propulsion of Trawlers*, Vol. 75, NECIES, Newcastle Upon Tyne, U.K., 1958–1959.

Edstrand, H., Freemanis, E. and Lindgren, H., Systematic tests with models of ships with CB-0.525, Publication No. 38, SSPA, 1956.

Edstrand, H. and Lindgren, H., Experiments with tanker models, Publication No. 23, 26, 29, 36 and 37, SSPA, Goteborg, Sweden, 1955–1956.

Eppler, R. and Shen, Y.T., Wing sections for hydrofoils – Part I, symmetrical profiles, *Journal of Ship Research*, 23(3), 46, 1979.

Eswara, A.K., Misra, S.C. and Ramesh, U.S., Introduction to natural gas: A comparative study of its storage, fuel costs and emissions for a harbour tug, in *Proceedings of the Annual Conference*, SNAME, Bellevue, WA, pp 201–217, 2013.

Executive Control Board, National shipbuilding research program, Benchmarking of US Shipyards – Industry Report, South Carolina, January 2001.

Faltinsen, O.M., *Sea Loads on Ships and Offshore Structures*, Cambridge University Press, Cambridge, U.K., 1990.

Fletcher, R., *Practical Methods of Optimisation*, Wiley, Chichester, U.K., 1987.

Freemanis, E. and Lindgren, H., Systematic tests with ship models, Publication No. 39, 41, 42 and 44, SSPA, 1957–1959.

Gallin, C., Inventiveness in ship design, *Transactions on NECIES*, 94(1), 17–32, 1977.

Gawn, R.W.L., Effect of pitch and blade width on propeller performance, *Transactions on Royal Institution of Naval Architects*, 95, 157–193, 1953.

Gerritsma, J., Keuning, J. and Onnink, R., The Delf Systematic yacht Hull (Series II) experiments, in *The 10th Chesapeake Sailing Yacht Symposium*, SNAME, Annapolis, MD, 1991, pp 27–40.

Gerritsma, J., Moeyes, G. and Onnink, R., Test results of a systematic yacht hull series, *International Shipbuilding Progress*, 25(278), 163–180, July 1978.

Gerritsma, J., Onnink, R., Versluis, A., Gemoetry, Resistance, stability of the Delft systematic yacht hull series, report, Technological University of Delft, Delft, the Netherlands.

Gertler, M., A reanalysis of the original test data of the Taylor standard series, DTMB report 806, Washington, DC, 1954.

Ghose, J.P. and Gokarn, R.P., *Basic Ship Propulsion*, Allied Publishers Pvt. Ltd., New Delhi, India, 2004.

Gonzalez, J.M., Specific hydrodynamic aspects of fast displacement ships and planing Craft, in *13th WEGMT School on Design Techniques for Advanced Marine Vehicles and High Speed Displacement Ships*, Delft, the Netherlands, 1989.

Graff, W., Kracht, A. and Weinblum, G.P., Some extensions of D.W. Taylor standard series, *Transactions on Society of Naval Architects and Marine Engineers*, 72, 374–401, 1964.

Guldhammer, H.E. and Harvald, S.A., *Ship Resistance: Effect of Form and Principal Dimensions*, Akademisk Forlag, Copenhagen, Denmark, 1974.

Hadler, J.B., The prediction of power performance on planing craft, *Transactions of the Society of Naval Architects and Marine Engineers*, 74, 563–610, 1966.

Harrington, L.R., *Marine Engineering*, SNAME, Jersey City, NJ, 1992.

Holtrop, J., A reanalysis of resistance and propulsion data, *International Shipbuilding Progress*, 31, 1984, pp 272–276.

Holtrop, J. and Mennen, G.G.J., A Statistical power prediction method, *International Shipbuilding Progress*, Vol. 25, no. 290, pp 253–256, October 1978.

Holtrop, J. and Mennen, G.G.J., An approximate power prediction method, *International Shipbuilding Progress*, Vol. 29, no. 235, pp 166–170, July 1982.

Hughes, O. and Pike, J.K., *Ship Structural Analysis and Design*, SNAME, Jersey City, NJ, 2010.

International Association of Classification Societies, Shipbuilding and repair quality standard, IACS recommendation 47, 1966, revised 1999, 2004, 2006, 2008, 2010, 2012 and 2013.

International Maritime Organisation (IMO), Reduced/ No B.W. ship designs, GloBallast Monograph 20, IMO, London, U.K., 2011.

International Towing Tank Conference (ITTC), 26th Conference in Rio De Janeiro, Brazil, 2011.

Jacobi, F., Modular ship design, *Naval Forces*, 1(2003), 36–40, 2003.

Keuning, J.A. and Gerritsma, J., Resistance tests of a series of planing hull forms with 25 degrees deadrise angle, *International Shipbuilding Progress*, 29, 289–296, September 1982.

Kuo, C., *Computer Applications in Ship Technology*, Heyden, London, U.K., 1977.

Kuo, C., Recent advances in marine design and applications, in *ICCAS 91*, Vol. IV, Rio de Janeiro, Brazil, September 1991.

Kuo, C., *Safety Management and Its Maritime Application*, 2nd in the Series of Maritime Futures, The Nautical Institute, London, U.K., 2007.

Lackenby, H. and Parker, M.N., The BSRA methodical series – An overall presentation – Variation of resistance with breadth-draught ratio and length-displacement ratio, *Transactions of the Royal Institute of the Naval Architects*, 108, 363–388, 1966.

Lamb, T., Ed., *Ship Design and Construction*, Vol. 1, SNAME, Jersey City, NJ, 2003a.

Lamb, T., Ed., *Ship Design and Construction*, Vol. 2, SNAME, Jersey City, NJ, 2003b.

Lamb, T., Worlwide shipbuilding productivity status and trends, in *Pan American Conference on Naval Engineering Maritime Transport and Port Engineering*, University of Michigan, Ann Arbor, MI, pp 253–256, October 2007.

Larsson, L. and Eliasson, R., *Principles of Yacht Design*, Adlard Coles, London, U.K., 2000.

Larsson, L., Raven, H.C. and Pauling, J.R., Ed., *Ship Resistance and Flows*, SNAME, Jersey City, NJ, 2010.

Latorre, R. and Ashcroft F., Recent developments in barge design, towing and pushing, *Marine Technology*, 18, 1981, pp 10–21.

Lavy, A.B., *The Basis of Practical Optimisation*, Society for Industrial and Applied mathematics, PI, 2009.

Lerbs, H.W., Moderately loaded propellers with a finite number of blades and an arbitrary distribution of circulation, *Transactions of the Society of Naval Architects and Marine Engineers*, 60, 73–123, 1952.

Lewis, E.V., Ed., *Principles of Naval Architecture*, Vol. 1, SNAME, Jersey City, NJ, 1989a.

Lewis, E.V., Ed., *Principles of Naval Architecture*, Vol. 2, SNAME, Jersey City, NJ, 1989b.

Lewis, E.V., Ed., *Principles of Naval Architecture*, Vol. 3, SNAME, Jersey City, NJ, 1989c.

Longuet-Higgins, M.S., Statistical properties of wave groups in a random sea state, *Philosophical Transactions of Royal Society of London*, 312, 219–250, 1984.

MAN-Diesel&Turbo, *Two-Stroke Low Speed Diesel Engines for Independent Power Producers and Captive Power Plants*, MAN-Diesel&Turbo, Copenhagen, Denmark, 5510-0067-00, May 2009a.

MAN-Diesel&Turbo, *Exhaust Gas Emission Control Today and Tomorrow Application on MAN B&W Two-stroke Marine Diesel Engines*, MAN-Diesel&Turbo, Copenhagen, Denmark, 5510-0067-00, August 2009b.

Mansoor, A., Liu, D. and Pauling, J.R. Ed., *Strength of Ships and Ocean Structures: The Principles of Naval Architecture Series*, SNAME, Jersey City, NJ, 2008.

Marchaj, C.A., *Sail Performance – Theory and Practice*, Adlard Coles Nautical, London, U.K., 1996.

Maruo, H. and Tokura, J., Prediction of hydrodynamic forces and moments acting on ships in heaving and pitching oscillations by means of an improvement of the slender ship theory, *Journal of the Society of Naval Architects of Japan*, 143, 104–111, 1978.

McLellan, H.J., *Elements of Physical Oceanography*, Pergamon Press, Oxford, U.K., 1965.

Miroyannis A., Estimation of ship construction cost, MSc dissertation, MIT, Cambridge, MA, 2006.

Misra, S.C., Gokarn, R.P., Sha, O.P., Banerjee, N., Suryanarayana, Ch. and Suresh, R.V., Development of a four-bladed surface piercing propeller series, *Naval Engineers Journal*, 124(4), 111–141(31), December 2012.

Misra, S.C., Sha, O.P. and Gokarn, R.P., Modular ship design for competitive construction in India, *Transactions of the Society of Naval Architects and Marine Engineers*, 110, 279–299, 2002.

Mistree, F., Decision based design: A contemporary paradigm for ship design, *Transactions of the Society of Naval Architects and Marine Engineers*, 98, 565–597, 1990.

Mohan Ram, N.S., Design, Lecture given at OE&NA Department, Bengal, India, November 2007.

Molland, A.F., Turnock, S.R. and Hudson, D.A., *Ship Resistance and Propulsion*, Cambridge University Press, Cambridge, U.K., 2011.

Moor, D.I., Longitudinal bending moments on models in head seas', *Transactions of the Royal Institution of Naval Architects*, 109, 117–165, 1967.

Moor, D.I., The effective horsepower of single screw ships: Average and optimum standards of attainment 1969, BSRA Report 317, BSRA, U.K., 1971.

Moor, D.I. and Murdey, D.C., Motion and propulsion of single screw models in head seas, *Transactions of the Royal Institution of Naval Architects, London*, 110, 403, 1968.

Moor, D.I. and Murdey, D.C., Motion and propulsion of single screw models in head seas (part 2), *Transactions of the Royal Institution of Naval Architects, London*, 112, 121, 1970.

Moor, D.I. and Patullo, R.N.M., The effective horsepower of twin screw ships, BSRA Report 192, BSRA, U.K., 1968.

Moor, D.I. and Small, V.F., The effective horsepower of single screw ships, *Transactions of the Royal Institution of Naval Architects*, 102, 315–365, 1960.

Morelli & Melvin, Designers, beginning of something new – 'Alpha 42', *Multihull Magazine*, pp 28–35, May/June 2012.

Morgan, N., *Marine Technology Reference Book*, Buttersworth, London, U.K., 1990.

Muckle, W., *Strength of Ship's Structures*, Edward Arnold, London, U.K., 1967.

Munchmeyer, F.C., Schubert, C. and Nowacki, H., Interactive design of fair hull surfaces using B-splines, in *Proceedings of the Third ICCAS*, North-Holland, Glasgow, Scotland, pp. 66–76, June 1979.

Munro-Smith, R., *Merchant Ship Design*, Hutchinson, London, U.K., 1964.

Munro-Smith, R., *Elements of Ship Design*, Marine Media Management, London, U.K., 1975.

Munro-Smith, R., *Ships and Naval Architecture*, Institute of Marine Engineers, London, U.K., 1988.

Newman, J.N., *Marine Hydrodynamics*, MIT Press, Cambridge, MA, 1977.

Newton, R.N. and Radar, H.P., Performance data for propellers of high speed craft, *Transactions of the Royal Institution of Naval Architects*, 103, 93–129, 1961.

Nowacki, H., *Curve and Surface Generation and Fairing*, Lecture Notes in Computer Science No. 89, J. Encarnacaoled (Ed.). Springer Verlag, Berlin, Germany, 1980.

O'Brien, T.P. *The Design of Marine Screw Propellers*, Hutchinson Scientific & Technical, London, U.K., 1962.

Okumoto, Y., Takeda, Y., Mano, M. and Okada, T., *Design of Ship Hull Structures*, Springer, Berlin, Germany, 2009.

Oosterveld, M.W.C. and Van Oossanen, P., Further computer analysed data of the wageningen B screw series, *International Shipbuilding Progress*, 22(251), 3–14, 1975.

Pike, J.K. and Frieze, P.A., Ship structural safety and reliability, in M. Erki, M.A. Bradford, R.E. Melchers (Eds.), *Progress in Structural Engineering and Materials*, John Wiley & Sons, New York, pp. 198–210, 2001.

Price, W.G. and Bishop, R.E.D., *Probabilistic Theory of Ship Dynamics*, Chapman & Hall, London, U.K., 1974.

Proceedings of the 15th to 26th International Towing Tank Conference, 2011.

Rausand, M. and Arnljot, H., *System Reliability Theory: Models, Statistical Methods and Applications*, 2nd edn., Wiley, Hoboken, NJ, November 2003.

Rawson, K.J., Maritime system design methodology, *Advances in Marine Technology*, 1, 25–41, 1979.

Ray, T., Gokarn, R.P. and Sha, O.P., A global optimisation model for ship design, *Computers in Industry*, 26, 175–192, 1995.

Ridgely-Nevitt, C., The development of parent hulls for a high displacement-length series of trawler forms, *Transaction of the Society of Naval Architects and Marine Engineers*, 71, 5–30, 1963.

Rodrigue, J.-P., *The Geography of Transportation Systems*, Hofstra University, New York, 2013.

Rommel, G., Bruck, F., Diederichs, R., Kempis, R.D. and Kluge, J., *Simplicity Wins*, Harvard Business School Press, Boston, MA, 1995.

Roseman, D.P., *The MARAD Systematic Series of Full Form Ships*, Vol. 85, SNAME, Jersey City, NJ, 1987.

Rowen, A.L., Gardner, R.F., Femenia, J., Chapman, D.S. and Wiggins, E.G., *Introduction to Practical Marine Engineering*, Vols. 1 and 2, Jersey City, NJ, 2005.

RoyChoudhury, S., Nagarajan, V. and Sha, O.P., Prediction of ship manoeuvering characteristics using CFD, in *24th International Ocean and Polar Engineering Conference, ISOPE-2014*, Busan, Korea, pp 470–477, 2014.

Sabit, A.S., Regression analysis of the resistance results of the BSRA series, *International Shipbuilding Progress*, 18, 3, 1971.

Sabit, A.S., An analysis of the series 60 results, Pt. 1. Analysis of forms and resistance results, *International Shipbuilding Progress*, 19, 71–95, 1972a.

Sabit, A.S., The SSPA Cargo liner series. Regression analysis of the resistance and propulsion coefficients, *International Shipbuilding Progress*, 23, 213–217, 1972b.

Saha, A. and Ray, T., Practical robust design optimization using evolutionary algorithms, *American Society of Mechanical Engineers*, 133, 101012–101022, 2011.

Saunders, H.E., Ed., *Hydrodynamics in Ship Design*, Vol. 2, SNAME, New York, 1957.

Savitsky, D., Hydrodynamic design of planing hulls, *Marine Technology*, 8(4), 381–400, October 1964.

Savitsky, D., Planing craft, special edition, Ellsworth, W.M., ed., Naval Engineers Journal, Vol. 97, no. 2, The American Society of Naval Engineers Inc., Alexandria, VA, pp 113–141, February 1985.

Savitsky, D. and Koelbel, J.R., Seakeeping considerations in design and operation of hard chine planing hulls, *Naval Architect*, 55–59, March 1979.

Schneekluth, H., *Ship Design for Efficiency and Economy*, Buttersworth, London, U.K., 1987.

Schneekluth, H. and Bertram, V., *Ship Design for Efficiency and Economy*, Butterworth-Heinemann, Oxford, U.K., 1998.

Schneider, T., *How We Know Global Warming is Real – The Science Behind Human-induced Climate Change*, Skeptic Magazine, Vol. 14, No. 1, pp 31–37, The Skeptics Society, Altadena, CA, 2008.

Sea-Europe Ships and Marine Equipment Association, Market and policy development on shipbuilding, Special session on market distorting factors, OECD, WP6, Paris, France, June 2012.

Sha, O.P., CFD applications in ship design, in *Workshop on EFD and CFD Applications in Ship Design*, IIT, Madras, India, December 2014.

Sha, O.P., Misra, S.C. and Gokarn, R.P., Ship hull form design – A modular approach, in *Ninth Symposium PRADS*, Lubeck, Germany, 2004.

Sha, O.P., RoyChoudhury, S., Dash, A.K. and Nagarajan, V., Development of shallow water manoeuvring mathematical model of a large tanker using RANS solver, in *European Inland Water Navigational Conference*, Budapest, Hungary, pp 402–413, September 2014.

Singer, D.J., Doerry, N. Capt. and Buckley, M. E., What is set-based design? *ASNE Naval Engineers Journal*, 121(4), 31–43, 2009.

Smith, T.C. and Thomas III, W.L., A survey of ship motion reduction devices, Report No. DTRC/SHD-1338-01, David Taylor Research Centre, Bathesda, MD, September 1990.

Sobek, D.K., II, Ward, A.C. and Liker, J.K., Toyota's principles of set-based concurrent engineering, *Sloan Management Review*, 40(2), 67–84, January 1999.

Soding, H. and Rabien U., Hull surface design by modifying an existing hull, in *Proceedings of the First International Symposium on Computer-Aided Hull Surface Definition*, Annapolis, MD, pp 19–29, 1977.

St. Denis, M., On the empiric design of seakindly ships, in *Second International Conference on Practical Design in Shipbuilding PRADS*, Tokyo, Japan, pp 1–9, 1983.

St. Denis, M. and Pierson, W.J., On the motion of ships in confused seas, *Transactions of the Society of Naval Architects and Marine Engineers*, NJ, pp 280–357, 1953.

Stopford, M., Forecasting – An impossible job? in *Trade Winds Norshipping Conference*, Oslo, Norway, June 2009.

Storch, R.L., Hammon, C.P. and Bunch, H.M., *Ship Production*, Cornell Maritime Pr/Tidewater, Centreville, MD, 1988.

Taggart, R.L., Ed., *Ship Design and Construction*, SNAME, Jersey City, NJ, 1980.

Taha, H.A., *Operations Research*, MacMillan, New York, 1971.

Talley, L.D., Pickard, G.L., Emery, V.J. and Swift, J.H., *Descriptive Physical Oceanography*, Elsevier, Amsterdam, the Netherlands, 2011.

Tarjan, G., *Catamarans*, Kindle Edition, Multihulls magazine, October 2007.

Taylor, D.A., *Introduction to Marine Engineering*, Buttersworth Heinmann, Oxford, U.K., 1990.

Taylor, D.W., *The Speed and Power of Ships*, 2nd revision, U.S. Maritime Commission, Washington, 1943.

The Shipbuilders' Association of Japan, *Shipbuilding Statistics*, The Shipbuilders' Association of Japan, March 2013.

Todd, F.H., Further model experiments on the resistance of mercantile ship forms – Coaster vessels, INA, London 1931 and 1938, NECIES, Scotland 1934 and 1942 and IME, London 1940.

Todd, F.H., *Ship Hull Vibration*, Edward Arnold, London, U.K., 1961.

Todd, F.H., *Series 60—Methodical Experiments with Models of Single Screw Merchant Ships*, DTMB report 1712, Maryland, 1963.

Townsin, R.L. and Kwon, Y.J., Approximate Formulae for the speed loss due to Added Resistance in Wind and Waves, RINA, 1982.

Tupper, E.C., *Introduction to Naval Architecture*, Butterworth-Heinemann, Amsterdam, the Netherlands, 2004.

Ulstein X-bow brochure, Published by the Ulstein group (shipbuilders, ship design and system solution providers), Ulsteinvik, Norway, 2005–2006.

UNCTAD report, *Review of Maritime Transport*, United Nations Publication UNCTAD/RMT/2013 and UNCTAD/RMT/2014.

van Manen, J.D. and van Lammeren, W.P.A., The design of wake adapted screws and their behaviour behind a ship, *International Embedded Systems Symposium*, 99, 113, 1955–1956.

van Oortmerssen, G., A power prediction method and its application to small ships, *International Shipbuilding Progress*, 18, pp 397–412, 1971.

Van Oortmerssen, G., A power prediction method for motor boats, NSMB, Netherlands Publication 429, 1973.

Vassalos, D., Risk based ship design, in A. Papanikolaou (Ed.), *Risk Based Ship Design – Methods, Tools and Applications*, Springer, Berlin, Germany, 2009.

Watson, D.G.M., *Practical Ship Design*, Elsevier, Amsterdam, the Netherlands, 2009.

Watson, D.G.M. and Gilfillan, A.W., Some ship design methods, *Transaction of the Royal Institute of the Naval Architects*, pp 279–303, 1977.

Werenskiold, P., Powering of planing chine hulls, *Ship and Boat International*, April 1970.

Yeh, H.Y.H., Series 64 resistance experiments on high speed displacement forms, *Marine Technology*, 2, pp 248–272, 1965.

Zakaria, G., Iqbal, K.S. and Hussain, K.A., Performance evaluation of contemporary shipbuilding industries in Bangladesh, *Journal of Naval Architects and Marine Engineers*, Vol. 7, Bangladesh, 2010, pp 73–82, December 2010.

Index

Printed in the United States
by Baker & Taylor Publisher Services